D1328840

KU-762-778

ESSENTIAL REPRODUCTION

To Bob Edwards and Joe Herbert
who first stimulated our interest
in the science of reproduction

FOURTH EDITION

ESSENTIAL REPRODUCTION

Martin H. Johnson

MA, PhD
Professor of Reproductive Sciences
Department of Anatomy, University of Cambridge
Fellow Christ's College, Cambridge

Barry J. Everitt

MA, BSc, PhD
Reader in Neuroscience
Department of Experimental Psychology,
University of Cambridge
Fellow and Director of Studies in Medicine
Downing College, Cambridge

b

Blackwell
Science

© 1980, 1984, 1988, 1995 by
Blackwell Science Ltd
Editorial Offices:
Osney Mead, Oxford OX2 0EL
25 John Street, London WC1N 2BL
23 Ainslie Place, Edinburgh EH3 6AJ
238 Main Street, Cambridge
   Massachusetts 02142, USA
54 University Street, Carlton
   Victoria 3053, Australia

Other Editorial Offices:
Arnette Blackwell SA
   224, Boulevard Saint Germain
   75007 Paris, France

Blackwell Wissenschafts-Verlag GmbH
Kurfürstendamm 57
10707 Berlin, Germany

Zehetnergasse 6
A-1140 Wien
Austria

All rights reserved. No part of
this publication may be reproduced,
stored in a retrieval system, or
transmitted, in any form or by any
means, electronic, mechanical,
photocopying, recording or otherwise,
except as permitted by the UK
Copyright, Designs and Patents Act
1988, without the prior permission
of the copyright owner.

First published 1980
Reprinted 1983,
Second edition 1984
Third edition 1988
Fourth edition 1995
Reprinted 1996 (twice)

DISTRIBUTORS
Marston Book Services Ltd
PO Box 87
Oxford OX2 0DT
(*Orders*: Tel: 01865 791155
          Fax: 01865 791927
          Telex: 837515)

USA
Blackwell Science, Inc.
238 Main Street
Cambridge, MA 02142
(*Orders*: Tel: 800 215-1000
          617 876-7000
          Fax: 617 492-5263)

Canada
Copp Clark, Ltd
2775 Matheson Blvd East
Mississauga, Ontario
Canada, L4W 4P7
(*Orders*: Tel: 800 263-4374
          905 238-6074)

Australia
Blackwell Science Pty Ltd
54 University Street
Carlton, Victoria 3053
(*Orders*: Tel: 03 9347 0300
          Fax: 03 9349 3016)

Set by EXPO Holdings, Malaysia
Printed and bound in Great Britain
at the Alden Press Ltd.,
Oxford and Northampton

A catalogue record for this title
is available from the British Library

ISBN 0-632-03525-0

Library of Congress
Cataloging-in-Publication Data

Johnson, M. H.
   Essential reproduction/Martin
   H. Johnson, Barry J. Everitt.—4th ed.
      p.   cm.
   Includes bibliographical references
   and index.
   ISBN 0-632-03525-0
   1. Mammals—Reproduction.
   I. Everitt, Barry J.
   II. Title.
QL739.2.J64   1995
599'.016—dc20   94-45116
                                      CIP

Contents

10 · Maternal Recognition and Support of Pregnancy, 181

8 · Coitus and Fertilization, 143

11 · The Fetus and its Preparations for Birth, 190

9 · Implantation and the Establishment of the Placenta, 161

12 · Parturition, 208

13 · Lactation and Maternal Behaviour, 217

14 · Fertility, 233

Preface to the Fourth Edition

In the few years since the third edition, several areas of reproductive science have moved forwards considerably. This progress has been due, in part, to the application to reproductive studies of the newer techniques of molecular biology and single cell physiology and biochemistry. But it is also undoubtedly the case that two major health problems have propelled the study of certain aspects of reproduction. First, considerable impetus has been provided by the continuing clinical developments in assisted conception for the alleviation or circumvention of subfertility. Second, the threat posed by the human immunodeficiency virus has put reproduction and sexual behaviour high on the agenda of medical research. Two social changes, namely the changing role of women in developed societies and a more general acknowledgement of the variety of sexual behaviour, have also influenced attitudes to the types of research considered acceptable or important. We have restructured all chapters, some of them substantially, to take these new developments into account. In addition, we have tried to simplify the text, and have re-organized the sequence of discussion, omitted one or two more marginal topics, speculated a little on the likely development of some new topics, and moved some topics between chapters. In particular, this has led to a completely new Chapter 2, which deals with the subject of reproductive messengers. We have again received many helpful comments and letters of advice from our students and from teachers all over the world, and have attempted, in this fourth edition, to provide a compact and comprehensive text for students of reproduction.

How to Use this Book

This book represents an integrated approach to the study of reproduction. There can be few subjects that so obviously demand such an approach. During our teaching of reproduction at Cambridge University, the need for a book of this kind was clear to both ourselves and our colleagues. We hope this volume goes some way towards filling this need.

We have written the book for medical, veterinary and science students of mammalian reproduction. Throughout, we have attempted to draw out the general, fundamental points common to reproductive events in all or most species. However, a great range of variation in the *details* of reproduction is observed amongst different species, and, in some respects, very *fundamental* differences are also observed. Where the details differ, we have attempted to indicate this in the numerous tables and figures, rather than clutter the general emphasis and narrative of the text. Where the fundamentals differ, an explicit discussion is given in the text. These fundamental differences should not be ignored. For example, preclinical medical students may consider the control of luteal life in the sheep or ovarian cyclicity in the rat to be irrelevant to their future interests. However, as a result of inappropriate extrapolation between species, the human female has been treated both as a ewe and as a rat in the past (much to her discomfort and detriment). If, on finishing this book, the student appreciates the dangers of wholesale extrapolations between species, we will have achieved a major aim.

Confinements of space have unfortunately necessitated omission from the text of the subject of embryonic development. We felt that to give only passing reference to this subject would be an injustice, to treat it fully would require a text of similar length to the present one. We recommend that the interested student seeks this information elsewhere.

We suggest that you first read through each chapter with only passing reference to tables and figures. In this way, we hope that you will grasp the essential fundamentals of the subject under discussion. Then re-read the chapter, referring extensively to the tables and figures and their legends, in which a great deal of detailed or comparative information is located. Finally, because we have adopted an integrated approach to the subject, the book needs to be taken as a whole, as it is more than the sum of its constituent chapters.

M.H.J., B.J.E.

Acknowledgements

For help at many stages of the preparation of this edition we owe particular thanks to many people: to our present and former students for their interest, stimulation and responsiveness; to our colleagues at Blackwell Science for their help and advice; to John Bashford and Roger Liles for their advice and help with photographic illustrations; to Rachel Chesterton for help with illustrations; to Professor Peter Braude, Dr Graham Burton, Dr Ruth Kleinfeld, Professor Tomas Hökfelt, J. Moeselaar, Dr Bernard Maro, Dr Orla McGuinness, Dr Tony Plant, Dr J. M. Tanner and Dr Pauline Yahr for allowing us to use their original photographs and data; and to our many colleagues who read and criticized our drafts and encouraged us in the preparation of this edition, especially: John Aitken, Graham Burton, Toby Bush, Chris Carne, Fran Ebling, Lynn Fraser, Michael Hastings, Barry Keverne, Chi Wong and Francis Woodman. We owe a special debt of thanks to Professor Peter Braude for his advice on clinical aspects. Finally, the stimulating environment generated by the research workers in our laboratories has been extremely important in providing us with the help and motivation to write this book, and to them we are very grateful.

Figure Acknowledgements

Chapter 1

Figs 1.3, 1.4, 1.5, 1.6 redrawn from Langman J (1969) *Medical Embryology*. Williams & Wilkins.

Fig. 1.7a reproduced with permission from Money J & Ehrhardt A (1972) *Man and Woman, Boy and Girl*, Johns Hopkins University Press.

Fig. 1.7b redrawn with permission from Overzier C (1963) *Inter-sexuality*. Academic Press.

Fig. 1.9 provided by Dr Pauline Yahr, University of California, Irvine.

Figs 1.10, 1.11 redrawn from Goy RW (1970) *Philosophical Transactions of the Royal Society of New York* **259**, 149–162.

Figs 1.14, 1.18 from Tanner JM (1962) *Growth at Adolescence*. Blackwell Scientific Publications, Oxford.

Figs 1.15, 1.16, 1.17 from van Wieringen JC, Wafelbakker F, Verbrugge HP & de Haas JH (1971) *Growth Diagrams 1965 Netherlands: Second National Survey on 0–24 year olds*. Netherlands Institute for Preventative Medicine TNO, Leiden and Wolters Noordhoff.

Chapter 2

Fig. 2.6 redrawn from Nikolics K *et al.* (1983) *Nature* **316**, 511–517.

Fig. 2.7 redrawn from Robinson ICAF (1986) *Neuroendocrinology* (Eds Lightman SL & Everitt BJ), pp. 154–176. Blackwell Scientific Publications, Oxford.

Fig. 2.11 redrawn from Steinetz BG *et al.* (1982) Advances in Experimental Medicine and Biology (New York) **143**, 79–104.

Chapter 4

Fig. 4.1 redrawn from Netter FH (1954) *The Ciba Collection of Medical Illustrations*. Ciba.

Fig. 4.3 drawn from original data from Baker T.

Chapter 5

Fig. 5.4a,b redrawn from Heimer L (1983) *The Human Brain and Spinal Cord*. Springer-Verlag.

Fig. 5.6 redrawn from Clarke IJ & Cummins JT (1982) *Endocrinology* **111**, 1737–1739.

Fig. 5.7 redrawn from Belchetz PE *et al* (1978) *Science* **202**, 631–633.

Figs 5.8, 5.9, 5.12, 5.13, 5.17, 5.23, 5.24, 5.26 from Yen SSC & Jaffe R (Eds) (1978) *Reproductive Endocrinology*. WB Saunders & Co.

Fig. 5.10 redrawn from Martin GB *et al.* (1986) *Journal of Endocrinology* (London) **111**, 287–296.

Fig. 5.11 redrawn from Rivier C *et al.* (1986) *Science* **234**, 205–208.

Fig. 5.14 from Clayton RN (1986) *Neuroendocrinology* (Eds Lightman SL & Everitt BJ). Blackwell Scientific Publications, Oxford.

Fig. 5.19 redrawn from Moore RY (1978) *Reproductive Endocrinology* (Eds Yen SSC & Jaffe R). WB Saunders & Co.

Figs 5.20, 5.27 from Lincoln GA (1979) *British Medical Bulletin (London)* **35(2)**, 167–172.

Fig. 5.21 redrawn from Dorner G (1979) *Sex, Hormones and Behaviour*. Ciba Foundation Symposium 62. Elsevier.

Fig. 5.28 redrawn from Bittman EL *et al.* (1984) *Biology of Reproduction (New York)* **30**, 585–593.

Figs 5.29, 5.30 from Keverne EB (1979) *Sex, Hormones and Behaviour*. Ciba Foundation Symposium 62. Elsevier.

Figs 5.31, 5.32, 5.33 from Everitt BJ & Keverne EB (1986) *Neuroendocrinology* (Eds Lightman SL & Everitt BJ). Blackwell Scientific Publications, Oxford.

Chapter 6

Figs 6.1, 6.7 from Tanner JM (1962) *Growth at Adolescence*. Blackwell Scientific Publications, Oxford.

Fig. 6.2 redrawn from Grumbach M (1978) *Reproductive Endocrinology* (Eds Yen SSC & Jaffe R). WB Saunders & Co.

Fig. 6.3 reproduced with permission from Weitzmann ED (1975) *Recent Progress in Hormone Research*. Academic Press.

Fig. 6.4 redrawn from Boyer RM *et al.* (1974) *Journal of Clinical Investigation (New York)* **54,** 609.

Fig. 6.5 from Grumbach M (1978) *Control of the Onset of Puberty* (Eds Grumbach M, Grave GD & Mayer FE). John Wiley & Sons, Chichester.

Fig. 6.6a from Wildt L, Marshall G & Knobil E (1980) *Science* **207,** 1373–1375.

Fig. 6.6b from Plant T (1980) *Endocrinology* **106,** 1451–1454.

Fig. 6.8 from Frisch RE (1972) *Pediatrics* **50,** 445–450.

Chapter 7

Fig. 7.3 redrawn from Netter FH (1954) *The Ciba Collection of Medical Illustrations*. Ciba.

Figs 7.7, 7.8 redrawn with permission from Bermant G & Davidson JM (1974) *Biological Bases of Sexual Behaviour*. Harper & Row.

Fig. 7.9 redrawn from Bancroft J (1983) *Human Sexuality and its Problems*. Churchill Livingstone.

Figs 7.10, 7.16 redrawn from Bancroft J (1989) *Human Sexuality and its Problems*, 2nd edition. Churchill Livingstone.

Fig. 7.13 from Udry JR, Morris NM & Waller I (1973) *Archives of Sexual Behavior (New York)* **2(3),** 205–214.

Fig. 7.15 data from Sherwin BB & Gelf MM (1987) The role of androgen in the maintenance of sexual functioning in oophorectomized women. *Psychosomatic Medicine* **49,** 397–409.

Fig. 7.17 data adapted from Bermant G & Davidson JM (1974) *Biological Bases of Sexual Behaviour*. Harper & Row.

Chapter 8

Figs 8.3, 8.4 redrawn from Netter FH (1954) *The Ciba Collection of Medical Illustrations*. Ciba.

Figs 8.7, 8.8 provided by courtesy of O McGuinness.

Fig. 8.9 from Maro B, Johnson MH, Pickering S & Flach G (1984) Changes in action distribution during fertil-

ization of the mouse egg. *Journal of Embryology and Experimental Morphology (Cambridge Eng)* **81,** 211–237.

Chapter 9

Fig. 9.1 Photography by courtesy of Professor PR Braude.

Fig. 9.4b,c redrawn from Hamilton WJ, Boyd JD & Mossman HW

Fig. 9.6 redrawn from Steven DH (1975).

Fig. 9.8 redrawn from Hamilton WJ, Boyd JD & Mossman HW (1972) *Human Embryology*.

Fig. 9.10 Photography by courtesy of Habashi S, Burton GJ & Steven DH.

Chapter 11

Fig. 11.2 redrawn from Metcalfe *et al.* (1967) *Physiological Reviews* **47,** 782.

Table 11.2, Figs 11.1, 11.3, 11.4, 11.5 after Biggers JD (1979) *Medical Physiology* (Ed. Mountcastle V). Mosby.

Chapter 12

Fig. 12.4 reproduced with kind permission of KR Niswander.

Chapter 13

Fig. 13.1 redrawn from Netter FH (1954) *The Ciba Collection of Medical Illustrations*. Ciba.

Figs 13.2, 13.4 redrawn from Cowie AT (1972) *Reproduction in Mammals* (Eds Austin CR & Short RV). Cambridge University Press, Cambridge.

Fig. 13.5 redrawn and adapted with kind permission of A McNeilly.

Fig. 13.9 redrawn with permission from Hinde RA (1974) *Biological Bases of Human Social Behaviour*. John Wiley & Sons.

Chapter 14

Figs 14.1, 14.8 from Professor Peter Braude.

Figs 14.5, 14.6 based on figures in Guillebaud J (1993) *Contraception*. Churchill Livingstone.

Fig. 14.7 data adapted from Hull M.

CHAPTER 1

Sex

The reproduction of mammals involves sex. The essential feature of mammalian *sexual reproduction* is that the new individual receives its genetic endowment in two equal portions: half carried in a *male gamete*, the *spermatozoon*, and half carried in a *female gamete*, the *oocyte*. These gametes come together at *fertilization* to form the new *zygote*. In order to reproduce itself subsequently, the individual must transmit only half its own chromosomes to the new zygotes of the next generation. In sexually reproducing species, therefore, a special population of *germ cells* is set aside. These cells undergo a *reduction division* known as *meiosis*, in which the chromosomal content of the germ cells is halved and the genetic composition of each chromosome is modified as a result of chromosomal exchange (Fig. 1.1). The increased genetic diversity that is generated within a population that is reproducing sexually may offer a richer and more varied source of material on which natural selection can operate. The population would, therefore, be expected to show greater resilience in the face of environmental challenge.

However, sex is not by any means an essential component of reproductive processes. Thus, *asexual* (or *vegetative*) reproduction occurs continuously within the tissues of our own bodies as individual cells grow, divide *mitotically* (Fig. 1.1) and generate two offspring that are genetically identical to each other and to their single parent. Many unicellular organisms reproduce themselves mitotically just like the individual cells of the body. Among multicellular organisms, including some complex vertebrates, such as lizards, several reproduce themselves by setting aside a population of germ cells that can differentiate in the absence of a fertilizing spermatozoon in order to generate a complete new organism that is genetically identical or very similar to its parent. This asexual process of reproduction, often called *parthenogenetic* development, is simply not available to mammals. Although it is possible to stimulate a mammalian oocyte (including human oocytes) in the complete absence of a spermatozoon, such that it undergoes the early processes of development and may even implant in the uterus, these parthenogenetic embryos always fail and die eventually. It seems that a complete set of chromosomes from a father and a complete set from a mother are a requirement for normal and complete development to occur in mammals (see Chapter 8 for further discussion).

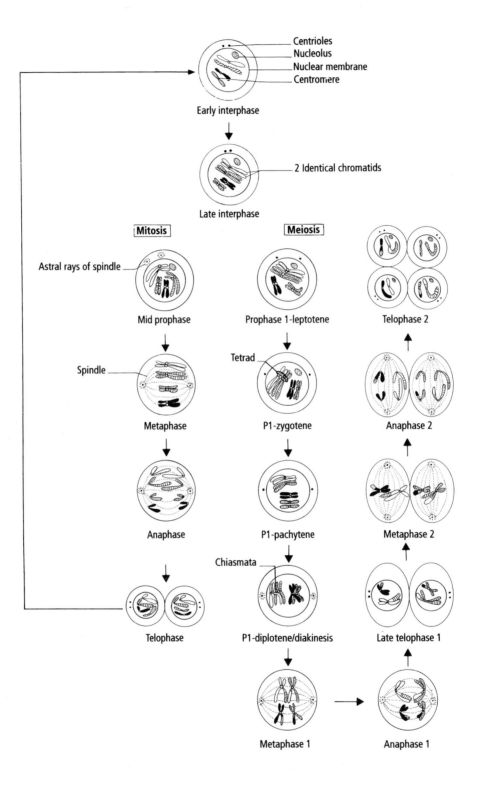

Centrioles
Nucleolus
Nuclear membrane
Centromere

Early interphase

2 Identical chromatids

Late interphase

Mitosis

Meiosis

Astral rays of spindle

Mid prophase

Prophase 1-leptotene

Telophase 2

Spindle

Tetrad

Metaphase

P1-zygotene

Anaphase 2

Anaphase

P1-pachytene

Metaphase 2

Chiasmata

Telophase

P1-diplotene/diakinesis

Late telophase 1

Metaphase 1

Anaphase 1

While the advantages of sex can be sought in terms of the increased responsiveness to environmental variation that it offers, the consequences of this mode of reproduction permeate all aspects of mammalian life. At the core of the process lies the creation and fusion of the two types of gamete.

The gametes take distinctive male or female forms (to prevent self-fertilization) and are made in distinctive male and female *gonads:* the *testis* and *ovary*, respectively. In addition, the gonads elaborate a group of hormones, notably the *sex steroids*, which modify the tissues of the body to generate distinctive male and female somatic phenotypes suited to maturing and transporting their respective gametes. The steroids also affect the behaviour and physiology of the individuals of each sex to ensure that, in most mammals, mating will only occur between opposite sexes at times of maximum fertility. Finally, in mammals, not only do the steroids provide conditions to facilitate the creation of new individuals, they also prepare the female to carry the growing embryo for prolonged periods (*viviparity*), and to nurture it after birth through an extended period of *maternal lactation* and *parental care*. Thus, sex has come to occupy a central position in mammalian biology, shaping not just anatomy and physiology, but also aspects of behaviour and social structure. This ramification of sex throughout a whole range of biological and social aspects of mammalian life is mediated largely through the actions of the gonadal hormones. However, in humans and in other higher primates, social learning also plays an important role in generating sex differences. Children are taught how to behave as women or men, what is *feminine* and what is *masculine*. In this way they acquire a sense of their *gender*.

Thus, although studies on mammals in general are relevant to humans, they are not in themselves sufficient. In order fully to understand human reproduction and sexuality, humans must be studied too.

THE GENESIS OF TWO SEXES

Genetic determinants

The genesis of two sexes has a genetic basis in mammals. Examination of human chromosomes reveals a consistent difference between the sexes in *karyotype* (or pattern of chromosomal morphologies). Thus, the human has 46 chromosomes, 22 pairs of *autosomes* and one pair of *sex chromosomes* (Fig. 1.2). Human females, and indeed all female mammals, are known as the *homogametic sex* because the sex chromosomes are both *X chromosomes* and all the gametes (oocytes) are similar to one another in that they each possess one X chromosome. Conversely, the male is termed the *heterogametic sex*, as his pair of sex chromosomes consists of one X and one Y, so producing two distinct populations of spermatozoa, one bearing an X and the other a Y chromosome (Fig. 1.2). Examination of a range of human patients with chromosomal abnormalities has shown that if a Y chromosome is present then the individual develops the male gonads (testes). If the Y chromosome is absent the female gonads develop (ovaries). The number of X chromosomes or autosomes present does not affect the primary determination of gonadal sex (Table 1.1). Similar studies on a whole range of other mammals show that Y-chromosome

---

Fig. 1.1 (*opposite*) Mitosis and meiosis in human cells. Each human cell contains 23 pairs of homologous chromosomes, making 46 chromosomes in total. Each set of 23 chromosomes is called a *haploid* set. When a cell has two complete sets, it is described as being diploid. In this figure, we show a single schematicized cell with just two of the 23 homologous pairs of chromosomes illustrated. Prior to division, the cell is in interphase, during which it grows and duplicates the DNA in each of its chromosomes. As a result, each chromosome consists of two identical *chromatids* joined at the *centromere*. Interphase chromosomes are not readily visible, being long, thin and decondensed (but are shown in this figure in a more condensed form for simplicity of representation).

In *mitotic prophase*, the two chromatids become distinctly visible under the light microscope as each shortens and thickens by a spiralling contraction; the nucleoli and nuclear membrane break down. In *mitotic metaphase*, microtubules form a *mitotic spindle* between the two *centrioles* and the chromosomes lie on its *equator*. In *mitotic anaphase*, the centromere of each chromosome splits and each chromatid from one chromosome migrates to opposite poles of the spindle. *Mitotic telophase* sees: the reformation of nuclear membranes and nucleoli; division of the cell into two daughters (known as *cytokinesis*); breakdown of the spindle; and decondensation of chromosomes so that they are no longer visible under the light microscope. Two genetically identical daughter cells now exist where one existed before. Mitosis is a non-sexual or vegetative form of reproduction.

Meiosis involves two sequential divisions. The *first meiotic prophase* (prophase 1) is extended and divided into several sequential steps: (1) *leptotene* chromosomes are long and thin; (2) during *zygotene*, homologous pairs of chromosomes from each haploid set come to lie side by side along parts of their length; (3) in *pachytene*, chromosomes start to thicken and shorten and become more closely associated in pairs along their entire length at which time *synapsis*, *crossing-over* and *chromatid exchange* take place and nucleoli disappear; (4) in *diplotene* and *diakinesis*, chromosomes shorten further and show evidence of being closely linked to their homologue at the *chiasmata* where crossing-over and chromatid exchange has occurred, giving a looped or cross-shaped appearance. In *meiotic metaphase 1*, the nuclear membrane breaks down, and homologous pairs of chromosomes align on the equator of spindle. In *meiotic anaphase 1*, homologous chromosomes move in opposite directions. In *meiotic telophase 1*, cytokinesis occurs; the nuclear membrane may reform temporarily, although this does not always happen, yielding two daughter cells each with half the number of chromosomes (only one member of each homologous pair), but each chromosome consisting of two unique chromatids. In the *second meiotic division*, these chromatids then separate as in mitosis, to yield a total of four haploid offspring from the original cell, each one containing only one complete set of chromosomes. Due to chromatid exchange and the random segregation of homologous chromosomes, each haploid cell is genetically unique. At fertilization, two haploid cells will come together to yield a new diploid zygote.

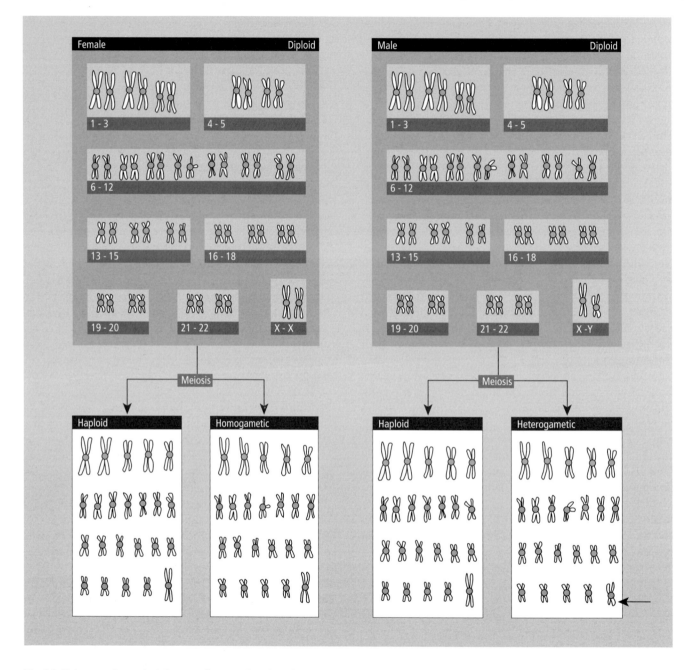

Fig. 1.2 Karyotypes of two mitotic human cells: one male and one female. Each cell was placed in colchicine, a drug which arrested them in mitotic metaphase when the chromosomes were condensed and clearly visible (see Fig. 1.1). The chromosomes were stained and then arranged and classified according to the so-called 'Denver' system. While the 44 *autosomes* (22 pairs of homologues) are grossly similar in size in each sex, the pair of sex chromosomes are distinguishable by size, being XX in the female and XY in the male. After meiotic division, all four female cells (only two shown) contain one X chromosome: the homogametic sex. In contrast, two of the male cells contain X chromosomes and two contain Y chromosomes: the heterogametic sex. An arrow indicates the position of the SRY gene on the short arm of the Y chromosome.

activity alone is sufficient to determine gonadal sex. Thus the first step towards sexual dimorphism in mammals is the issuing of an instruction by the Y chromosome saying: 'make a testis'.

The Y chromosome itself is small. Moreover, most of its DNA is *heterochromatic* (that is, very condensed and incapable of synthesizing RNA). Therefore, the many structural genes required to make an organ as complex as the testis cannot be located on the Y chromosome alone. Indeed, these genes are known to lie on other autosomal chromosomes, and some even lie on the X chromosome. What the Y chromosome contains is a 'switching' or controller gene, which regulates the expression of all these other structural genes by deciding whether and when they should become activated. The identity and location on the Y chromosome of this gene has been discovered recently by the study of some rare and atypical individuals.

Clinicians have identified men with an XX sex chromosomal constitution and women with an XY chromosomal constitution. At first sight, these people appear to contradict all that has been said above (see Table 1.1). However, careful examination of the DNA sequences on the short arm of the Y chromosome of the XY females has revealed either that short pieces of DNA are missing (*chromosomal deletions*) or that there are *mutations* of one or more nucleic acid bases. By comparing the DNA sequences in a large number of such patients, it is possible to find one region of the Y chromosome common to all of them that is affected by deletion or mutation. This region is a likely locus for the testis determining gene. Supportive evidence comes from the XX males, who are found to have translocations of small pieces of the Y chromosome to one of their autosomes or X chromosomes. Again, the critical piece of Y chromosome that must be translocated to yield an XX male seems to come from the same region as is damaged in the XY females. This region contains a gene called *SRY* (in humans), which stands for 'sex-determining region on the Y chromosome'. The gene is located close to the end of the short arm of the human Y chromosome (see arrow in Fig. 2.2). Genes with a similar base sequence composi-

tion, which are also associated with the development of a testis, have been found in other mammals, including the mouse (in which the gene is called *Sry*).

The identification of the mouse homologue was important, because it enabled a critical experimental test of the function of this region of the Y chromosome to be performed. Thus, the *Sry* region of DNA was excised from the Y chromosome and injected into the nuclei of 1-cell XX mouse zygotes. The excised material can integrate into the chromosomal material of the XX recipient mouse, which now has an extra piece of DNA. If this piece of DNA is functional in issuing the instruction 'make a testis', the XX mouse should develop as a male. This is what happened, strongly supporting the idea that the region containing the controller gene had been identified. This region of the Y contains a gene encoding a DNA-binding protein, which is exactly what would be expected if indeed it is a controller gene, influencing other down stream genes. How and where does *SRY* act to cause a testis to be generated?

Gonadal dimorphism

The early development of the gonad proceeds indistinguishably in males and females. In both sexes the gonads are derived from two distinct tissues: *somatic mesenchymal tissues*, which form the matrix of the gonad, and the *primordial germ cells* (PGCs), which migrate into and colonize this matrix to form the gametes.

The *genital ridge primordia* are two knots of mesenchyme overlain by a columnar *coelomic (or germinal) epithelium*. The primordia develop at about 3.5–4.5 weeks in human embryos, on either side of the central dorsal aorta on the posterior wall of the embryo in the lower thoracic and upper lumbar region (Fig. 1.3b,c). The genital ridges are superficial and medial to the developing *mesonephric tissue* (Fig. 1.3). Columns of cells derived from proliferation and inward migration of both the mesonephros and the coelomic epithelium form the *primitive medullary* and *sex cords* respectively (Fig. 1.3d).

The primordial germ cells arise outside the genital ridge region, and are first identifiable in the human embryo at about 3 weeks in the epithelium of the yolk sac near the base of the developing allantois (Fig. 1.3a). There they may be recognized by their distinctive morphology and alkaline phosphatase activity. By the 13- to 20-somite stage, the PGC population, expanded by mitosis, can be observed migrating to the connective tissue of the hind gut and from there into the gut mesentery (Fig. 1.3b). From about the 25-somite stage onwards, 30 days or so after fertilization, the majority of cells has passed into the region of the developing kidneys, and

Table 1.1 Effect of human chromosome constitution on the development of the gonad

Chromosomal number		Gonad	Syndrome
Autosomes	Sex chromosomes		
44	XO	Ovary	Turner's
44	XX	Ovary	Normal female
44	XXX	Ovary	Super female
44	XY	Testis	Normal male
44	XXY	Testis	Klinefelter's
44	XYY	Testis	Super male
66	XXX	Ovary	Triploids (non-viable)
66	XXY	Testis	
44	XX^{sxr}	Testis	Sex reversed*

* An X^{sxr} chromosome carries a small piece of Y chromosome translocated onto the X: see text.

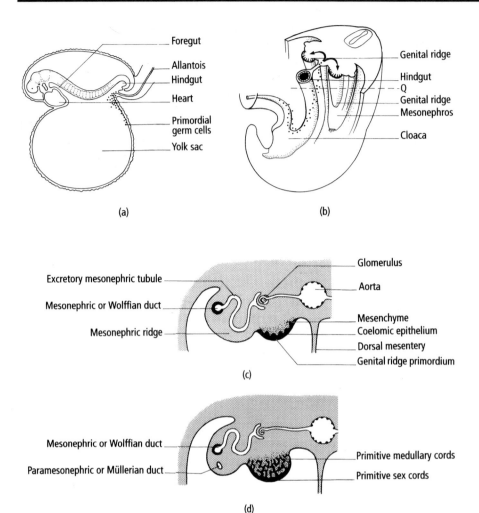

Fig. 1.3 Drawings of a 3-week human embryo showing: (a) the origin of the primordial germ cells; (b) the route of their migration. Section Q is the plane of transverse section through the lumbar region shown in (c) and (d) at 4 and 5 weeks of development: the 'indifferent gonad' stage.

thence into the adjacent gonadal primordia (Fig. 1.3) where they join the ingressing cells of the sex and medullary cords. This migration of PGCs occurs primarily by amoeboid movement. The genital ridges may produce a chemotactic substance to entice the PGCs in, as PGCs placed in a culture dish with a genital ridge specifically move towards it. Moreover, gonadal primordial tissue grafted into abnormal sites within the embryo attracts germ cells to colonize it.

Throughout this phase of PGC migration and early colonization it is not possible to discriminate between male and female gonads, which are therefore said to be *indifferent*. The Y-chromosomal determination of gonadal sex is visible only when PGC colonization is completed during the sixth week in the human embryo. At this time, and in male embryos only, there is a vigorous proliferation of the sex cord cells deep into the medullary region of the gonad, thereby establishing contact with ingrowing medullary cords of mesonephric tissue (the *rete blastema or rete testis cords*) and leading to the formation of the *definitive testis cords* (Fig. 1.4a,b). These cords incorporate most of the PGCs within their columns and secrete an outer basement membrane. They are then known as the *seminiferous cords*, and will give rise to the *seminiferous tubules* of the adult (Fig. 1.4b). Of the two cell populations within the cords, the PGCs, now known as *prospermatogonia*, will give rise to spermatozoa, and the mesodermal cord cells give rise to *Sertoli cells*. Between the cords, the loose mesenchyme vascularizes and develops as stromal tissue, within which cells condense in clusters to form specific endocrine units, the *interstitial glands of Leydig*. During this period of testis cord formation, the *SRY* gene is expressed for the first time, exactly as would be expected if it controlled the process. Its major site of expression is the Sertoli cells, and this has led to the proposal that the *SRY* gene actually issues the instruction 'make a Sertoli cell'. It is not expressed in the PGCs; indeed, testis cord formation occurs quite happily in the absence of any PGCs, confirming that they have no role to play in primary testis determination.

While the male gonad undergoes these marked *SRY*-directed changes in organization, the female gonad continues to appear indifferent and does not express *SRY*. The primitive sex cords of the human female are ill-defined, although in sheep and pigs well-formed cords do appear transiently. Unlike the situation in the male, the ingrowth of coelomic and mesonephric epithelial cells is followed by their condensation in the more cortical regions of the gonad (Fig. 1.4c). Small cell clusters then surround the germ cells to initiate formation of the *primordial follicles* characteristic of the ovary (Fig. 1.4d). In these follicles the mesenchymal cells secrete an outer basement membrane, the *membrana propria*. The mesenchymal cells will give rise to the *granulosa cells* of the follicle, while the germ cells, which have now ceased their mitotic proliferation, give rise to *oocytes* (see

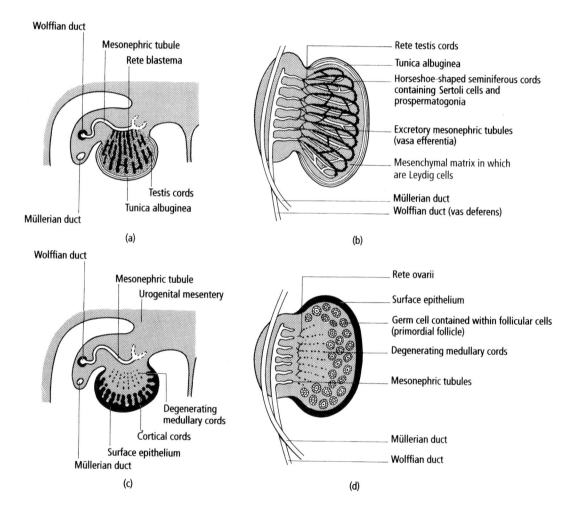

Fig. 1.4 Testicular development during (a) the eighth, and (b) the sixteenth to twentieth week of human fetal life. (a) The primitive sex cords proliferate into the medulla, establish contact with the medullary cords of the rete testis blastema, and become separated from the coelomic epithelium by the tunica albuginea (fibrous connective tissue), which eventually forms the testicular capsule. (b) Note the horseshoe shape of the seminiferous cords and their continuity with the rete testis cords. The vasa efferentia, derived from the mesonephric tubules, connect the seminiferous cords with the Wolffian duct (see text).

Comparable diagrams of ovarian development around (c) the seventh, and (d) the twentieth to twenty-fourth weeks of development. (c) Any primitive medullary sex cords degenerate to be replaced by the well-vascularized ovarian stroma. The cortical cells proliferate, and mesenchymal condensations later develop around the arriving PGCs to yield primordial follicles shown in (d). In the absence of medullary cords and a true persistent rete ovarii, no communication is established with the mesonephric tubules. Hence, in the adult, oocytes are shed from the surface of the ovary, and are not transported by tubules to the oviduct (compared to male, see Chapters 3 & 4).

Chapter 4). Clusters of *interstitial gland cells* form from the mesenchyme around the primordial follicles.

Although the initial decision as to whether to make an ovary or a testis depends simply on the presence or absence of the *SRY* activity in mesenchymal cells, subsequent development of the gonad, particularly of the ovary and its follicles, is dependent upon the presence of a population of normal germ cells. For example, women suffering from *Turner's syndrome* (see Table 1.1), who have a normal autosomal complement but only one X chromosome, develop an ovary. Subsequently, however, normal oocyte growth requires the activity of both X chromosomes, and the activity of only one X in individuals with Turner's syndrome leads to death of the oocyte. Secondary loss of the follicle cells follows, leading to *ovarian dysgenesis* (abnormal development), and a highly regressed or *streak* ovary. Conversely, men with *Klinefelter's syndrome* (see Table 1.1) have a normal autosomal complement of chromosomes but three sex chromosomes, two X and one Y. Testes form normally in these individuals. However, all the germ cells die when they enter meiosis and their death is the result of the activity of two rather than one X chromosomes. These syndromes provide us with two important pieces of clinical evidence. First, *initiation* of gonad formation can occur when *mesenchymal* cells have only *one* Y (testis) or *one* X (ovary) chromosome. Second, *completion* of normal gonad development requires that the *germ* cells have *two X chromosomes in an ovary* but *do not have more than one X chromosome in a testis*.

Primary hermaphroditism

We have established that *SRY* activity on the Y chromosome converts an indifferent gonad into a testis, whereas the absence of its activity results in an ovary. *Genetic maleness* leads to *gonadal maleness*. This primary step in sexual differentiation is remarkably efficient, and only rarely are individuals found to have both testicular *and* ovarian tissue. Such individuals are called *primary (or true) hermaphrodites* and arise, in most if not all cases, because of the presence of a mixture of XY and XX (or XO) cells.

The main role of the *SRY* gene in sexual determination is completed with the establishment of the fetal gonad, and the gene is no longer expressed in the fetus. From this point onwards, the gonads themselves assume the pivotal role in directing sexual differentiation both pre- and postnatally. Again, it is the male gonad, like the Y chromosome before it, that plays the most active role, taking over the 'baton of masculinity' in this sexual relay.

THE DIFFERENTIATION OF TWO SEXES

Endocrine activity in the ovaries is *not* essential for sexual differentiation during fetal life. In contrast, the testes actively secrete two *essential* hormones. The mesenchymal interstitial cells of Leydig secrete steroid hormones, the *androgens*, and the Sertoli cells within the seminiferous cords secrete a dimeric glycoprotein hormone called *Müllerian inhibiting hormone* (MIH). These hormones, which are discussed in more detail in Chapter 2, are the messengers of masculine sexual differentiation sent out by the testis. In their absence feminine sexual differentiation occurs. Thus, sexual differentiation must be actively diverted along the male line, whereas differentiation along the female line reflects an inherent trend requiring no active intervention.

The internal genitalia

Examination of the primordia of the male and female *internal genitalia* (see Figs 1.4 & 1.5) shows that instead of one indifferent but bipotential primordium, as was the case for the gonad, there are two separate sets of primordia, each of which is *unipotential*. These are termed the *Wolffian* or *mesonephric* (male) and *Müllerian* or *paramesonephric* (female) ducts. In the female, the Wolffian ducts regress spontaneously and the Müllerian ducts persist and develop to give rise to the *oviducts, uterus, cervix* and possibly the *upper vagina* (Fig. 1.5). None the less, if a female fetus is *castrated* (its gonads removed), internal genitalia develop in a typical female pattern. This observation demonstrates that ovarian activity is not required for development of the female tract.

In the male, the two testicular hormones prevent this spontaneous development of female genitalia. Thus androgens, secreted in considerable amounts by the testis, actively maintain the Wolffian ducts, which develop into the *epididymis, vas deferens and seminal vesicles*. If androgen secretion by the testes should fail, or be blocked experimentally, then the Wolffian duct system regresses and these organs fail to develop. Conversely, exposure of female fetuses to androgens causes the development of male internal genitalia.

Testicular androgens have no influence on the Müllerian duct system, however, and its regression in males is under the control of the second testicular hormone, MIH. Thus, incubation of MIH *in vitro* with the primitive internal genitalia of female embryos provokes abnormal regression of the Müllerian ducts.

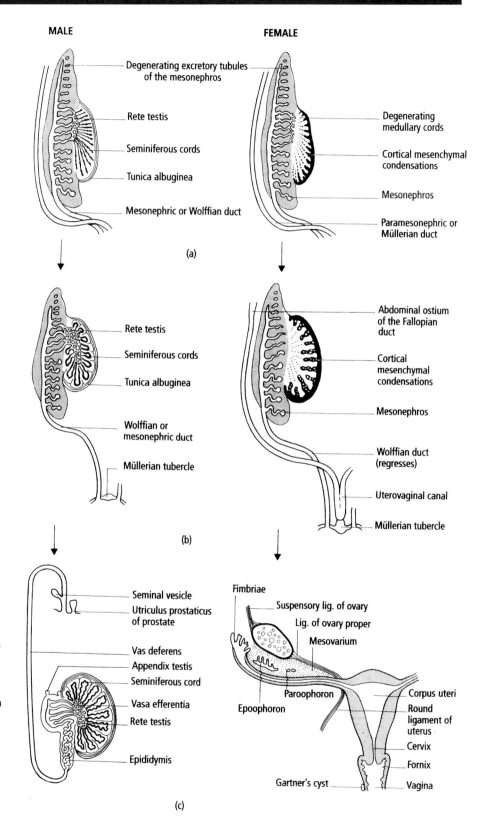

MALE

FEMALE

Degenerating excretory tubules
of the mesonephros

Rete testis

Seminiferous cords

Tunica albuginea

Mesonephric or Wolffian duct

Degenerating
medullary cords

Cortical mesenchymal
condensations

Mesonephros

Paramesonephric or
Müllerian duct

(a)

Rete testis

Seminiferous cords

Tunica albuginea

Wolffian or
mesonephric duct

Müllerian tubercle

Abdominal ostium
of the Fallopian
duct

Cortical
mesenchymal
condensations

Mesonephros

Wolffian duct
(regresses)

Uterovaginal canal

Müllerian tubercle

(b)

Seminal vesicle
Utriculus prostaticus
of prostate

Vas deferens
Appendix testis
Seminiferous cord
Vasa efferentia
Rete testis

Epididymis

Fimbriae
Suspensory lig. of ovary
Lig. of ovary proper
Mesovarium

Paroophoron
Epoophoron

Corpus uteri
Round
ligament of
uterus
Cervix
Fornix

Gartner's cyst   Vagina

(c)

Fig. 1.5 Differentiation of the internal genitalia
in the human male and female at:
(a) week 6; (b) the fourth month; and (c) the
time of descent of testis and ovary. Note the
Müllerian and Wolffian ducts are present in both
sexes early on, the former eventually regressing
in the male and persisting in the female, while
the converse is true of the latter. The appendix
testis and utriculus prostaticus in the male, and
epoophoron, paroophoron and Gartner's cyst in
the female are remnants of the degenerated
Müllerian and Wolffian ducts, respectively.

The external genitalia

The primordia of the external genitalia, unlike those of the internal genitalia, are bipotential (Fig. 1.6). In the female, the *urethral folds* and *genital swellings* remain separate, thus forming the *labia minora* and *majora*, while the *genital tubercle* forms the *clitoris* (Fig. 1.6). If the ovary is removed, these changes still occur, indicating their independence of ovarian endocrine activity. In contrast, androgens secreted from the testes in the male cause: the urethral folds to fuse (so enclosing the *urethral tube* and contributing, together with cells from the genital swelling, to the *shaft of the penis*); the genital swellings to fuse in the midline (so forming the *scrotum*); and the genital tubercle to expand (so forming the *glans penis*) (Fig. 1.6). Exposure of female fetuses to androgens will 'masculinize' their external genitalia, while castration, or suppression of endogenous androgens, in the male results in 'feminized' external genitalia.

Secondary hermaphroditism

Failure of proper endocrine communication between the gonads and the internal and external genital primordia

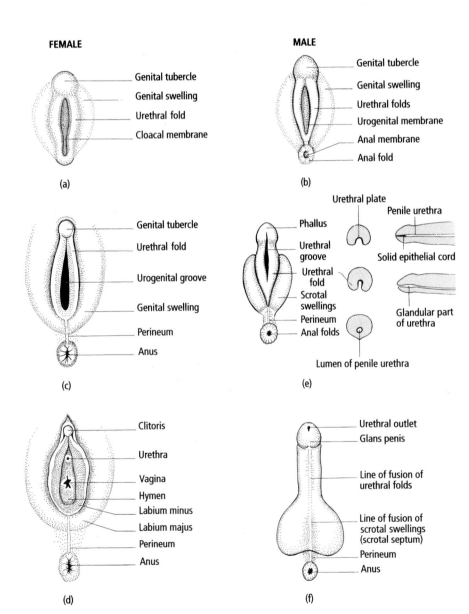

Fig. 1.6 Differentiation of the external genitalia in the human female (left) and male (right) from common primordia shown at: (a) 4 weeks; and (b) 6 weeks. (c) In the female, the labia minora form from the urethral folds and the genital tubercle elongates to form the clitoris. (d) The definitive external genitalia of the female at birth. (e) Subsequent changes by the fifth month are more pronounced in the male, with enlargement of the genital tubercle to form the glans penis and fusion of the urethral folds to enclose the urethral tube and the shaft of the penis (the genital swellings probably contribute cells to the shaft). (f) The definitive external genitalia of the male at birth.

can lead to a dissociation of gonadal and genital sex. Such individuals are called *pseudo* or *secondary hermaphrodites*. For example, in the syndrome of *testicular feminization* (*Tfm*) the genotype is XY (male), and testes develop normally and secrete androgens and MIH.

However, the fetal genitalia are genetically insensitive to the action of androgens (see detailed discussion in Chapter 2), which results in complete regression of the androgen-dependent Wolffian ducts and in the development of female external genitalia. Meanwhile, the MIH

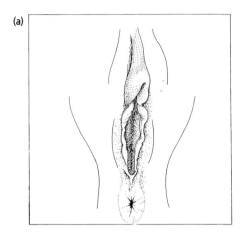

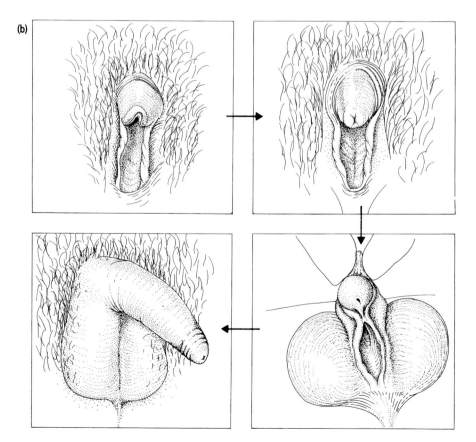

Fig. 1.7 (a) External genitalia of an XY adult with testicular feminization syndrome and a testis. The genitalia are indistinguishable from those of a female, and at birth the child would be classified as a girl (see Chapter 2 for more details). (b) The external genitalia from XX girls with adrenogenital syndrome to varying degrees of severity, from an enlarged clitoris to development of a small penis and (empty) scrotum. Ovaries are present. The adrenal cortex has inappropriately secreted androgens at the expense of glucocorticoids during fetal life and directed development of the genitalia along the male line. Clearly, the more severe cases could lead to sex assignment as a boy, or to indecision.

secreted from the testes exerts its action fully on the Müllerian ducts, which regress. Thus, this genetically male individual bearing testes appears female with labia, clitoris and a vagina, but totally lacks other components of the internal genitalia (Fig. 1.7a).

A naturally occurring counterpart to testicular feminization is the *adrenogenital syndrome* in female fetuses, in which the XX female develops ovaries as usual. However, the fetal adrenal glands are hyperactive and secrete large quantities of steroids, some with strong androgenic activity (see Chapter 2). These stimulate development of the Wolffian ducts, and also cause the external genitalia to develop along the male pattern. The Müllerian system remains, as no MIH has been secreted. Thus, the individual appears male with penis and scrotum, but is, none the less, genetically and gonadally *female* and possesses the internal genitalia of *both* sexes (Fig. 1.7b).

Individuals with *persistant Müllerian duct syndrome* present as genetic males in whom either MIH production, or responsiveness to it, is inadequate. They therefore have testicular androgens that stimulate external genitalia and Wolffian ducts, but retain Müllerian duct structures. These men are thus genetically and gonadally *male* but possess the internal genitalia of *both* sexes.

Apart from the problems of immediate clinical management raised by diagnosis of these syndromes, abnormalities of development of the external genitalia may have important long-term consequences. The single, most important event in the identification of sex of the new-born human is examination of the external genitalia. These may be unambiguously male or female, regardless of whether the genetic and gonadal constitutions correspond. They may also be ambiguous as a result of partial masculinization during fetal life. Sex assignment at birth is one important step that contributes to the development of an individual's *gender identity*, so errors at this early stage can have major consequences for an individual's self perception later in life as a man or a woman. We shall now look at this aspect of gender development in more detail.

THE BRAIN, BEHAVIOURAL DIMORPHISM AND GENDER

Males and females differ from each other during childhood, adolescence and adulthood not only physically, but also behaviourally. How does *sexually dimorphic behaviour* develop? Two major influences have been proposed, and the balance of their relative importance depends on the species. It is clear that exposure to sex steroids during critical periods of early life is a major determinant of the development of sexually dimorphic behaviour patterns, just as it determines the development of sexually dimorphic genitalia. However, there is also evidence in primates for an influence of social interactions during childhood and adolescence. We will consider the evidence for both primates and non-primates that bears on the relative significance of the *endocrine* and *social determinants* of *behavioural dimorphism*.

Non-primates

In non-primates, exposure to sex hormones, particularly androgens and their metabolites, during critical periods of early life is specifically associated with the subsequent display of sexually dimorphic behaviour in adulthood. This critical period occurs either in late fetal life (e.g. guinea-pig, sheep) or neonatally (e.g. rat, mouse, hamster). Sexually dimorphic behaviour displayed in adulthood includes, for example, the distinctive urination patterns shown by the dog (cocked leg) and bitch (squatting). However, the most intensively studied behaviour patterns are those associated with copulation. Thus, in adult males, *courtship behaviour* (e.g. *following* and *anogenital investigation*), *mounting*, *intromission* and *ejaculation* may be evident during sexual interaction with females (Fig. 1.8). Conversely, adult females display equally distinctive patterns of courtship behaviour (e.g. *soliciting*) and *receptive postures* (e.g. *lordosis*) (Fig. 1.8). These behaviours are *predominantly*, *but not exclusively*, typical for each sex. Thus, normal males will occasion-

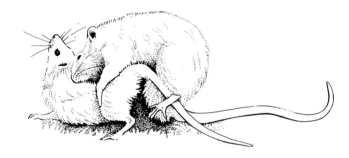

Fig. 1.8 Sexually dimorphic patterns of behaviour in male and female rats. Note the immobile lordosis posture shown by the receptive female which enables the male to mount and achieve intromissions which will result in ejaculation. Lordosis is predominantly shown by females, while mounting, intromission and ejaculation patterns of behaviour are predominantly shown by males.

ally solicit and even accept mounts by other males, i.e. show patterns of behaviour usually associated with females, while females in heat will often mount one another. *The differences in behaviour are not absolute but quantitative.*

Treatment of female rats with testosterone during the first 5 days of life enhances their capacity to display masculine patterns of sexual behaviour in adulthood and reduces their capacity to display feminine patterns. Conversely, castration of male rats to remove the influence of androgens during this same critical period results in an enhanced capacity to display feminine patterns of sexual behaviour in adulthood while reducing the capacity for masculine patterns of behaviour. Thus, 'masculinization' in the rat (and many other non-primate mammals) is accompanied by 'defeminization'.

It is clear that the impact of steroids on the development of sex differences in behaviour is reflected by their actions on the developing brain of male and female animals during these critical periods of late fetal or early neonatal life. Indeed, implantation of androgens directly into a region of the brain called the *anterior hypothalamus* (see Chapter 5 for more details) results in adult behaviour patterns that are both defeminized and reciprocally masculinized.

It is now well established that there are many *sex differences in the structure of the brain*, some of which may be closely associated with the altered propensity to show masculine or feminine patterns of behaviour in adulthood. To what extent are these differences in brain structure caused by the same perinatal hormonal events that affect the development of sex differences in behaviour? Most attention has focused on the *anterior hypothalamus* and *adjacent medial preoptic area* (anatomy discussed in more detail in Chapter 5), as both are intimately involved with the control of sexual behaviour (see Chapter 7). Thus, the medial preoptic area is sexually dimorphic, and experiments in several species have made it clear that development of a male-like medial preoptic area is the direct result of neonatal exposure to testosterone or its metabolites (Fig. 1.9), just as it is for the development of sexually dimorphic external genitalia. Thus, neonatal treatment of female rodents with testosterone not only masculinizes their adult reproductive behaviour, but also results in a male-like medial preoptic area (Fig. 1.9). However, it would be wrong to imply that each of the sub-regions of this and other sexually dimorphic areas of the brain is reliably associated with specific reproductive behavioural functions. As we will see later, such assumptions have led prematurely to assertions of such structure–function relationships in humans.

Non-human primates

To what extent do androgens exert the same effect on the development of sexually dimorphic behaviour in primates? Results from experiments on rhesus monkeys suggest some similarities, as young females exposed to high levels of androgens during fetal life display levels of sexually dimorphic behaviour in childhood patterns of play that are intermediate between normal males and females (Fig. 1.10). Moreover, although both male *and* female infant monkeys will mount other infants (Fig. 1.11a), only males progressively display mounts of a mature pattern (Fig. 1.11b). Androgenized females, however, do develop this mounting pattern. Moreover, as adults they attempt to mount other females at a higher frequency than do non-androgenized monkeys. Thus, neonatal androgenization does produce persistent 'masculinization' of behaviour. However, the androgenized female monkeys, unlike androgenized female rats, may as adults show normal menstrual cycles and become pregnant. Presumably, therefore, they display patterns of adult feminine sexual behaviour at least adequate for them to interact successfully with males, suggesting that they are not totally or permanently 'defeminized'.

These results suggest a less complete or persistent effect of androgens on the development of sexually dimorphic behaviour in primates than in non-primates. Why might this be? One explanation lies in the timing of the critical androgen-sensitive period. In rats, this is neonatal, after genital phenotype is established, so making it easily accessible to manipulation. If a critical period exists in primates, it occurs during fetal life, and may be prolonged. Attempts to androgenize primate fetuses *in utero* often lead to abortion if doses of administered androgens are too high and the genitalia also tend to be masculinized, which might affect the subsequent social interactions and learning of the infant. Thus, a clear and specific effect of a high dose of androgen on the brain may not yet have been achieved. Alternatively, it is possible that in non-human primates, a hormonal 'fixing' of sexually dimorphic behaviour simply does not occur. Androgens may predispose to masculine patterns of behaviour, but social learning may also influence the degree to which they are expressed.

Humans

Not surprisingly, the difficulty in studying primates is exacerbated further when the human is considered.

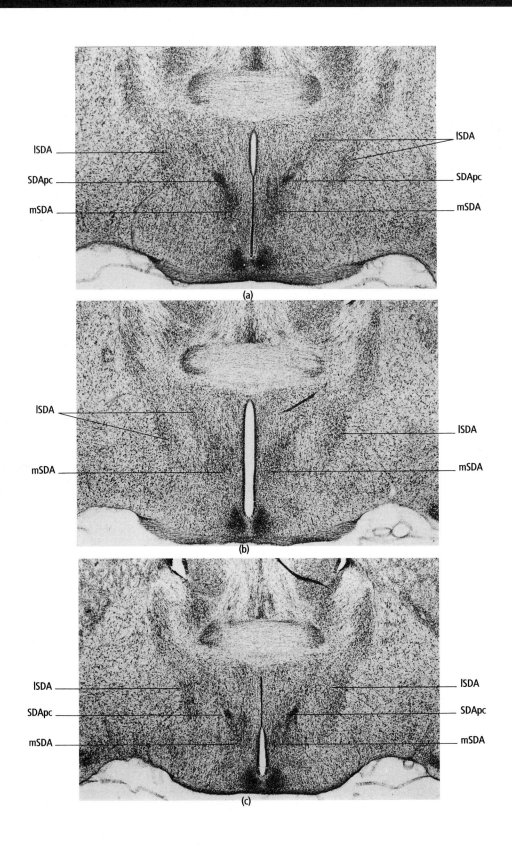

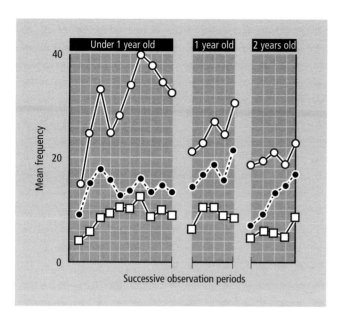

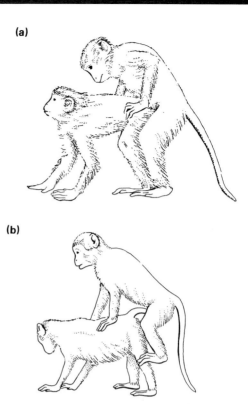

Fig. 1.10 Frequency of 'rough-and-tumble play' during the first, second and third years of life of a rhesus monkey male (○), female (□), and pseudohermaphroditic female (●) that had been treated with testosterone prenatally. Note that males display this behaviour at a higher frequency than females and that pseudohermaphroditic females are intermediate.

Fig. 1.11 Sexually dimorphic patterns of mounting behaviour in young rhesus monkeys. Early in life, both males and females show immature mounts (a) by standing on the cage floor. During development, males show progressively more mature mounts in which (b) they clasp the female's calves so that she supports his weight entirely. Pseudohermaphroditic (androgen treated) females display more of the latter type of mature mounts than do untreated females.

Sexually dimorphic behaviour patterns that might be considered analogous to those studied in monkeys can be observed and quantified in humans. For example: (a) energy expenditure during play is higher for boys than girls; (b) parental rehearsal patterns in children, such as doll playing and fantasies about adulthood, differ; (c) attention to personal appearance such as clothes, hair and body decoration differ in ways that may be analogous to grooming differences in monkeys; and (d) explicitly sexual behaviour patterns also differ. Humans may also be studied uniquely by interview or questionnaire techniques in order to determine attitudes and assump-

tions about *gender identity* and *role* (Table 1.2). It is important to stress again that by all these criteria, the sexes differ quantitatively and not exclusively, and that the criteria may change in value over time or with different cultures. For example, the gender specificity of the wearing of trousers by females or of ear-rings or 'skirts' by men has varied historically in Western culture and does so now between different cultures. Moreover, even within a particular culture at a particular time, some individuals will not conform to all the criteria associated with their gender. For example, a genetically (XY) and gonadally male individual who may be, and consider

Fig. 1.9 (*opposite*) Photomicrographs of coronal sections through the preoptic area of three, 21-day-old gerbils (*Meriones unguiculatus*). Section A is taken from a male, B from a female, and C from a female treated neonatally with testosterone propionate (50 μg on the day of birth and 50 μg the next day). The sexually dimorphic area (SDA) can be divided into several regions: medial (mSDA), lateral (lSDA) and pars compacta (SDApc). The SDA differs between males and females in a number of aspects: prominence (not necessarily size); acetylcholinesterase histochemistry; steroid binding; and various other neurochemical characteristics, but most obviously in the presence or absence of the SDApc. Thus, the SDApc is virtually never found in

females (compare A with B). Note that in females treated neonatally with testosterone, there is a clear SDApc (compare C with B). These pictures provide clear evidence of the impact of hormones during a critical period of early life on the differentiation of this part of the brain. The medial preoptic area in general is closely involved with the regulation of sexual behaviour (see Chapter 7), and some progress has been made in relating specific aspects of sexual behaviour to subdivisions of the SDA. It is also important to note that such sex differences in the structure of the preoptic area are found in many species, from rats to humans, but the precise details of the dimorphism varies considerably.

Table 1.2 Definition of some elements of human sexuality and behaviour

Gender identity	The self-awareness of one's identity or self-image as male, female, or ambivalent. It is a personal experience, but can be expressed socially as gender role.
Gender role	Everything an individual says and does to convey (consciously or subconsciously) to other people his/her gender identity as male, female or ambivalent. It may be taken socially to include the gender of sexual arousal, preference and response but need not be, and is not restricted to them. Thus, many individuals who show a sexual preference for the same sex may none the less have a strong and clear gender identity.
Sexual preference	The preference shown in a sexual context by an individual for members of the same or the opposite sex. A heterosexual woman shows sexual preference for a man. A homosexual man also shows partner preference for a man. A bisexual person has no one preference.

himself to be, typically male by the above criteria, nevertheless may have as the object of his sexually orientated fantasy and/or activity another man (*homosexual*). Or an XY individual with testes and male phenotype may none the less feel that he is female (*male to female transsexual*) and even request surgical and endocrine reassignment of *his* physical appearance to match *her* psychological state. Such transsexuals may have as the object of their sexually oriented activity either men or women. In the face of such complexity, it is both naïve and quite unjustifiable to extrapolate, as some have done, from the demasculinized, genetically male rat that readily shows feminine sexual behaviour patterns to the homosexual or transsexual man in well-meaning, but misguided, attempts to provide a biological 'explanation' of such variations in human sexual behaviour.

However, the influence of prenatal hormones on subsequent behaviour has been investigated in humans. The two clinical conditions we encountered in the section on genital development have proved particularly useful. Genetic females with the adrenogenital syndrome (AGS) are nature's counterpart to experimental animals treated exogenously with androgens during the critical period of neural differentiation. Two sub-groups of these subjects have been studied: those exposed to continuing high levels of androgens after birth, so-called 'late-treated' (this group dates back to the 1950s before the simple and effective therapy for this condition was discovered); and 'early-treated' subjects, in whom the hypersecretion of androgens was controlled by cortisol therapy soon after birth. The converse of this syndrome is testicular feminization (Tfm), which occurs in genetic males who are unable to respond to their own androgens, a counterpart of experimental castration of neonatal males.

Studies of girls with AGS suggest *increased* levels of energy expenditure and athletic interests more characteristic of boys, and a *decreased* incidence of 'rehearsals'

of maternal behaviour and doll-play activities, together with diminished interest in dresses, jewellery and hairstyles. This spectrum of behaviour, termed '*tomboyism*', is well recognized and accepted in Western culture, and provides few if any problems for children so affected. This tomboyism seems likely to represent, directly or indirectly, a consequence of the effects of androgens on the fetal brain, rather like the changes in rough-and-tumble play in infant monkeys exposed prenatally to androgens. However, further study of these human subjects as adults, revealed little evidence of enduring behavioural consequences of the abnormal fetal endocrine events. Thus, they were reported to have boyfriends (although they began dating a little later than controls), some were married, had children and proved to be unexceptional mothers. They apparently showed no evidence of a higher incidence of dissatisfaction with their female gender identity nor of homosexuality than did controls. Thus, although the early androgen exposure did affect childhood play, gender identity in the adult was not affected to any great extent. However, these results were only observed if one important condition was fulfilled, namely that the affected individuals were *assigned as girls at birth and reared unambiguously as such*. If the 'masculinization' of their external genitalia was severe enough for them to have been assigned and reared as boys, then their adult gender identity was reported to be masculine; they were attracted to and dated girls and had no desire for sex reassignment. A 'masculinizing' puberty can be achieved in such individuals only by giving exogenous androgens after removal of their ovaries. Surgical reconstruction of masculine external genitalia is possible, but difficult.

Evidence from subjects with the syndrome of complete or partial testicular feminization leads to similar conclusions. In the complete form of Tfm, in which external genitalia were unambiguously female at birth, leading to rearing as girls, subjects displayed a clearly feminine gender identity, and were attracted to, dated and married men. Conversely, Tfm subjects in which the syndrome was incomplete, and who had only partially masculinized external genitalia such that they were reared as boys, developed a masculine gender identity. They faced a major problem at puberty as the androgen insensitivity, which characterizes the syndrome, ensures that a masculinizing puberty is impossible; indeed, these subjects develop breasts and female body form under the action of their own oestrogens.

These clinical examples have been taken as evidence that sex of rearing is the most important single event determining subsequent differentiation of gender identity, as it establishes the framework for sexually specified social interactions during childhood and adolescence. Clinical psychologists and psychiatrists generally believe that this process of differentiation of masculine or feminine gender identity and sexual orientation is probably completed by about 2–3 years of age, after which it is an irrevocable part of a boy's or girl's personality. Ambiguity on the part of parents about the sex of their child may affect the subsequent development of gender identity and sexuality. Thus, it has been suggested, but not universally agreed, that cases of transsexualism are associated with a sexually ambiguous childhood; for example, parents treating their son more as a girl, clothing him in dresses and not reinforcing 'boyish' activities in play and sport. If this is indeed the case, much depends on the genitalia at birth!

However, there are two lines of evidence against a sole or major role for social factors in determining gender and sexuality. The first has come from the recognition of a syndrome first identified among a sub-group of the population of the Dominican Republic, but now identified in Europe also. These individuals are genetic males with normal functional testes. However, they are born with external genitalia sufficiently diminutive and ambiguous as to be classifiable as 'female', and consequently are reared as girls. At puberty, the 'clitoris' enlarges rapidly to form a penis (as a result of which they are called 'Guevodoces' or 'penis-at-twelves'), the 'labia' enlarge to receive the testes, the body hair pattern and phenotype assume male characteristics, and their psychosexual identity is reported to switch from feminine to masculine, such that they function behaviourally as men. The syndrome is caused by the presence of an autosomal recessive gene that results in the lack of a specific enzyme, 5α-reductase (see Chapter 2 for details). This enzyme normally converts the weak androgens secreted by the testes to a very potent form that is more active in the tissues of the external genitalia. Thus, although the individuals have only weak androgens

during development, at puberty the output of testicular androgens rises dramatically to overcome the lack of enzyme and thereby to complete the full process of masculinization. However, as they are known as Guevodoces, it is clear that these individuals are not reared unambiguously as females. Thus, it is difficult to determine the degree to which their adult masculine sexual behaviour, gender identity and role reflects the prevailing effect of having brains that were exposed to androgens *in utero*, or whether their early life had taught them to expect the rather dramatic change from Guevodoces to men. However, there are several reports of such individuals in the Western world, who, although treated during early childhood unambiguously as girls, assumed a masculine gender identity at or after puberty, suggesting that exposure to prenatal hormones might be an important determinant of adult sexual behaviour, as in non-primate species.

The second line of evidence comes from recent studies on the structure of parts of the anterior hypothalamus which are, homologous to those areas that in rats are influenced by neonatal testosterone. These studies have suggested size differences between the sexes and ambiguity in the brains of self-declaring homosexual men. However, the studies are small and there is a great deal of overlap of the size values between both men and women and between gay men and either of these groups. The validity of these much publicized claims needs to be established carefully, and they certainly do not form the basis of useful predictors of either gender or sexual orientation. There are no experimental data to suggest a direct relationship between this area of the brain and something as complex as homosexual orientation and behaviour. Indeed, studies in animals (see Chapter 7) suggest this is very unlikely, as the preoptic area appears to be much more concerned with copulatory reflexes than with the expression of partner preference.

It will be evident from the foregoing discussion that the postulated social determination of sexual behaviour can only be a tentative conclusion. Clearly, only the less severely androgenized females with AGS will be brought up as girls, and only the more severely as boys. The converse applies with Tfm syndrome. The Guevodoces are clearly definable as distinct from girls even when they are immature (hence their special name) and thus perhaps are brought up more as 'not typical but perhaps special boys' rather than truly as 'girls'. Thus, a clear separation of endocrine and social factors has not been achieved. It might be suggested that such a rigid separation is also unlikely, endocrine and social reinforcement being the norm.

Summary

It is perhaps not surprising that the relatively simple rules governing the development of sex differences in the behaviour of rodents cannot easily be applied to primates and humans. The finding that exposure to androgens during a critical period of early life both alters the structure of the brain and is also reflected subsequently in the masculinization, with attendant defeminization, of patterns of sexual behaviour in adulthood, does not find a simple counterpart in monkeys or humans. In both the latter species, behavioural evidence of the effects of exposure of the fetal brain to androgens is abundant in the form of sexually dimorphic childhood behaviour, but affected individuals can display apparently typical patterns of feminine sexual behaviour as adults. It remains to be determined whether and how sex assignment at birth and subsequent gender-specific patterns of social behaviour interact with endocrine factors during fetal life to influence behavioural dimorphism, gender identity and sexual orientation.

PRE- AND POSTNATAL GROWTH OF THE GONADS

We have seen how phenotypic and behavioural features of the male and female develop. The male arises by active genetic and then endocrine activities, the female develops passively in the absence of these activities. These same rules operate during the remainder of prenatal life and during postnatal life up to puberty. Over this period, further sexual divergence of physical phenotypes occurs only at a very slow pace. In the male, this process is dependent upon low levels of gonadal androgens. The female grows and retains a feminine phenotype in the absence of such androgens or indeed of any gonadal steroids.

The testis

The genital ridge develops in the upper lumbar region of the embryo, yet in the adults of most species of mammals the testes have *descended* through the abdominal cavity and over the pelvic brim to arrive in the scrotum (Fig. 1.12). Evidence of this extraordinary migration is found in the nerve and blood supplies to the testis, which retain their lumbar origins and pass on an extended course through the abdomen to reach their target organ. Testicular migration may, in certain pathological conditions, be arrested at some point on the migratory route resulting in *cryptorchidism* (hidden gonad). Examples of 'evolutionary cryptorchidism' are provided by a few species in which testes: normally do not descend at all from the lumbar site (e.g. monotremes, elephants and hyraxes); migrate only part of the route to the rear of the lower abdomen (e.g. armadillos, whales and dolphins); lodge in the inguinal canal (e.g. hedgehogs, moles and some seals); retain mobility in the adult, migrating in and out of the scrotum to and from inguinal or abdominal retreats (e.g most rodents, wild ungulates).

The transabdominal descent of the testis towards the inguinal canal occurs independently of androgens but a role for MIH has been suggested. Failure of the testes to descend is associated with lack of MIH. The MIH may act on a fibrous structure called the *gubernaculum*, which attaches the testis to the posterior abdominal wall (Fig. 1.12). As the body increases in size, the gubernaculum does not elongate in the male (but does in the female), and thus in the male the relative position of the testis becomes increasingly caudal. Cryptorchid testes are frequently associated clinically with residual Müllerian duct structures. The passage of the testis through the inguinal canal into the scrotum is probably controlled by the action of androgens.

The consequences of cryptorchidism in those mammals with scrotal testes demonstrate that a scrotal position is essential for normal testicular function. Although endocrine activity is not affected in any major way, spermatogenesis is arrested, testicular metabolism is abnormal and the risk of testicular tumours increases. These effects can be simulated by prolonged warming of the scrotal testis, either experimentally or by wearing thick tight underwear. The normal scrotal testis functions best at temperatures 4–7°C lower than abdominal 'core' temperature. Cooling of the testis is improved by *copious sweat glands in the scrotal skin* and by the circulatory arrangements in the scrotum. The *internal spermatic arterial* supply is coiled, or even forms a rete in marsupials, and passes through the *spermatic cord* in close association with the draining venous *pampiniform plexus*, which carries peripherally cooled venous blood (Fig. 1.13). Therefore, a heat exchange is possible, cooling the arterial and warming the venous blood. However, although the scrotal testis clearly requires a lower ambient temperature, this requirement may be a secondary *consequence* of its scrotal position rather than the original evolutionary *cause* of its migration. Those species in which testes remain in the abdomen survive, flourish and reproduce despite the high testicular temperature. It remains unclear as to why testes are scrotal in so many species.

By the sixteenth to twentieth week of human fetal life, the testis consists of an outer *fibrous tunica albuginea* enclosing vascularized stromal tissue, which contains

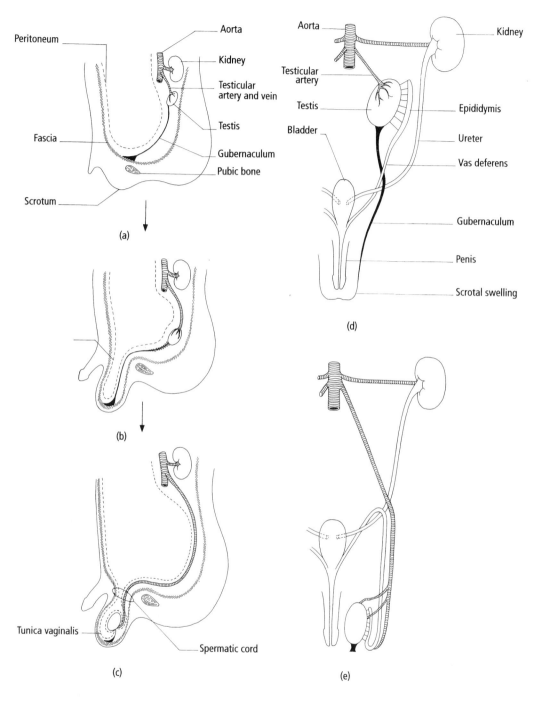

Fig. 1.12 (a)–(c) Parasagittal sections through a developing male abdomen. The initial retroperitoneal, abdominal position of the testis shifts pelvically between 10 and 15 weeks, extending the blood supply (and Wolffian duct derivatives, not shown) as the gubernaculum becomes shorter. A musculo-fascial layer evaginates into scrotal swelling accompanied by peritoneal mem-brane, which forms the processus vaginalis. Between 25 and 28 weeks of pregnancy in the human, the testis migrates over pubic bone behind the processus vaginalis (which wraps around it forming a double-layered sac), reaching the scrotum by 35–40 weeks. The fascia and peritoneum become closely apposed above the testis, obliterating the peritoneal cavity leaving only a tunica vaginalis around the testis below. The fascial layers, obliterated stem of processus vaginalis, vas deferens and testicular vessels and nerves form the spermatic cord. In some species, and pathologically in man, the closure may not be complete and the testis may be able to move back up into a pelvic location. (d) and (e) Front view of the migration, showing the extended course ultimately taken by the testicular vessels and vas deferens.

condensed Leydig cells and solid seminiferous cords comprised of a basement membrane, Sertoli cells and prospermatogonial germ cells. These cords connect to the cords of the *rete testis, the vasa efferentia* and thereby to the *epididymis* (Fig. 1.13). The Leydig cells in the human testis actively secrete testosterone from at least 8–10 weeks of fetal life onwards, with blood levels peaking at 2 ng/ml at around 13–15 weeks. Thereafter, blood levels decline and plateau by 5–6 months at a level of 0.8 ng/ml (compared with the adult male level of *c.* 9 ng/ml). This transient *prenatal peak* of blood testosterone is a feature of many species, although in some, for example the rat and sheep, the peak may approach, or span, the period of parturition, and only begin its decline postnatally. This testosterone secretion is, as we

saw, essential for establishing the male phenotype. In addition, the males of some primate species, including man, show a second *postnatal peak* in plasma testosterone, concentrations reaching 2–3 ng/ml by 2 months postpartum, but declining to around 0.5 ng/ml by 3–4 months. The Sertoli cells continue to produce MIH throughout fetal life up until puberty when levels drop sharply.

Throughout pregnancy and early postnatal growth, testis size tends to increase slowly but steadily, but the germ cells do not contribute to this by proliferating, and indeed, in some species, many of them die. At puberty, however, there is a sudden increase in testicular size: endocrine secretion by the Leydig cells increases sharply; the solid seminiferous cords canalize to give rise to

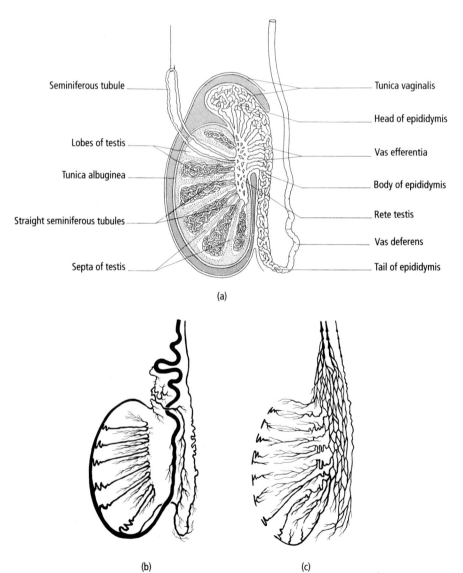

Seminiferous tubule — — — Tunica vaginalis

Head of epididymis

Lobes of testis — — Vas efferentia

Tunica albuginea — Body of epididymis

Rete testis

Straight seminiferous tubules — Vas deferens

Septa of testis — Tail of epididymis

(a)

(b)   (c)

Fig. 1.13 Section through adult human testis to show: (a) general structure; (b) arterial supply; and (c) venous drainage.

tubules; the Sertoli cells increase in size and activity; and the germ cells resume mitotic activity. The causes of this sudden growth at puberty will be discussed in detail in Chapter 6. The consequences are the subject of much of the remainder of this chapter.

The ovary

The ovary, unlike the testis, retains its position within the abdominal cavity, shifting slightly in some species, such as the human, to assume a pelvic location. It is attached to the posterior abdominal wall by the *ovarian mesentery or mesovarium*. The ovary, like the testis, grows slowly but steadily in size during early life. Over this period its output of steroids is minimal and, indeed, removal of the ovary does not affect prepubertal development. However, at puberty marked changes in both the structure and endocrine activity of the ovary occur, and for the first time the ovary becomes an essential and positive feminizing influence on the developing individual.

PUBERTY

Puberty is best regarded as a collective term that encompasses all the physiological, morphological and behavioural changes that occur in the growing individual as the gonads change from an infantile to an adult condition. Puberty has been studied in most detail in the human. A fairly definitive sign of its occurrence in girls is *menarche*, the first *menstrual bleeding*. An indication of a similar stage of maturity in boys is the first *ejaculation*, which often occurs nocturnally and is thus much more difficult to date precisely. The first menstruation and ejaculation do not signify fertility. Indeed, in early puberty the ovary does not ovulate and the ejaculate consists of small quantities of seminal plasma lacking spermatozoa. Rather, these two dramatic events are signs that the gonads have been awakened and are beginning to assume adult levels of activity.

Some 2–4 years prior to these obvious signs of sexual maturation, a series of other changes in most of the organs and in the structure of the body is initiated. These changes are dependent on, and orchestrated by, the increasing level of sex steroids from the gonads and also from the adrenal glands. This sequence of maturational changes does not begin at the same chronological age or take the same length of time to reach completion in all children, even though the sequence in which these changes occur varies but little.

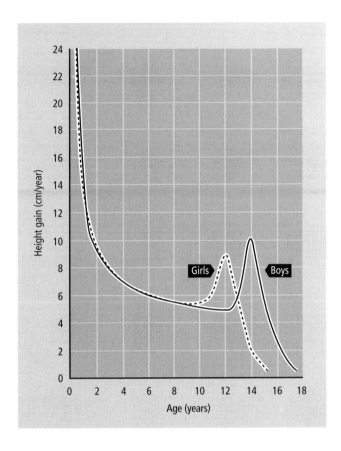

Fig. 1.14 Growth velocity curves for boys and girls. Note the later time of 'take-off' in boys, which generally ensures a greater height at the start of the adolescent growth spurt. Also note that average peak height velocity (PHV) is 9 cm/year for girls and 10 cm/year for boys.

Physical changes

The *adolescent growth spurt* is an acceleration, followed by a deceleration, of growth in most skeletal dimensions, and can be divided into three stages: (a) the *time of minimum growth velocity* (or 'age at take-off'); (b) the time of *peak height velocity* (PHV); and (c) the time of decreased growth velocity and cessation of growth at *epiphyseal fusion*. Figure 1.14 shows that boys commence their growth spurt about 2 years later than girls. They are therefore taller at the age of take-off and reach their PHV 2 years later. The height gain of boys and girls between take-off and cessation of growth is similar, about 28 cm and 25 cm respectively, indicating that the 10 cm difference in mean height between adult males and adult females is due more to the height difference at take-off than gain during the spurt. Age at take-off and

PHV are poor indicators of adult height, and, in addition, they show a poor correlation with the rate of passage through the various stages of puberty described below. Virtually every muscular and skeletal dimension is involved in the adolescent growth spurt; however, sex differences in the growth rate of different regions occur, which enhance sexual dimorphism in the adult, in, for example, the shoulders (greater in boys) and hips (greater in girls). This dynamic phase of growth is dependent not only on sex steroids but also on *growth hormone* from the *anterior pituitary*. Thus, patients with poorly functional pituitaries (*hypopituitarism*) must be given both growth hormone and steroids if a pubertal growth spurt is to occur.

As well as growth, considerable changes in body composition occur during puberty. Lean body mass and body fat are virtually identical in prepubertal girls and boys, but in adulthood, men have about 1.5 times the lean body mass of women, while women have twice as much body fat as men. In addition, the skeletal mass of men is 1.5 times that of women. These alterations in body mass commence at about 6 and 9 years of age in girls and boys respectively, and represent the earliest changes in body composition at puberty. The greater average strength of men compared to women reflects a greater number of larger muscle cells.

Secondary sexual characteristics

In addition to growth and changes in body composition, development of the *secondary sexual characteristics* occurs at puberty, for example breasts (Fig. 1.15), genitalia and pubic hair (Figs 1.16 & 1.17), and beard growth and voice change. Ovarian oestrogens regulate growth of the breast and female genitalia, but *androgens* from both the ovary and the adrenal gland (see Chapter 6) control the growth of female pubic and axillary hair. Testicular androgens not only control development of the genitalia and body hair in boys, but also, by enlarging the *larynx* and *laryngeal muscles*, lead to *deepening* of the voice.

These various characteristics develop at very different chronological ages in different individuals (see for example Fig. 1.18). However, the *sequence* in which the changes occur is quite characteristic for each sex. This is important for the clinician, who has *staging criteria* (summarized in Figs 1.15, 1.16 & 1.17) by which abnormalities can be detected and comparisons made between individuals, populations and cultures. For example, a boy with advanced penile and pubic hair growth but small testes must have a non-gonadal source of excess androgen,

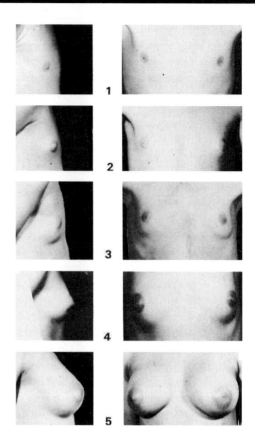

Fig. 1.15 Stages of breast development. 1: Pre-adolescent stage during which the papilla (nipple) alone is elevated. 2: Breast bud stage in which the papilla and breast are elevated as a small mound. 3: Continued enlargement of the breast and areola, but without separation in their contours. 4: Further breast enlargement but with the papilla and areola projecting above the breast contour. 5: Mature stage in which the areola has become recessed, and forms a smooth contour with the rest of the breast. Only the papilla is elevated. Classification of this stage is independent of breast size, which is determined principally by genetic and nutritional factors.

such as congenital adrenal hyperplasia or an adrenal tumour.

Puberty in other mammals

All mammals undergo pubertal changes equivalent to those described in detail for the human. However, the details differ as do the timings. For instance, whereas in the human female the first evidence of fertility occurs between 12 and 14 years postnatally, in most other species it occurs much earlier, for example in the ewe (6–7 months), cow (12 months), pig (7 months), and mouse (30–35 days). Prepubertal development is prolonged in higher primates.

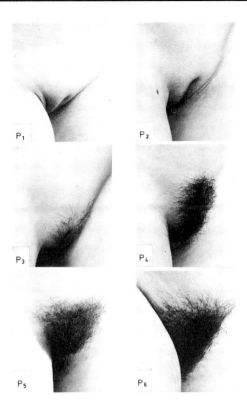

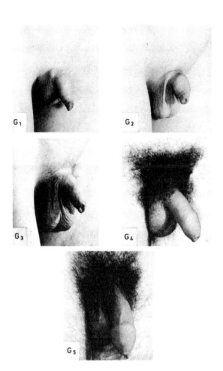

Fig. 1.16 Stages of pubic hair development in girls. 1: No pubic hair is visible. 2: Sparse growth of long, downy hair which is only slightly curled and situated primarily along the labia. 3: Appearance of coarser, curlier and often darker hair. 4: Hair spreads to cover labia. 5: Hair spreads more over the junction of the pubes and is now adult in type but not quantity. No spread to the medial surface of the thighs. 6: Adult stage in which the classical 'inverse triangle' of pubic hair distribution is seen, with additional spread to the medial surface of the thighs.

Fig. 1.17 Stages of external genitalia development in boys. 1: Pre-adolescent stage during which penis, testes and scrotum are of similar size and proportion as in early childhood. 2: Scrotum and testes have enlarged; texture of the scrotal skin has also changed and become slightly reddened. 3: Testes and scrotum have grown further but now the penis has increased in size: first in length and then in breadth; facial hair appears for the first time on upper lip and cheeks. 4: Further enlargement of the testes, scrotum (which has darkened in colour) and penis; the glans penis has now begun to develop. 5: Adult stage; facial hair has now extended to lower lip and chin.

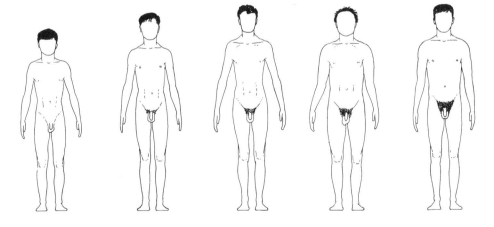

Fig. 1.18 Five boys all aged 14, illustrating the marked differences in maturation at the same chronological age.

SUMMARY

In this chapter we have seen that sexual differentiation is an enduring process of divergence, which begins with the expression of a genetic message that establishes the structure and nature of the fetal gonad, and then extends from the gonad, via its hormonal secretions and over a period of several years, to many tissues of the body. Thus, sex may be defined at several levels and by several parameters. Concordance at all levels is often incomplete, and the medical, social and legal consequences of this 'blurring' of a clear, discrete sexual boundary may pose problems. However, in this chapter we have been able to define, by a broad set of criteria, how the two sexes are established and attain sexual maturity. In subsequent chapters, we will look at the physiological processes regulating fertility and sexual behaviour that result in conception, pregnancy, parturition, lactation and maternal care.

FURTHER READING

Austin CR & Edwards RG (1981) *Mechanisms of Sex Differentiation in Animals and Man.* Academic Press, London.

Bancroft J (1989) *Human Sexuality and its Problems.* Churchill Livingstone, London.

Byskov AG (1986) Differentiation of mammalian embryonic gonad. *Physiological Reviews* **66**, 71–117.

Green R (1987) *The 'Sissy Boy Syndrome' and the Development of Homosexuality.* Yale University Press, New Haven.

Hughes IA (1983) Precocious puberty and its management. *British Medical Journal* **66**, 664–665.

Josso N (1994) Anti-Müllerian hormone: a masculinizing relative of TGF-β. *Oxford Reviews of Reproductive Biology* **16**, 139–164.

Koopman P *et al.* (1991) Male development of chromosomally female mice transgenic for *Sry. Nature* **351**, 117–121.

McLaren A (1991) Development of the mammalian gonad: the fate of the supporting cell lineage. *BioEssays* **13**, 151–156.

Money J (1972) *Man and Woman, Boy and Girl.* Johns Hopkins University Press, Baltimore.

Tanner JM (1966) *Growth at Adolescence.* Blackwell Scientific Publications, Oxford.

Yahr P (1988) Sexual differentiation of behavior in the context of developmental psychobiology. In: *Handbook of Behavioral Neurobiology* (Ed. Blass EM) **9**, 197–243. Plenum Press, New York.

CHAPTER 2

Reproductive Messengers

Two major conclusions were drawn in Chapter 1. First, the gonads are pivotal organs in the reproductive process, translating genetic sex into phenotypic and behavioural sex. Second, this translation is mediated by chemical messengers (steroids and the Müllerian inhibiting hormone (MIH)), which have pervasive effects on reproductive maturation and function. These messengers, which will be referred to by the generic name *hormones*, form only part of a complex network of communication within the reproductive system, some of the main routes of which are shown in Fig. 2.1. Each of these routes will be considered in more detail in later chapters. In this chapter, the main hormones will be introduced and some general principles underlying their activities examined.

PATTERNS OF SECRETION

Coordination of function within the body requires effective communication over a range of distances. A variety of hormonal secretion and transport patterns have developed to meet these needs. For example, we saw that the steroid hormones produced by the gonads influence the structure and function of the brain, the external genitalia and breasts, and the distribution of body hair. To achieve these pervasive effects, the steroids are secreted into the bloodstream, in which they are distributed rapidly throughout the tissues of the body. This secretory process is called *endocrine*, meaning literally 'secreted inwards'. However, not all steroid actions are achieved in this way. For example, steroids within the testis also pass into the fluids of the seminiferous tubules, in which they are carried into the epididymis and vas deferens (see Chapters 3 & 8). Similarly, some of the ovarian hormones pass into and along the oviduct with the newly ovulated egg and its secretions (see Chapters 4 & 8). In both cases, these hormones exert more localized actions within the duct systems themselves. This sort of secretion is termed *exocrine*, meaning 'secreted outwards'.

Hormones can also act very close to their site of secretion on adjacent cells and tissues. For example, both MIH and testosterone influence the differentiation of the internal genital primordia by local diffusion from the fetal testis. In this case, the secretion is said to be *paracrine*, or

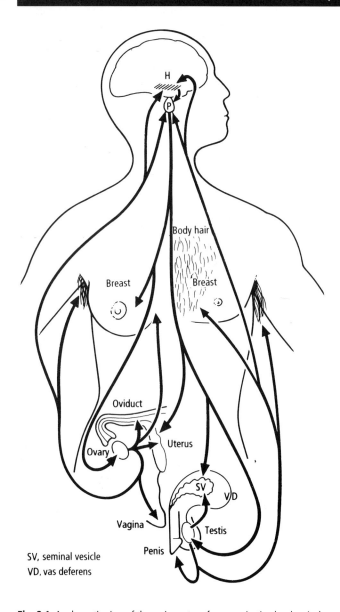

Fig. 2.1 A schematic view of the main routes of communication by chemical messengers within the reproductive system. The hypothalamic region (H) of the brain influences (→) the pituitary gland (P), which in turn influences the breasts (Chapter 13), the uterus, cervix and seminal vesicle (Chapters 7 & 12) and the gonads: ovary and testis (Chapters 3 & 4). The gonads themselves exert influences back on the pituitary and the brain (see Chapters 5 & 6 for details). The gonads also influence multiple sites in the internal and external genitalia (Chapters 7–13), and at least some of these sites send messages back to the gonads directly or indirectly. Finally, in a number of tissues or organs messages are passed around internally between different cell types or even fed back on the very cells that produced them (e.g. see Chapter 4).

SV, seminal vesicle
VD, vas deferens

acting 'close by'. Indeed, hormones secreted by a cell may act back on the same cell to influence its own function. This autostimulation is described as *autocrine* activity. Finally, some hormones, which in certain circumstances can be released from a cell to act on other cells at a distance, can also appear on the surface of the secreting cell, in which case they can act on adjacent cells only. This latter, very localized action has been termed *juxtacrine*.

Clearly, from what we have seen already, the same hormone can use a variety of routes. Whether it acts locally or pervasively will depend not only on the *nature* and *direction* of its secretion, but also on the *blood supply* of its secreting tissue and the *solubility* of the hormone itself in the fluids bathing this tissue. These latter points will be considered after the main hormones have been introduced.

TYPES OF HORMONE

The hormones involved in reproductive activity can be subdivided broadly into lipids, proteins, peptides and monoamines. In this section, the main family members of each of these classes of hormone will be introduced and their general properties described. Detailed discussion of their actions and the control of their production will follow in later chapters, as indicated in the text and tables.

Lipids

There are two classes of lipid-based reproductive hormones, the steroids and the *eicosanoids*.

Steroids

The steroid hormones comprise a large group of molecules all derived from a common sterol precursor: *cholesterol* (Fig. 2.2). Cholesterol is synthesized from acetate in many tissues of the body, and is an important structural component of cell membranes. Indeed, if the cholesterol content of mammalian cells is manipulated to abnormal levels experimentally, or varies pathologically, then the cell membranes malfunction, destabilize and rupture more easily. In steroidogenic tissues, most of the steroid output is derived from acetate with cholesterol as an intermediate product (Fig. 2.2). Cholesterol itself can also be used as a starting substrate, being gained either from intracellular stores of esterified cholesterol or from cholesterol circulating in the blood.

The pathway for steroid biosynthesis from cholesterol is illustrated in Fig. 2.3, which also provides a visual framework of the molecular relationships of the different steroid family members. The conversion of cholesterol to *pregnenolone* marks the first and common step in the formation of all the major steroid hormones. This conversion is rate-limiting and therefore an important point of regulation. The conversion occurs on the inner mitochondrial

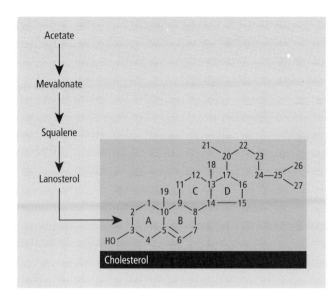

Fig. 2.2 The basic structure of the cholesterol molecule. Each of the 27 carbon atoms is assigned a number, and each ring a letter. The individual carbon atoms are simply carrying hydrogen atoms unless otherwise indicated (e.g. C1 and C19 are —CH_2— and —CH_3 residues, whereas C3 is —CHOH—). Cholesterol is converted to pregnenolone by cleavage of the terminal six carbons leaving a steroid nucleus of 21 carbons. This conversion occurs within the mitochondria, and requires NADPH and oxygen.

membrane and requires NADPH, oxygen, and cytochrome P-450. Pregnenolone is then converted to the sex steroids in the adjacent smooth endoplasmic reticulum. There are four main families of steroid: the *progestagens, androgens, oestrogens* (American spelling estrogens) and *corticosteroids* (Fig. 2.3), of which only the first three are regarded as the sex steroids. Although there are three distinctive classes of sex steroids, they are related structurally to each other. Indeed, they can be seen as different generations of a biosynthetic family, the progestagens being 'parental' to the androgens and the oestrogens being 'filial' to the androgens (Fig. 2.3). Interconversion from one class of steroid to another is undertaken by a series of enzymes arranged together as a 'biosynthetic unit', taking in substrate and passing the molecule along a production line with little 'leakage' of any intermediates. For example, the enzymes, 17α-hydroxylase, 17,20-desmolase, 17-ketosteroid reductase and 3β-hydroxy-steroid dehydrogenase, would form an enzyme package for the synthesis of testosterone from pregnenolone.

The close relationship of the different classes of steroids means that an enzymic defect at one point in the synthetic pathway may have far-reaching effects. For example, quite common genetic deficiencies in the fetal adrenal are reduced activity of 21-hydroxylase (which converts 17α-hydroxyprogesterone to 11-desoxycortisol;

Fig. 2.3) or of 11β-hydroxylase (which converts 11-desoxycortisol to cortisol). In each case, the resulting deficiency of corticosteroids feeds back to stimulate the corticosteroid biosynthetic path (*congenital adrenal hyperplasia*), which leads to the accumulation of high levels of 17α-hydroxyprogesterone. This steroid is then converted by 17,20-desmolase to androgens, which masculinize female fetuses (see Chapter 1; adrenogenital syndrome). Conversely, genetic deficiency of 17,20-desmolase itself results in depressed androgen output, and the failure of male fetuses to masculinize.

Within each class of sex steroids there are several natural members. There are two criteria for membership of each class: similarity of chemical structure and sharing common functional properties. Tables 2.1–2.3 summarize the principal natural progestagens, androgens and oestrogens, together with some of the functional characteristics of each class. It is largely, but by no means absolutely, true to say that: progestagens are associated with the preparations for pregnancy and its maintenance; androgens with the development and maintenance of masculine characteristics and fertility; and oestrogens with the development and maintenance of feminine characteristics and fertility. These activities will be examined more closely in later chapters.

The functional properties shared by members of each steroid class are reflected in common structural features. Natural progestagens are characterized by 21 carbons (C21 steroids), a double bond between C4 and C5, a β-acetyl at C17 and a β-methyl at C13. Natural androgens are characterized by 19 carbons (C19 steroids); strong androgenic activity is associated with a β-hydroxylated C17 and a ketone structure (—C=O) at C3. Natural oestrogens have 18 carbons (C18 steroids) with an aromatized A ring, hydroxylated at C3 and with a β-hydroxyl group at C17.

Eicosanoids

The eicosanoids comprise two major classes of messenger: the *prostaglandins* (PGs), which are of principal interest for reproductive processes, and the *leukotrienes,* which play a relatively minor role. The pathway of their biosynthesis is shown in Fig. 2.4. The essential polyunsaturated fatty acid, *arachidonic acid,* is the common precursor. It is derived from glycerophospholipids in the cell membranes by the actions of two enzymes: *phospholipase A_2* (acting primarily on phosphatidyl-choline and -ethanolamine) and *phospholipase C* (acting on phosphatidyl-inositol). Arachidonic acid is converted to the leukotrienes by the action of cytosolic *5-lipoxygenase,* and to the prostaglandins by the microsomal *cyclooxygenase* (Fig. 2.4). The rate-limiting step appears to be the availability of arachidonic acid, which thus forms the principal control point. Control is exerted mainly by

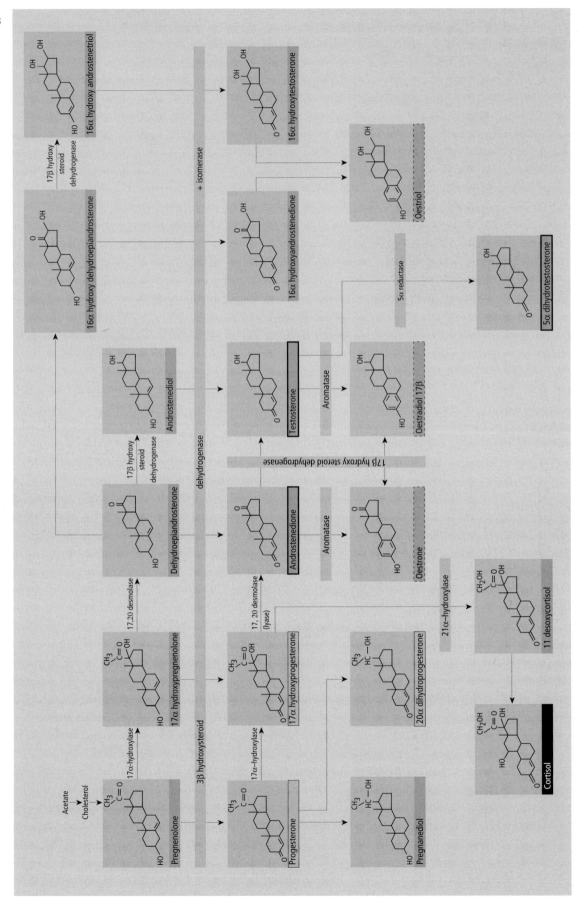

Fig. 2.3 Pathway of interconversion of steroids. Some of the principal enzymes involved are indicated. Note that extensive interconversions are possible; however, in any given tissue, absence of certain enzymes will mean that only parts of the matrix will be completed. Clearly, the enzymes present will therefore determine which steroids can and cannot be made from which available substrates. The members of the four major classes of sex steroids are shown distinctively: progestagens, faint solid surround; androgens, bold solid surround; oestrogens, dashed surround; and corticosteroids, solid background. Note that in steroid terminology the suffixes denote the following: -ol: hydroxyl group; -diol: two hydroxyl groups; -one: ketone group; -dione: two ketone groups. An unsaturated —C═C— link is indicated by -ene; two such links are indicated by -diene. Note also that aromatase denotes 19-hydroxysteroid dehydrogenase + C10, 19 lyase.

Table 2.1 Principal properties of natural progestagens and their receptors

Progestagens (relative potency)*	Properties
Progesterone (P) (100%)	1 Prepare uterus to receive conceptus
	2 Maintain uterus during pregnancy
17α-Hydroxyprogesterone (=17αOHP) (40–70%)	3 Stimulate growth of mammary glands, but suppress secretion of milk
	4 Mild effect on Na$^+$ loss via distal convoluted tubule of kidney
20α-Hydroxyprogesterone (20α-Dihydroprogesterone or 20α-OHP) (5%)	5 General mild catabolic effect
	6 Regulate secretion of gonadotrophins

The progestagen receptor is a 903 amino acid single subunit protein with three binding domains: a C-terminal progesterone-binding domain (amino acids 645-933), a central DNA-binding domain (amino acids 568–644) and an N-terminal transcription factor-binding domain (amino acids 1–567).

* In Tables 2.3–2.5 the relative potencies are only approximate since they vary with species and with the assay used. This variation is due partly to differences in the relative affinity of receptors in different tissues, partly to differences in local enzymic conversions of steroids within tissues and partly due to differences in systemic metabolism: see text for discussion of these factors.

Common abbreviations or alternative names encountered in the literature are also recorded in Tables 2.1–2.3.

Table 2.2 Principal properties of natural androgens

Androgens (relative potency)*	Properties
5α-Dihydrotestosterone (DHT) (100%)	1 Induce and maintain differentiation of male somatic tissues
	2 Induce secondary sex characters of males (deep voice, body hair, penile growth) and body hair of females
Testosterone (T) (50%)	3 Induce and maintain some secondary sex characters of males (accessory sex organs)
Androstenedione (A 4) (8%)	4 Support spermatogenesis
	5 Influence sexual and aggressive behaviour in males and females
Dehydroepiandrosterone (DHA) (4%)	6 Promote protein anabolism, somatic growth and ossification.
	7 Regulate secretion of gonadotrophins (testosterone)

The androgen receptor is a 918 amino acid single subunit protein with three binding domains: a C-terminal androgen-binding domain (amino acids 628–918), a central DNA-binding domain (amino acids 559–627) and an N-terminal transcription factor-binding domain (amino acids 1–558).

* As for Table 2.1.

varying the activity of phospholipase A_2. This enzyme is present in an inactive membrane-bound form in lysosomes, and its release and activation depends primarily on reducing the stability of the lysosomal membranes. The type of prostaglandin synthesized varies in different tissues at different times, presumably due to variation in the relative activities of the down-stream enzymes.

PGs are synthesized in most tissues of the body, including the uterine myometrium, cervix, ovary, placenta and fetal membranes (see Chapters 4, 10 & 12). They have half-lives of the order of 3–10 minutes, and so act mainly as local hormones, either paracrinologically or after a short passage through the local bloodstream. They are inactivated totally by a single passage through the systemic circulation, and particularly through the lungs.

Proteins

The protein hormones can be sub-grouped into: *gonadotrophic glycoproteins, somatomammotrophic polypeptides, cytokines* and *small peptides*.

Gonadotrophic glycoproteins

There are three gonadotrophins, so called because they stimulate the gonads, namely: *follicle stimulating hormone (FSH), luteinizing hormone (LH)* and *chorionic gonadotrophin (CG)*. They are closely related to a fourth hormone, only indirectly involved in reproductive

Table 2.3 Principal properties of natural oestrogens

Oestrogens (relative potency)*	Properties
Oestradiol 17β (Estradiol or E_2) (100%)	1 Stimulate secondary sex characters of female
	2 Prepare uterus for spermatozoal transport
	3 Increase vascular permeability and tissue oedema
Oestriol (Estriol or E_3) (10%)	4 Stimulate growth and activity of mammary gland and endometrium
	5 Prepare endometrium for progestagen action
Oestrone (Estrone or E_1) (1%)	6 Mildly anabolic; stimulates calcification
	7 Active during pregnancy
	8 Regulate secretion of gonadotrophins
	9 Associated with sexual behaviour in some species

The oestrogen receptor is a 595 amino acid single subunit protein with three binding domains: a C-terminal oestrogen-binding domain (amino acids 251–595), a central DNA-binding domain (amino acids 186–250) and an N-terminal transcription factor-binding domain (amino acids 1–185).

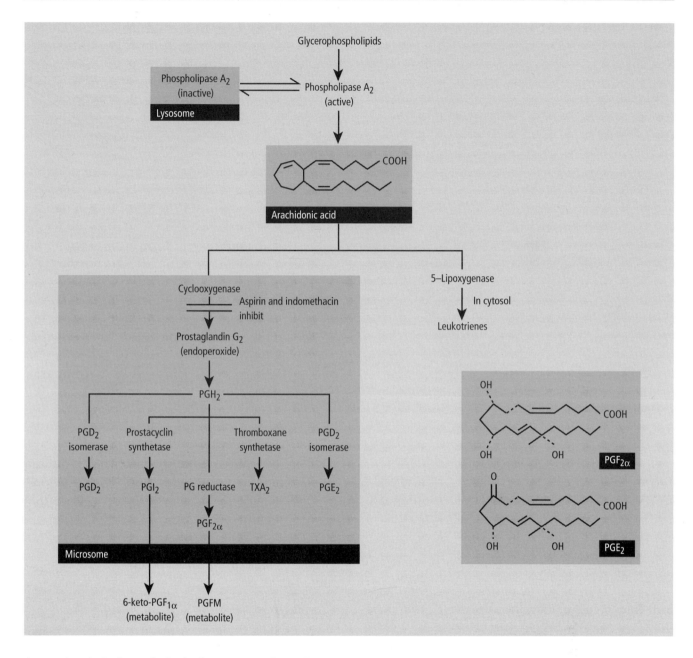

Fig. 2.4 Biosynthesis of prostaglandins (PGs): PGD$_2$, PGF$_{2\alpha}$ and PGE$_2$. The rate-limiting step is arachidonic acid availability. Factors, for example steroids, which affect the activity of phospholipase A$_2$ are therefore critical determinants of the rate of PG synthesis. Leukotrienes stimulate myometrial contractility; PGE$_2$ stimulates myometrial activity, cervical ripening and autocrine output of PGE$_2$ from the amnion; PGF$_{2\alpha}$ stimulates myometrial activity, cervical ripening; PGI$_2$ may inhibit myometrial activity. PGI$_2$: prostacyclin; TXA$_2$: thromboxane A$_2$.

processes, called *thyroid stimulating hormone* (TSH or *thryotrophin*). FSH, LH and TSH are produced in the pituitary gland, while CG is produced in the placenta. Each of these hormones is a globular protein consisting of two glycosylated polypeptides (α and β chains) linked non-covalently. The same α chain is used in all four hormones, but the β chain is unique to each, conferring

most of the specific functional properties of each hormone. Additionally, the carbohydrate side-arm structures are unique for each α and β chain of each hormone. If these carbohydrates are modified by removal of sugars, the stability of the hormones in the circulating blood falls dramatically, as does the biological activity of any hormone that survives to reach its target cells.

Table 2.4 Properties of human gonadotrophins

	Luteinizing hormone (interstitial cell stimulating hormone) LH (ICSH)	Follicle stimulating hormone (FSH)	Chorionic gonadotrophin (CG)
Secreted from	Anterior pituitary gonadotrophs	Anterior pituitary gonadotrophs	Cytotrophoblast, then syncytiotrophoblast of placenta in pregnancy
Acts upon	Leydig cells (Chapter 3) Thecal cells–antral follicles (Chapter 4) Granulosa cells–pre-ovulatory follicles (Chapter 4) Luteal cells–corpus luteum (Chapter 4) Interstitial glands of ovary (Chapter 4)	Sertoli cells (Chapter 3) Granulosa cells–follicles (Chapter 4)	Luteal cells (Chapters 10 & 11)
Molecular weight	c. 28 000	c. 28 000	c. 37 000
Composition	α chain* 92 amino acids and two carbohydrate chains; ß chain † 121 amino acids and two carbohydrate chains	α chain* as LH and two carbohydrate chains; ß chain † 111 amino acids and two carbohydrate chains	α chain* as LH and two carbohydrate chains; ß chain † 145 amino acids (a 24 carboxy addition) and six carbohydrate chains (Four additional on carboxy terminal tail)
Receptor	85–92 kDa glycoprotein; G protein coupled	146 kDa multimeric glycoprotein; G protein coupled	As for LH

* The common or backbone subunit chain—encoded in a single gene with four exons on chromosome 6.
† The subunit chain conferring specificity. For LH and FSH, single genes on chromosome 19. For CG, three genes expressed and three with low activity, all on chromosome 19; three exons in each. There is 80% homology between the common parts of CG and LH ß chains, but only 36% homology between LH and FSH ß chains.

Analysis of plasma and urinary gonadotrophins reveals an enormous size heterogeneity for each hormone that reflects slight modifications to the glycosylation patterns of each. Many believe that these heterogeneities in glycosylation are reflected in subtle differences in physiological properties, but there is no universal agreement on what these might be. The main features of each human gonadotrophin are summarized in Table 2.4.

Somatomammotrophic polypeptides

This family is named for the pervasive effects its members have on tissue growth and function, including effects on the mammary gland. The family has three main members: *prolactin* (PRL), *placental lactogen* (PL; also called *placental somatomammotrophin*) and *growth hormone* (GH; also called *somatotrophin*). Each consists of a single polypeptide chain. PRL and PL are particularly concerned with lactation, while GH plays a role in puberty, and a placental variant of GH is active in preg-

nancy (Table 2.5). However, in addition to these very specific reproductive functions, these polypeptides have a widespread, generally supportive role, helping other hormones to achieve their effects more fully. In this way, they resemble cytokines, with which they also share a similar mode of interaction with their target cells.

Cytokines

Cytokines are also polypeptides, having one or two chains, and molecular weights usually less than 100 kDa. Conventionally, they are distinguished from the 'classical' hormones (gonadotrophic glycoproteins and somatomammotrophic polypeptides, described above) by several criteria: (a) they are made in a variety of cell types, not a defined gland; (b) they act on a multiplicity of target cell types; (c) different cytokines interact to modulate each other's effects; (d) several cytokines may have identical or overlapping effects, that is, they show considerable molecular redundancy; (e) they tend to act

Table 2.5 Properties of human somatomammotrophic polypeptides

	Prolactin (PRL)	Placental lactogen (PL) (chorionic somatomammotrophin)	Growth hormone (GH) (somatotrophin)
Secreted from	Anterior pituitary lactotrophs and placental decidua	Cytotrophoblast to week 6, then syncytiotrophoblast of placental villi. Also invasive mononuclear trophoblast	Anterior pituitary somatotrophs (GH-N) Syncytial trophoblast placental variant (GH-V)
Acts upon	Leydig cells (Chapter 3) Seminal vesicle and prostate (Chapter 7) Ovarian follicles (Chapter 4) Corpus luteum (Chapter 4) Mammary gland (Chapter 13) Amnion (Chapter 11)	Maternal intermediary metabolism (Chapters 10 & 11) Mammary gland (Chapter 13) Fetal growth (Chapter 11)	In puberty (Chapter 1) Breast development (Chapter 13)
Molecular weight	22 500	22 278	GH-N 22 000 (20 and 17KDa variants by alternative splicing) GH-V 22 000 (93% homology with GH-N)
Composition	Single polypeptide chain of 199 amino acids. Post-translationally modified to give multiple forms (cleavage, glycosylation, phosphorylation and dimerization)	Single non-glycosylated polypeptide chain of 191 amino acids (67% homology to PRL, 96% homology to GH)	191 amino acids (single glycosylation site in GH-V) 96% homology with PL
Genes	A single 15kb gene on chromosome six with six exons. Exon 1 encodes non-translatable sequence and is only expressed in decidua. Exon 2 encodes 5′- untranslated sequence and part of the leader sequence	Two genes (each with five exons) encoding same polypeptide on chromosome 17	Two genes (five exons each) on chromosome 17 in PL cluster
Receptors	Binds PRL receptor	Binds PRL-R (same affinity) and GH-R (affinity 1/2000 < GH)	Binds GH-R and PRL- R (GH-V binds PRL-R >GH-N and GH-R < GH-N)

mainly in an autocrine, paracrine and/or juxtacrine way, and less often in an endocrine way; (f) they often seem to function to modulate or mediate the actions of more conventionally endocrine hormones, such as LH and FSH. However, as more is learnt about both cytokines and classical hormones, these distinctions become increasingly blurred. For example, inhibin is a cytokine that functions in both endocrine and paracrine roles in reproduction (see below), and we have already mentioned the resemblance that the somatomammotrophins have to cytokines.

Like the steroids, cytokines can be grouped in families on the basis of both their molecular structure and their overlapping activities. The main cytokine families and their members relevant for reproductive processes are recorded in Table 2.6. We already encountered MIH in Chapter 1. The *inhibins* and *activins* are in the same family as MIH, and form a series of heterodimers from

three basic subunits made on three genes, one encoding an α subunit and two encoding β subunits as shown in Fig. 2.5.

Small peptides

There are four small peptides that have important roles in reproduction. Each of them is made in the form of a larger polypeptide precursor, the active hormone (or hormones) being cleaved out just prior to their secretion.

Gonadotrophin releasing hormone (GnRH) is a member of a large family of peptides, produced in neurosecretory cells in a part of the brain called the *hypothalamus* (Fig. 2.1 and Chapter 5). The family also includes releasing hormones for thryotrophin, corticotrophin and somatotrophin, and, as their names imply, they are concerned with release of these hormones from cells in the *anterior pituitary*. GnRH is a decapeptide with the structure (Pyro)-Glu-His-Trp-Ser-Tyr-Gly-Leu-Arg-Pro-Gly-

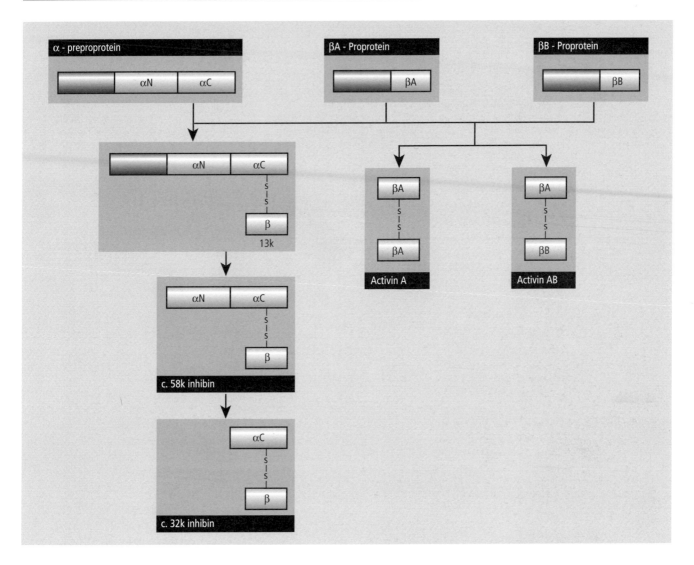

Fig. 2.5 The biosynthesis of inhibins and activins occurs from three genes producing an α preproprotein and two β proproteins. In each case, the N-terminus (hatched) is cleaved off and the subunit peptides are linked in different combinations. Activin BB has been made in the test-tube but not identified in an animal. Two forms of inhibin exist due to a further N-terminal cleavage. The presence of glycosylation sites on the peptides means that the molecular weights of the two forms of inhibin, particularly the larger one, can vary considerably.

NH₂ and it is derived by cleavage from a larger precursor called *prepro-GnRH* (Fig. 2.6).

Oxytocin is a nonapeptide, which is also mainly produced by neurons in the hypothalamus. Like *arginine vasopressin* (*AVP*: also called *antidiuretic hormone: ADH*), a similar hormone with a less direct involvement in reproductive processes, oxytocin is transported along long neuronal processes from the hypothalamus into the *posterior pituitary*, where it is released into the blood to act on the uterus (see Chapter 12) and the mammary gland (see Chapter 13). Like GnRH, oxytocin and AVP are each derived from a larger precursor consisting of an N-terminal signal peptide (which is removed before packaging occurs), followed by the hormone itself and then a *neurophysin* sequence (Fig. 2.7). The neurophysin is able to bind oxytocin, and in this way the hormone is packaged for transport along the axon to the terminals of the neurons in the posterior pituitary. Neurophysin is released, along with oxytocin, into the bloodstream, but it has no clear function in the periphery.

Recently, production of both GnRH and oxytocin has been identified at several other sites; for example, GnRH has been found in the placenta, ovary and other regions of the brain, and oxytocin has been found in the ovary, testis, uterus and widespread sites in the forebrain and brainstem. These findings have suggested a more diverse role for these hormones, or neurotransmitters, and this will be discussed later in the book.

Table 2.6 Cytokine families (Alternative names and/or abbreviations shown in brackets)

Family members	General structure	Amino acid no. of membrane bound (juxtacrine) form (MW) kDa	Amino acid no. of secreted (paracrine) form (MW) kDa	Receptor (R) type	Reproductive involvement
Epidermal growth factor (EGF)					
EGF	Single chain glycoprotein	1207aa	53aa (6kDa)	EGF-R (tyrosine kinase)	Follicular, pre- and post-implantation development (Chapters 4, 9 & 10); breast development (Chapter 13)
TGF-α (transforming GF; 44% homology EGF)	Single chain glycoprotein	160aa	50aa	EGF-R (tyrosine kinase)	Pre- and post- implantation development (Chapter 9); breast development (Chapter 13)
HB-EGF (heparin binding EGF-like GF; 41% homology EGF)	Single chain glycoprotein	204aa	(22kDa)	EGF-R + heparan sulphate	Implantation (Chapter 9)
Insulin					
Insulin	Disulphide-linked A and B chains	–	A = 24 and B = 29aas	I- and IGF-1-Rs	Pre-implantation development (Chapter 9); fetal growth and development (Chapter 11);
IGF-1 and 2 (insulin like GFs or somatomedins)	Single chains	–	70 and 67aas respectively	IGF-1- and 2-Rs	breast development (Chapter 13)
Relaxin	Disulphide-linked A and B chains (two forms)	–	A = 24 and B = 26 to 29aa	Relaxin-R	Parturition (Chapter 12)
Transforming Growth Factor β					
TGFs β1, β2, β3	Homo-dimers, disulphide-linked	–	(25kDa)	TGFβ R types I,II and III	Ovarian function (Chapter 4); implantation (Chapter 9); pre-implantation development (β1), Extra-embryonic membranes (β2), fetal tissues especially heart (β1,2,3) (Chapter 9)
Inhibins and activins	Disulphide-linked A and B chains	–	see Fig 2.5	Inhibin-Rs (two low affinity described) Activin-Rs (one high and one low affinity)	Intra-ovarian activity (Chapter 4); intratesticular activity (Chapter 3); gonad–pituitary interactions (Chapter 5); post-implantation development (Chapter 10)
Müllerian inhibitory hormone (MIH)	Homodimer +13.5% carbohydrate	–	(140kDa)		Development of internal genitalia; descent of testis (Chapter 1)
Fibroblastic growth factors (FGFs) (seven members) aFGF bFGF, kFGF, int-2	Single chain glycoprotein	–	(17kDa)	Flg/bek R (four classes)	Follicles and corpus luteum of ovary (Chapter 4); pre-implantation development (kFGF); organogenesis (Chapter 9)
Leukaemia Inhibiting Factor (LIF)	Single chain glycopeptide	–	(32–67kDa)	LIF-R	Implantation (Chapter 9)
Colony stimulating factors (CSFs)					
CSF.1 (MSF)	Glycoproteins (membrane = monomers; secreted = homodimer)	256/554aa	(86kDa)	CSF-1-R (c-fms)	Implantation (Chapter 9)
GM.CSF				GM-CSF-R(IL-3 subfamily)	
Tumour necrosis Factor α (TNFα or cachetin)	Membrane = monomer; secreted = trimer	233aa (26kDa)	157aa (3 x 17kDa)	TNF-Rs (2 receptors)	Implantation (Chapter 9)
Platelet derived growth factors (PDGFs)					
PDGF	Homo/heterodimeric glycoproteins AA, AB, BB. Disulphide bonded	–	(25kDa)	PDGF-α and β class	Pre-implantation development; fetal and placental development (Chapter 9)
Placental growth factor (PlGF)	Glycosylated dimeric peptide (two splice sites)	–	(46–50kDa) 149 or 170aa	Vascular endothelial GF-R	Implantation (Chapter 9)
Type 1 interferons					
IFN-τ(trophoblastin or trophoblast protein 1)		–	172aa	Type I IFN-R	Anti-luteolysin (Chapter 10)

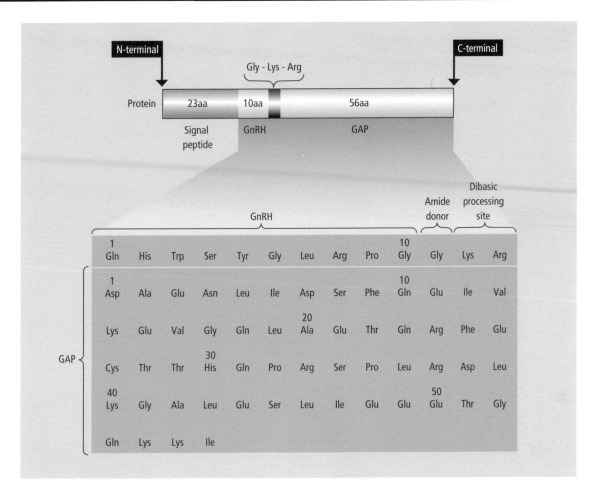

Fig. 2.6 Structure and encoded amino acid sequence of human cDNA for prepro-GnRH (molecular weight 10 000), which comprises the decapeptide GnRH preceded by a signal sequence of 23 amino acids and followed by a Gly-Lys-Arg sequence necessary for enzymatic processing and C-terminal amidation of GnRH. The C-terminal region of the precursor is occupied by a further 56 amino acids that constitute the so-called GnRH-associated peptide or *GAP*. GAP has been shown capable of releasing LH and FSH equipotentially from the anterior pituitary and of inhibiting the basal secretion of prolactin, but whether these actions are physiologically important has not been established (see Chapter 5).

The other important peptides, which we will encounter, are *β-endorphin* and *vasoactive intestinal peptide* (*VIP*), which is a member of the large *glucagon–secretin* peptide family and may be important in controlling the release of prolactin from the pituitary (see Chapter 5). As with GnRH and oxytocin, both are also derived from large precursor molecules within neurons in the hypothalamus and elsewhere. The synthesis of β-endorphin is especially interesting. Its precursor molecule, *pre-pro-opiomelanocortin*, undergoes extensive intracellular post-translational processing to yield β-endorphin (which is an opioid peptide), as well as several other biologically active hormones, especially *adrenocorticotrophic hormone* (*ACTH*; also called *corticotrophin*). Figure 2.8 illustrates the processing of pre-pro-opiomelanocortin and emphasizes the complex nature of post-translational processing that occurs during peptide secretion, and which sometimes results in neurons producing and releasing several, biologically important messengers. The effects of β-endorphin on reproduction are mediated entirely via intrahypothalamic interactions (see Chapter 5).

Monoamines

The catecholamines, *dopamine*, *noradrenaline* and *adrenaline*, as well as the indolamine, *N*-acetyl-5-methoxytryptamine, or *melatonin*, have all been implicated in reproductive neuroendocrine control mechanisms. Dopamine is synthesized from tyrosine and is also the precursor of noradrenaline, which in turn is the precursor of adrenaline (Fig. 2.9(a). These catecholamines are

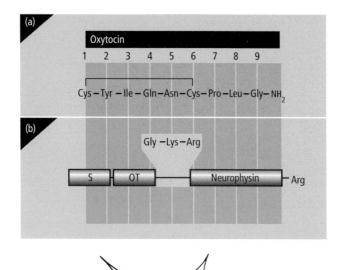

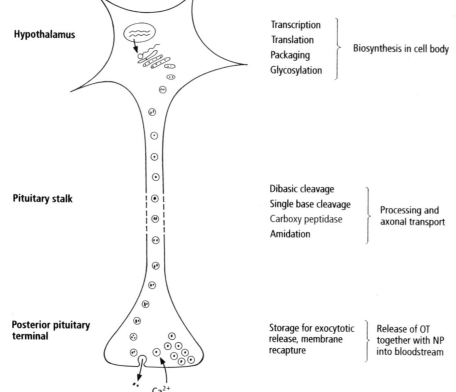

Fig. 2.7 (a) The nine amino acids comprising oxytocin contain a hexapeptide 'ring' and a tripeptide 'tail'. Oxytocin (OT) is synthesized as a precursor molecule, with: a leader sequence (S); a Gly-Lys-Arg linker sequence, serving the same function as described in the legend to Fig. 2.6 for GnRH; and a neurophysin (NP) sequence. (b) A schematized structure of a hypothalamic neurosecretory neurone indicating the cellular location of the various synthetic and processing stages that result in the release of oxytocin (OT; or ADH/arginine vasopressin) from neurohypophyseal terminals.

found in many neurons in the brainstem and, in the case of dopamine, in the hypothalamus, and they have a well-established role as neurotransmitters. Dopamine is also an important hormone; it is released from terminals in the hypothalamus and reaches the anterior pituitary where it is an important modulator of prolactin secretion (see Chapter 5). Melatonin (Fig. 2.9(b) is synthesized from serotonin (5-hydroxytryptamine) in the *pineal gland* and is under environmental control. It is released into the bloodstream and exerts important actions within the hypothalamus in order to regulate reproductive activity in seasonally breeding mammals (see Chapter 5).

Having introduced most of the main messengers used in relaying information in the reproductive system, we will now consider some of the general properties of messengers that influence how we interpret and understand their activities.

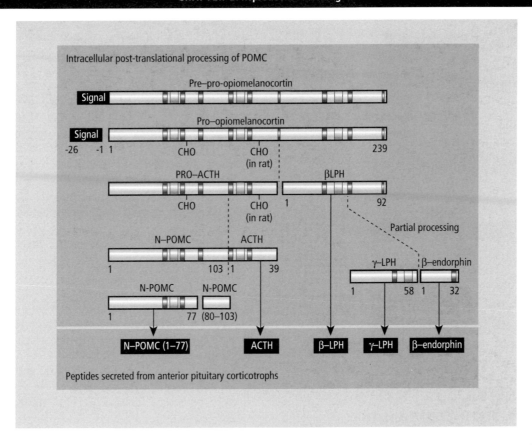

Fig. 2.8 The processing of pre-pro-opiomelanocortin and pro-opiomelanocortin (POMC). Note that there are several products of post-translational processing, including the opioid peptide β-endorphin, the non-opioid peptide ACTH and lipoproteins (LPH). CHO: carbohydrate side chains; stippled areas: melanocyte stimulating hormone core sequences; solid areas: sites for proteolytic cleavage.

HOW DO HORMONES ACT?

Tissue receptors

In the text above, we introduced the wide variety of molecules acting as reproductive messengers. Despite this diversity, we can draw some useful and important *general* conclusions about how these chemical messengers act. The hormones are grouped according to their general chemical structure and their functional activities. These two systems of classification are directly interrelated, as each of these messengers exerts its influence only by combining with a specific *receptor*. This receptor has a molecular conformation that matches some 3-dimensional physical feature(s) on the messenger molecule (which is often described as *the ligand*, in the context of its receptor). The interaction between the two, like a key in a lock, then leads to secondary changes that alert the cell to the arrival of the ligand. The information can then be incorporated by the cell into its pattern of behaviour. Thus, the cell responds to the hormone via the mediation of its receptor. As ligand–receptor interaction depends upon a good stereochemical fit between the two molecules, it is not surprising that the general molecular structure of messengers correlates with their general biological activities.

We have already grouped and sub-grouped hormones according to their structural relatedness. The more closely related molecules are, the more likely they are to bind to each other's receptors, albeit perhaps not as well as to their own. For example, human GH and prolactin each have and bind to their own specific receptors (GH-R and PRL-R). Placental lactogen binds to both these receptors; additionally, placental GH (GH-V) binds much better to PRL-R than does pituitary GH. These binding patterns reflect structural variations at the level of their amino acid sequences.

Sometimes these effects can appear paradoxical; thus, progesterone will also bind to the glucocorticoid receptor and so function as a glucocorticoid agonist! Similarly, several closely related members of a family may all work through the same receptor, but they may not be equally effective. For example, the 'strong' oestrogen, oestradiol

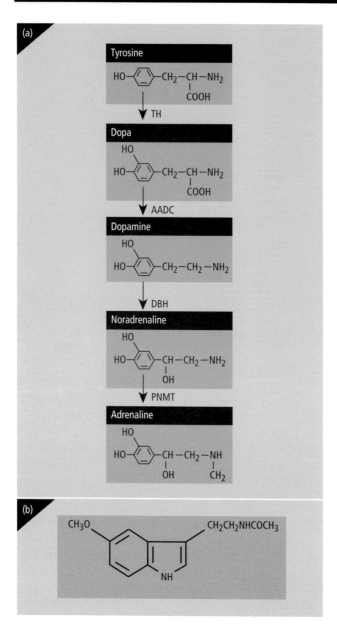

Fig. 2.9 (a) The biosynthetic pathway and structures of the catecholamines, dopamine, noradrenaline and adrenaline. (b) The structure of the indolamine, melatonin TH: tyrosine hydroxylase AADC: aromatic-L-amino acid decarboxylase; DBH: dopamine β-hydroxylase; PNMT: phenylethanolamine N-methyltransferase.

17β, is much more effective at binding and activating the oestrogen receptor than are oestriol and oestrone, which are therefore said to be 'weak' oestrogens. In Tables 2.1–2.3, the natural members of each steroid class are ranked in order of decreasing potency, and details of the receptors with which they interact are also given. This concept of *strong and weak hormone activity* is often encountered, and can be understood in terms of patterns of interaction with receptors.

Finally, some molecules may have the required structure to bind to a given receptor, but may not be able to activate it. This can happen if one part of the natural ligand is involved in receptor binding, but a second structurally and spatially distinct domain of the ligand is required for receptor activation. If a molecule is generated that retains only the first domain, it can occupy the receptor without activating it. In so doing, it may prevent the natural ligand(s) from binding and thus act as an *antagonist*. For example, we saw earlier that gonadotrophins from which sugar residues had been removed were inactive biologically. They can, however, bind to receptors. Thus, deglycosylation has affected selectively the domain(s) concerned with activation. Similarly, progesterone will compete with dihydrotestosterone to bind the androgen receptor, and so function as an anti-androgen.

Pharmacologists have taken advantage of both the blocking and mimicking effects that can operate through receptors to develop drugs for therapeutic use, such as controlling fertility or hormone dependent tumour growth. Chemical synthesis of 'analogues' of natural hormones can be achieved most easily for steroids and small peptides, and these provide the best examples, some of which, in clinical or experimental use, are listed in Table 2.7.

Receptor regulation

As the tissue specificity of hormonal action depends on both the hormone and its receptor being present, hormonal activity can be regulated *either* by controlling the availability of the hormone *or* by controlling the expression of the receptor. The latter control is extremely important in reproduction. Thus, the maturation of the ovarian follicle (see Chapter 4), the changing uterine function during the menstrual cycle (see Chapters 5 & 7), and the changes in the female tract at birth (see Chapter 12) are all critically dependent on the acquisition or loss of particular receptors on relevant tissues at appropriate times. Details of these patterns of receptor regulation will be discussed later on. For the moment, it is important to note that the measurement of variation in hormone levels alone will not provide an adequate basis for understanding reproductive function. The receptor profile must also be known.

The importance of this conclusion is underlined by reference to the genetic deficiency condition of testicular feminization (Tfm) described in Chapter 1. You will recall that affected individuals are phenotypic females with normal external genitalia, breasts and a female gender identity. On examination, these women are

Table 2.7 Some synthetic agonists/antagonists in clinical or experimental use

Progestagens	1 Derivatives of 19-nortestosterone (testosterone lacking the C19-methyl group attached to C10 and with an ethynyl group -C=CH at C17) active as progestagens:
	Norethisterone (also called Norethindrone), norethisterone acetate
	Norethynodrel/Ethynodiol diacetate: converted to Norethisterone before active as progestagens
	Norgestrel (also called levonorgestrel)
	2 Derivatives of 17α-hydroxyprogesterone by esterification of the 17-hydroxyl group (have progestagenic activity):
	Medroxyprogesterone acetate
	Chlormadinone acetate
	Magestrol acetate
	Provera
	3 Desogestrel, gestodene, norgestimate (so-called third generation progestins which have little androgen activity and can be used at much lower doses)
	4 RU486 (mifepristone): anti-progestin
Oestrogens	1 Derivatives of oestradiol 17β having oestrogenic activity: Ethinyloestradiol with an ethinyl group at C17α, and Mestranol with an ethinyl group at C17α and an -OCH$_3$ group at C3
	2 Clomiphene citrate (Clomid): initial stimulation, then anti-oestrogenic
	3 Diethylstilboestrol: agonist
	4 Nafoxidine, tamoxifen, 4-hydroxytamoxifen, MER 25: all anti-oestrogens (nafoxidene binds to the receptor but the complex fails to bind to chromatin; in contrast, the tamoxifen-receptor complex binds to the chromatin but is ineffective in stimulating the acceptor site)
Androgens	Cyproterone and Cyproterone acetate: anti-androgens
Peptides	GnRH analogues (buserelin, nafarelin, histrelin, goserelin, Lupon) used to suppress gonadotrophin output during infertility treatment (see Chapter 14) or, in conjunction with selective sex steroid replacement, in the control of a range of steroid dependent pathologies.

found to have an XY chromosome constitution, abdominal testes and blood levels of androgens in the male range. Examination of the usual androgen target tissues of these women reveals an absence or deficiency of androgen receptors. Thus, although the androgens are produced and present, the body is blind to them, and masculinizing activity is lost.

Receptor stimulation

The coupling of a hormone and its receptor leads to activation of the complex. This activation can take different forms depending on the chemical structure of the hormone (Fig. 2.10).

Thus, a steroid, being lipid soluble, can pass *into* a target cell to combine with its free *intranucleoplasmic receptor*, thereby activating it. Activation involves phosphorylation of the receptor and a conformational change so that the complex, but not steroid or receptor alone, can then bind to specific DNA sequences in the chromatin: the so called acceptor sites. Binding of adjacent transcription factors also occurs, leading to: a rapid rise

in the activity of RNA polymerase II; production of mRNA; and, in the continuing presence of the complex, a more general stimulation of nucleolar and transfer RNA synthesis. Protein synthesis rises within 30 minutes of steroid stimulation, and includes proteins specifically and selectively induced by the stimulating steroid.

Protein hormones cannot enter cells, and so use a different sort of receptor mechanism. Their receptors are located at the surface of the cell. After binding of the hormone to its receptor, different sorts of transmembrane activation events can occur (summarized in Fig. 2.10). In each of these cases, the receptor is acting as a *transducer*, passing the information on the arrival of the hormonal ligand *outside* the cell to effect metabolic responses *inside* the cell. The intracellular molecular systems mediating these responses are called *second messenger systems*.

We have considered some of the ways in which the presence or activity of receptors can influence hormonal function. We will now look at the factors influencing the levels of hormone available to interact with these receptors.

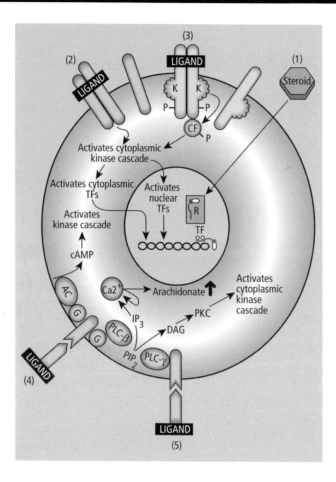

Fig. 2.10 Receptor activation and down-stream consequences. Ligands can influence cell function, and ultimately nuclear gene expression, in a variety of ways. Here we summarize in a very simplified form a very complex and rapidly evolving field of study. (1) Steroid hormones pass freely into cell nuclei where they bind receptors (R), displacing associated stabilizing proteins, such as HSP90, leading to activation by phosphorylation; the activated complex can then bind to both DNA and transcription factors (TF). Protein hormones bind to surface receptors which then act as transducers in various ways. (2) Some ligands (prolactin, GH, LIF, GM-CSF, TNFα) cross-link two receptor chains (as hetero- or homodimers). The cross-linked complex then activates a cytosolic kinase cascade, resulting in either the phosphorylation of TFs in the cytoplasm and their translocation to the nucleus, or the translocation of the kinases themselves to the nucleus where they phosphorylate and activate TFs. These ligands can cross-link overlapping spectra of receptor monomers to induce overlapping spectra of down-stream cascades, so accounting for some of the redundancy seen amongst cytokines (see text). (3) Other ligands (cytokines of the EGF, insulin and CSF-1 families) bind to, and homodimerize, receptors that are themselves kinases (K). Dimerization leads to kinase activation and both autophosphorylation (-P) and phosphorylation of cytoplasmic factors (CF-P), and thereby to a cascade of kinase activity culminating in TF activation. Again, there is some overlap of down-stream kinases and their targets both within this class of ligand–receptor interaction and between this class and class 2 above. (4) Some ligands (LH, FSH, CG, GnRH, oxytocin, arginine vasopressin) bind to receptors which then associate with G proteins (G). G proteins can then act in at least two ways: (a) to stimulate the phospholipase Cβ (PLCβ) mediated hydrolysis of phosphatidylinositol phosphate (PIP_2) to 1,4,5-triphosphate (IP_3) and diacylglycerol (DAG), which release Ca^{2+} and activate protein kinase C (PKC) respectively; or (b) to modulate activity of adenyl cyclase (AC) and so the output of cAMP. (5) Some ligands (EGF, spermatozoa?) may activate phospholipase Cγ (PLC-γ) directly, without G protein mediation.

HORMONE LEVELS IN TARGET TISSUES

The levels of a hormone in the fluids bathing potential target cells will depend upon the balance between its arrival and its removal. These in turn depend upon the local circulation, the nature and proximity of its secreting cells (local and paracrine or distant and endocrine), its circulating blood levels, and its stability. A high local blood flow will tend to dissipate autocrine and paracrine secretions but facilitate the local distribution of endocrine secretions. We will examine in more detail the factors influencing blood levels of endocrine secretions and the stability of hormones in tissues.

Blood levels of hormones

To a clinician, the hormones most directly visible are those measured in a sample of systemic blood. Measurements may be made by *bioassays*, which provide

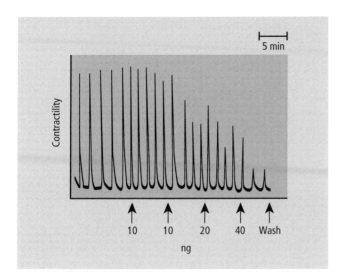

5 min

Contractility

10   10   20   40   Wash

ng

Fig. 2.11 Bioassay of hormones: various doses of a biological sample are compared with known hormonal standards for their effects in a biological system. In this example, increasing amounts (10, 20 and 40 ng) of a preparation of the cytokine, relaxin, are investigated for their inhibitory effects on the spontaneous contractility of a segment of oestrous mouse uterus. From this experiment a plot of inhibition versus concentration can be made. Unknown samples can then be used in the same system, and the concentration estimated from the standard plot. The disadvantage of bioassays is that the endpoints can be variable and susceptible to influence by extraneous factors, including the inherent variability of the biological material in the assay itself and contaminating agents that interfere with the assay. The advantage of bioassays is that they actually measure biological activity (see legend to Fig. 2.12).

a direct test of the hormone's functional activity (example in Fig. 2.11), or by use of a *protein-binding radioassay* (example in Fig. 2.12). What interpretation can be placed on such measurements?

Hormone levels may fluctuate with time

Observations on one sample of blood represent a single static measurement taken from a highly dynamic system. Figure 2.13 shows a series of values of blood testosterone taken from a man over a 24-hour period. Two important points emerge from inspection of these values. First, mean blood levels of testosterone, and of other hormones, such as prolactin, vary over a 24-hour period, showing *circadian rhythmicity*. Thus, samples should always be taken and compared at similar times. Second, testosterone (and many other hormones, such as gonadotrophins, GnRH and prolactin) is secreted not continuously but in a pulsatile manner at intervals of between 1 and 3 hours. The amplitude of these pulses is often large. A single blood sample taken at one point in time does not take this pulsing into account and may be misleading. This pulsatile secretion pattern can be

crucial to hormonal function. Figure 5.7 shows the effect of pulsatile and continuous infusions of GnRH on the output of LH from the pituitary. Clearly, continuous infusion suppresses output. It appears that when receptors are occupied constantly by ligand, they become 'exhausted' and are uncoupled from the internal second messenger systems that would normally operate. This uncoupling is called *receptor down regulation*. Examples of receptor down regulation in the face of continuous stimulation (both physiological and pharmacological) will be encountered later.

Thus, care must be taken in interpreting values for hormone levels in the blood. Moreover, even when a reliable measurement has been made, its interpretation depends on understanding that the value depends not simply on hormone production rates but on the balance between the rate of hormone secretion and its rate of clearance by various routes.

Hormone turnover

Most, but not all, hormones detectable in the blood are produced from a single major source, and so hormone levels will usually reflect the activity of that source. However, some hormones may come from multiple sources, either naturally (e.g. during pregnancy when fetal, placental and/or maternal sources of many hormones are present; see Chapters 10 & 11) or pathologically (e.g. tumours commonly secrete chorionic gonadotrophin and placental lactogen). Additionally, other hormones may be produced by a mixture of direct secretion and by interconversion from other hormonal substrates (e.g. oestrone, 10–30% from ovaries directly, the remainder derived from metabolic conversion of ovarian oestradiol 17β and of adrenal androstenedione by peripheral tissues, notably the liver). This distinction between primary secretion and metabolic conversion can be important in interpreting changes in the mean blood levels of a steroid, particularly in pathological conditions.

The rate at which a hormone is removed from the blood is reflected in its half-life. An awareness of hormonal half-lives is important in interpreting changes in hormone levels (Table 2.8). Thus, short half-lives mean that hormonal levels are more responsive to secretory changes. Changes in half-life may also inform about the state of the hormone itself (e.g. has it been modified to increase its rate of clearance) or about the functional competence of the organs effecting clearance. Removal of hormones from the blood is affected only marginally by utilization in receptor complexes, and mainly by metabolic conversion (e.g. in the liver: steroids, or lung: prostaglandins) to a range of biologically inactive or less active derivatives.

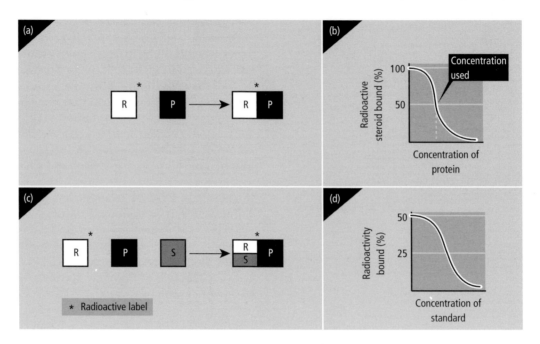

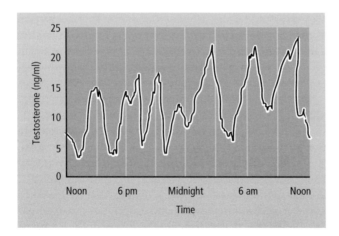

Fig. 2.12 Radioassay of hormones. (a) A sample of pure, radioactive hormone (R) is mixed with different amounts of a binding protein (P), which may be a specific antibody (radioimmunoassay) or an isolated plasma-binding protein or cell receptor (radioreceptor assay). The bound and free hormone are then separated. This may be accomplished in several ways: the protein can be used fixed to the side of the sample tube (solid phase assay), in which case the free hormone is easily washed away; alternatively, proteins in solutions may be precipitated by adding salts or a second antibody directed against the protein. The proportion of the hormone binding to the protein can then be plotted (b). Usually, protein levels binding about half the radioactive hormone are selected for use as shown in (c), in which the selected level of protein is incubated with radioactive hormone in the presence of various dilutions of a sample of plasma containing a known level of non-labelled standard hormone (S). The unlabelled standard hormone competes for the binding protein, and so the percentage of bound radioactive hormone declines. In this way a calibration curve can be constructed (d). Dilutions of samples of plasma containing unknown levels of hormone can now be used instead of the standard hormone, and the concentration of hormone thereby determined from the standard curve. Protein-binding assays are more reliably quantitative than bioassays, but may only measure one feature of the hormone, which is not necessarily equivalent to its activity. Thus, immunoassays are measuring the antigenic activity of the hormone not its biological activity; but, for example, quite large amounts of free gonadotrophin α chain are present in the circulation, as well as non-functional polymorphic gonadotrophins. Receptor assays measure binding to the receptor, but as we saw, deglycosylated LH binds but does not activate.

Fig. 2.13 Testosterone levels in blood samples taken from a male patient at 45-minute intervals over a 24-hour period. Note that the mean level is higher from midnight to noon and that there are pronounced oscillations.

Table 2.8 Half-lives of some hormones in the blood

Hormone	Half-life
Steroids	2–3 min
Prostaglandins	3–10 min
Gonadotrophins LH, FSH, CG	1–3 hours and 36 hours (biphasic) 6 hours and 36 hours (biphasic)
Prolactin and placental lactogen	10–20 min and 24 hours (biphasic)

Binding proteins

Finally, the mean blood level of some hormones is also influenced by the presence of binding proteins. Thus, the lipid hormones and their precursors are relatively insol-

Table 2.9 Steroid binding proteins in human plasma

Binding protein	Percentage of non-conjugated steroids bound*			
	Progestagens	Androgens	Oestrogens†	Cortisol
Albumin	48	32	63	20
Cortisol binding globulin§	50	1	-	70
Sex steroid binding globulin	-	66	36	-
Free steroid	2	1	1	10

* Steroids conjugated as sulphates or glucosiduronates bind weakly to albumin only.
† Oestrone and oestriol bind mainly to albumin.
§ Also called transcortin.
Note: Albumin is a low affinity/high capacity binding protein whilst the globulins are high affinity/low capacity binding proteins. There are sex differences, females in general binding proportionately more androgens/oestrogens to sex steroid binding globulin than to albumin.

uble in water. In order for them to reach their target cells, they must either bind to *carrier proteins* or be *chemically modified* to increase their solubility in plasma. Thus, most of the cholesterol and sex steroids in the blood are complexed with carrier protein molecules in equilibrium, with much lower levels of free steroid in aqueous solution, rather in the way that haemoglobin carries reservoirs of bound oxygen (Table 2.9). Steroids are also rendered more soluble in plasma by conjugation to give glucosiduronates and sulphates. In such a state, they show reduced biological activity and must be deconjugated again in order to become fully active.

In contrast, protein hormones are more freely soluble in the aqueous fluids of the body, but do not readily enter cells, tending to act at their surfaces. However, some protein hormones do complex with binding proteins in the blood, and this binding can affect their availability for binding to receptors. Thus, insulin and insulin-like growth factor 1 (IGF-1) can bind to six different insulin-binding proteins, and the levels of these proteins modulate both the plasma concentration and the activity of the cytokines, complicating the interpretation of their actions. Similarly, *follistatin* (a single chain glycopolypeptide of 32–35 kDa produced in the ovary) binds to activin and modulates its capacity to influence follicular development in the ovary (see Chapter 4) and to release FSH from the pituitary (see Chapter 5).

The presence of variable levels of binding proteins complicates the interpretation of hormone measurements. Thus, if the levels of the binding proteins themselves vary, the bound hormone levels will follow. For example, cases of androgenization in women can be associated with normal androgen levels but *reduced*

levels of sex steroid-binding globulin. In these cases, the proportion of androgen readily available to receptors has risen and androgenization occurs. Conversely, during pregnancy, the levels of steroid-binding proteins rise sixfold (see Chapter 10 for details), leading to a corresponding increase in total, measurable steroid levels, although not of free steroid. It is difficult to determine accurately the ratio of free to bound steroid in blood samples. However, samples of body secretions, such as saliva, contain levels of steroids that reflect primarily the levels of free blood steroid available for equilibration. Analysis of saliva for steroid levels is also useful because frequent sampling on an outpatient basis is possible.

Tissue levels of hormones

We have considered some of the factors affecting tissue levels of hormones, but one important factor not so far discussed is the ability of some tissues to activate a hormone or its precursor in the region of target cells carrying receptors. This ability to take a circulating hormone, transform it enzymatically and then utilize it locally will be illustrated with two examples here, and will be referred to again in other chapters.

The enzyme, 5α-reductase, is present in many of the androgen target tissues, for example the male accessory sex glands and the skin and tissues of the external genitalia. This enzyme converts testosterone, the main circulating androgen, to 5α-dihydrotestosterone (Fig. 2.3), which has a much higher affinity for the androgen receptor in these target cells, and thus is a much 'stronger' androgen than testosterone. In the prenatal and immature male, in which both the testicular and blood levels of testosterone are relatively low, testosterone is secreted and converted to 5α-dihydrotestosterone in the tissues possessing 5α-reductase. The low levels of this highly active androgen then exert local effects on the external genitalia, increasing phallic and scrotal size and rendering them clearly distinguishable from the clitoris and labia in females. The importance of this 5α-reductase activity is revealed in people with a genetic deficiency of it. Affected male infants have poorly developed male external genitalia, and at birth may be classed as females. At puberty, the external genitalia are suddenly exposed to much higher levels of circulating testosterone, and respond with a sudden growth to normal size. This so-called 'penis-at-twelve' syndrome illustrates vividly the role of local 5α-reductase activity in mediating many of the actions of testosterone (see also Chapters 1 & 7).

A second example of local steroid interconversion appears even more dramatic. The hypothalamus of male and female rodents, and indeed primates, is able to take circulating testosterone and convert it to oestradiol 17β by local aromatizing activity. The high local levels of oestradiol 17β then interact with a local oestrogen receptor to stimulate activity. Thus, paradoxically, the actions of testosterone to masculinize the neonatal rat brain (see Chapters 1 & 5) and to maintain masculine sexual behaviour in adulthood (see Chapter 7) are accomplished only after aromatization to oestradiol. This phenomenon emphasizes that the terms 'male' and 'female' hormones should be used with care.

SUMMARY

Although the hormones involved in reproduction are diverse, an understanding of the activities of each of them is helped if two fundamentally important general points are grasped. First, hormonal activity in the body (tissue fluids or blood) may reflect changing primary secretion, secondary interconversions, metabolic clearance or the levels of binding proteins. Second, hormone activity may also be regulated at the level of the target tissues by altering the level or activity of endogenous receptors. These important conclusions will be revisited in following chapters as we explore reproductive function in more detail.

FURTHER READING

Beato M (1989) Gene regulation by steroid hormones. *Cell* **56,** 335–344.

Lebeau M-C & Baulieu EE (1994) Steroid antagonists and receptor associated proteins. *Human Reproduction* **9,** 437–444.

Moore JW & Bulbrook RD (1988) The epidemiology and function of sex hormone-binding globulin. *Oxford Reviews of Reproductive Biology* **10,** 180–236.

Pardridge WM (1988) Selective delivery of sex steroid hormones to tissues by albumin and by sex hormone-binding globulin. *Oxford Reviews of Reproductive Biology* **10,** 237–292.

Robinson ICAF (1986) The magnocellular and parvocellular oxytocin and vasopressin systems. In: *Neuroendocrinology* (Eds Lightman SL & Everitt BJ), pp. 154–176. Blackwell Scientific Publications, Oxford.

Savouret JF, Misrahi M & Milgrom E (1988) Molecular action of progesterone. *Oxford Reviews of Reproductive Biology* **10,** 293–347.

Ying SY (1988) Inhibins, activins follistatins—Gonadal proteins modulating the secrection of FSH. *Endocrine Reviews* **9,** 267–293.

CHAPTER 3

Testicular Function

In Chapter 1, we described how the fetal testis formed from an indifferent genital ridge, and how endocrine function in the fetal testis was critical for the establishment of male phenotype and behaviour. We saw that postnatal growth of the testis was slow and the output of androgens low, albeit higher than in females. At puberty, however, a rapid growth and maturation of the testis to its adult form was accompanied by increases in androgen output (Fig. 1.13). In this chapter, we consider how the adult testis functions.

THE COMPARTMENTS OF THE TESTIS

The testis has two major products: spermatozoa, which transmit the male's genes to the embryo, and hormones, required for the maintenance of reproductive functions in adulthood. Androgens are the most important of these hormones, although oestrogens, inhibin and activin are also produced. The production of androgens and spermatozoa occurs in two discrete compartments within the testis. Spermatozoa develop *within* the tubules in close association with *Sertoli cells*, while androgens are synthesized *between* the tubules in the *Leydig cells* (Fig. 1.13).

These two compartments are not only structurally distinct but are also separated physiologically by cellular barriers, which develop during puberty and limit the free exchange of water-soluble materials. The precise cellular location of these barriers has been analysed by injection of dyes or electron-opaque materials into the blood, thus allowing the visualization of their distribution within the testis. Fairly free equilibration occurs between blood and the interstitial and lymphatic tissues (marked **I** in Fig. 3.1), but penetration through the peritubular wall into the basal compartment of the tubule (marked **B** in Fig. 3.1) is slower. Surprisingly, however, the major barrier to diffusion lies not at the peritubular boundary itself but between the basal compartment of the tubule (**B**) and an adluminal compartment (**A**). The physical basis for this barrier comprises multiple layers of both gap and tight junctional complexes completely encircling each Sertoli cell, and linking it firmly to its neighbours (Fig. 3.1, upper insert). Marker molecules are rarely seen penetrating through these junctional barriers between the adjacent Sertoli cells. The barrier constitutes the main element of the so-called *blood–testis barrier*. It is absent prepubertally, but develops prior to the initiation of spermatogenesis.

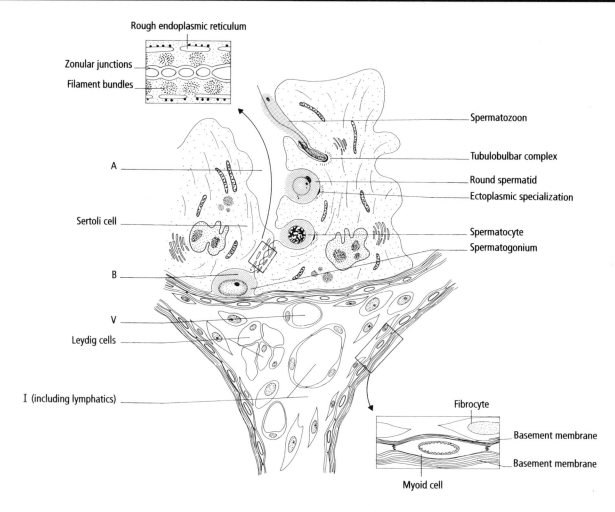

Fig. 3.1 Cross-section through part of an adult testis. The four compartments are: vascular (**V**); interstitial (**I**), including the lymphatic vessels and containing the Leydig cells; basal (**B**); and adluminal (**A**).The latter two compartments lie within the seminiferous tubules. An acellular basement membrane containing myoid cells and invested with a loose coat of interstitial fibrocytes separates the interstitial and basal compartments (see lower insert box). Myoid cells are linked to each other by punctate junctions. No blood vessels, lymphatic vessels or nerves traverse this boundary into the seminiferous tubule. Within the tubule, the basal and adluminal compartments are separated by rows of zonular tight and gap junctional complexes (see upper insert box), linking together adjacent Sertoli cells round their complete circumference. Internal to the region of junctional complexes are bundles of actin filaments running parallel to the surface around the 'waist' of the Sertoli cells, and internal to the filaments are cisternae of rough endoplasmic reticulum. Within the basal compartment are the spermatogonia, whilst spermatocytes, round spermatids and spermatozoa are in the adluminal compartment, in intimate contact with the Sertoli cells with which they form special junctions.

The presence of this barrier has two major functional consequences. First, it prevents intratubular spermatozoa from leaking out into the systemic and lymphatic circulations. This function is important because the body's immune system is not tolerant of spermatozoal antigens, which are capable of eliciting an immune response. Such a response can lead to *anti-spermatozoal antibodies* and even to an autoimmune inflammation of the testis (*autoallergic orchitis*), both of which are associated with subfertility. Second, the composition of *intratubular fluid* differs markedly from that of the *intertubular fluids*: blood, interstitial fluid and lymph. If radioactive 'marker molecules' are injected into the blood and their equilibration between blood, lymph and interstitial fluid is measured over time, it is found that ions, proteins and charged sugars enter the interstitial fluid and lymph rapidly, confirming the absence of a significant barrier at the capillary level. In contrast, these molecules do not gain free access to the tubular lumen, arriving there only as a result of selective transport. The compositions of the testicular lymph and tubular fluid reflect these dynamic processes (Fig. 3.2). The barrier function means that the later stages of spermatogenesis occur in a quite distinct and controlled chemical micro-environment.

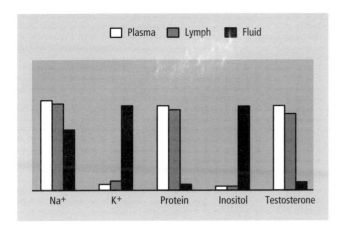

Fig. 3.2 Relative concentrations of substances in the venous plasma (vascular compartment), lymph (interstitial compartment) and fluid leaving the seminiferous tubules (adluminal compartment). Note that seminiferous tubule fluid differs markedly from plasma and lymph.

This intratubular micro-environment is not simply a consequence of passive equilibration across the blood–testis barrier, but is additionally influenced by Sertoli cell secretory products, such as androgen binding protein (ABP), testicular transferrin and sulphated glycoproteins 1 and 2 (SGP1&2); the functional significance of this will become clear later (see Chapter 8). These proteins are constituents of a fluid secretion, which can occur against considerable hydrostatic and diffusional gradients. Thus, if the outflow of fluid from the tubules towards the epididymis is blocked, secretion continues none the less, and the tubules dilate with fluid and spermatozoa, generating a hydrostatic pressure that will lead eventually to pressure necrosis and atrophy of intratubular cells.

In summary, the testis may be divided into two major compartments, each of which is further subdivided. The *extratubular* compartment consists of an *intravascular* component, in free communication with an *interstitial* component, including the lymphatics, and in which androgen synthesis occurs in Leydig cells. The *intratubular* compartment consists of a *basal* component, in restricted communication with the interstitium. A unique feature of the seminiferous tubule is that it also has a distinct *adluminal* intratubular compartment, which is effectively isolated from the other three compartments, and in which most of the spermatogenic events occur.

CYTODIFFERENTIATION OF SPERMATOZOA

The mature spermatozoon is an elaborate, highly specialized cell. *Spermatogenesis*, the process by which spermato-

zoa are formed, has three elements: *mitotic proliferation* to produce large numbers of cells; *meiotic division* to generate genetic diversity and halve the chromosome number; and *extensive cell modelling* to package the chromosomes for effective delivery to the oocyte. Large numbers of these complex cells are produced, between 300 and 600 sperm per gram of testis per second! How?

Mitotic proliferation

The interphase prospermatogonial germ cells of the immature testis are reactivated at puberty to enter rounds of mitosis in the basal compartment of the tubule. Henceforth they are known as *spermatogonial stem cells*. From within this reservoir of self-regenerating stem cells emerges, at intervals, groups of cells with a distinct morphology called *A1 spermatogonia*. Their emergence marks the beginning of spermatogenesis. Each of these A1 spermatogonia undergoes a limited number of mitotic divisions at about 42-hour intervals, thus producing a *'clone'* of cells. The number of divisions is characteristic for the species, and clearly will determine the number of cells in the clone. Thus, in the rat there are six divisions leading to a maximum clone size of 64 cells, although cell death during mitosis can reduce this number considerably. The morphology of the cells produced at each mitotic division can be distinguished from that of its parent, enabling us to subclassify spermatogonia (e.g. in the rat) as *'type A1–4'* during the first three mitoses, *'type intermediate'* after the fourth mitosis, and *'type B'* after the subsequent fifth division (Fig. 3.3). All the spermatogonia type B of the clone then divide to form *resting primary spermatocytes*.

A remarkable feature of this mitotic phase of spermatogenesis is that while nuclear division (*karyokinesis*) is completed successfully, cytoplasmic division (*cytokinesis*) is incomplete. Thus, all the primary spermatocytes derived from one type A1 spermatogonium are linked together by thin cytoplasmic bridges, constituting effectively a large syncytium. Even more remarkable is the fact that this syncytial organization persists throughout the further meiotic divisions, and individual cells are only released during the last stages of spermatogenesis as mature spermatozoa.

Meiosis

The proliferative phase of spermatogenesis takes place in the basal intratubular compartment of the testis. The clone of resting preleptotene primary spermatocytes formed duplicate their DNA content and then push their

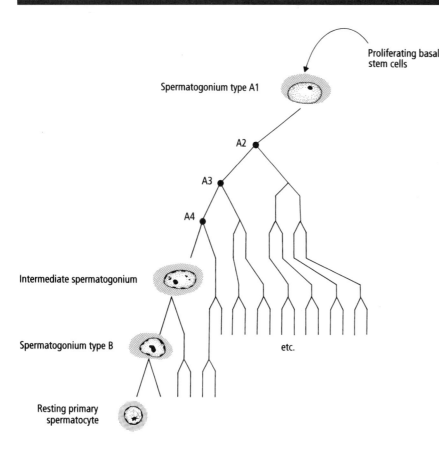

Fig. 3.3 Cells in the mitotic phase of spermatogenesis in the rat (present in the basal intratubular compartment). From the population of proliferating stem cells arise type A spermatogonia, which have large, ovoid, pale nuclei with a dusty, homogeneous chromatin. The intermediate spermatogonia have a more crusty or scalloped chromatin pattern on their nuclear membranes, and this feature is heavily emphasized in type B spermatogonia, in which the nuclei are also smaller and rounded.

way into the adluminal intratubular compartment by disrupting transiently the zonular junctions between adjacent Sertoli cells. They then enter the first prolonged meiotic prophase (Fig. 3.4; see Fig. 1.1 for details of meiosis). During prophase, the sister chromatid strands on the paired homologous chromosomes come together to form synaptonemal contacts at pachytene, during which the chromatids break, exchange segments of genetic material and then rejoin, thereby shuffling their genetic information, before pulling apart (Figs 1.1 & 3.4). Primary spermatocytes at different steps in this sequence can be identified by the characteristic morphologies of their nuclei, reflecting the state of their chromatin (Fig. 3.4). During the prolonged meiotic prophase, and particularly during pachytene, the spermatocytes are especially sensitive to damage, and widespread degeneration can occur at this stage.

The first meiotic division ends with the separation of homologous chromosomes to opposite ends of the cell on the meiotic spindle, after which cytokinesis yields, from each primary spermatocyte, two *secondary spermatocytes*, each containing a single set of chromosomes. Each chromosome consists of two chromatids joined at the centromere. The chromatids then separate, move to opposite ends of the second meiotic spindle, and the short-lived secondary spermatocytes divide to yield haploid *early spermatids* (Fig. 3.4). Thus, from the maximum of 64 primary spermatocytes that entered meiosis (in the rat), 256 early spermatids could result. Again, the actual number is much less than this, as, in addition to any losses at early stages of mitosis, the complexities of the meiotic process result in the further loss of cells. Yet again, the whole cluster of spermatids are linked syncytially via thin cytoplasmic bridges. With the formation of the early round spermatids, the important chromosomal reduction events of spermatogenesis are completed.

Packaging (spermiogenesis or spermeteliosis)

The major visible changes during spermatogenesis occur during the remarkable cytoplasmic remodelling of the spermatid, called spermiogenesis (Fig. 3.5). A *tail* is generated for forward propulsion; the *midpiece* forms, containing the mitochondria (energy generators for the cell); the *equatorial and post-acrosomal cap* region forms, and is important for sperm–oocyte fusion; the *acrosome* devel-

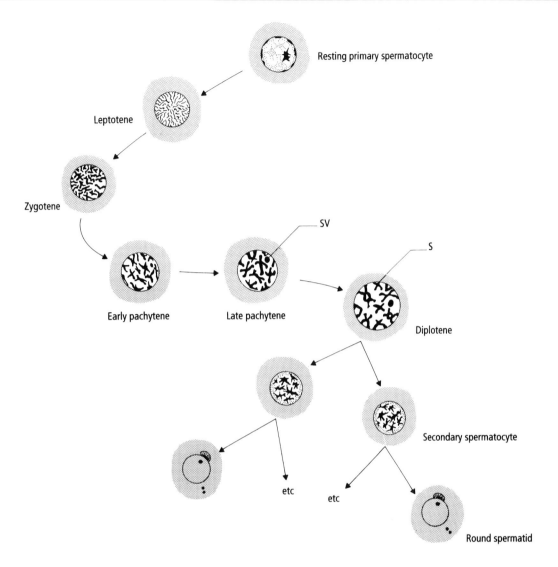

Fig. 3.4 Progress in the rat of a primary spermatocyte through meiosis in the adluminal intratubular compartment. DNA synthesis is completed in the resting primary spermatocyte although 'repair DNA' associated with crossing over occurs in late zygotene and early pachytene. In leptotene, the chromatin becomes filamentous as it condenses. In zygotene, homologous chromosomes thicken and come together in pairs (synapsis) attached to the nuclear membrane at their extremities, thus forming loops or 'bouquets'. In pachytene, the pairs of chromosomes (bivalents) shorten and condense, and nuclear and cytoplasmic volume increases. It is at this stage that autosomal

crossing over takes place (the sex chromosomes are paired in the 'sex vesicle', SV). The synapses (S) can be seen at light-microscopic level as chiasmata during diplotene and diakinesis, as the chromosomes start to pull apart and condense further. The nuclear membrane then breaks down, followed by spindle formation, and the first meiotic division is completed, with breakage of centromeric and chiasmatic contacts to yield two secondary spermatocytes each containing one set of chromosomes. These rapidly enter the second meiotic division, and the chromatids separate at the centromere to yield four haploid spermatids.

ops to function like an 'enzymic knife' when penetrating towards the oocyte; the *nucleus* contains the compact packaged haploid chromosomes; and the *residual body* acts as a dustbin for the residue of superfluous cytoplasm, and is phagocytosed by the Sertoli cell after the spermatozoon departs. Spermiogenesis is completed with the formation of a fully mature spermatozoon (Fig. 3.6). With the appearance of the spermatozoa, the

thin cytoplasmic bridges that make up the syncytium rupture, and the cells are released into the lumen of the tubule in a process called *spermiation*.

Genetic activity during spermatogenesis

Spermatogenesis is a complex and specialized process

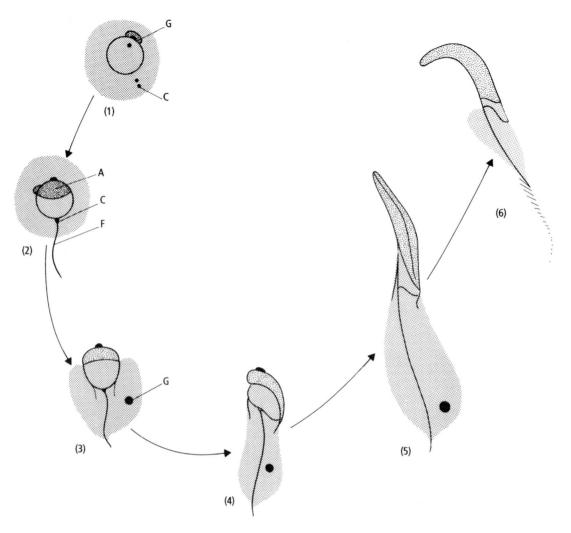

Fig. 3.5 Progress in the rat of a round spermatid through the packaging phase of the spermatogenic lineage. The Golgi apparatus (G) of the newly-formed round spermatid (1) gives rise to glycoprotein-rich granules, which coalesce to a single acrosomal granule (A) that apposes and grows over the nuclear surface to form a cap-like structure (2). Between the acrosome and the nucleus, a subacrosomal cytoskeletal element, the perforatorium, forms in many species. The nuclear membrane at this site loses its nuclear pores. The two centrioles (C) lie against the opposite pole of the nuclear membrane, and a typical flagellum (F) (9 + 2 microtubules) grows outwards from the more distal centriole (2), whilst from the proximal centriole the neck or connecting piece forms, linking the tail to the nucleus. The nucleus moves with its attached acrosomal cap towards the cytoplasmic membrane and elongation begins (3 & 4). Chromatin condensation commences beneath the acrosomal cap, generating a nuclear shape, which is characteristic for the species (3–6), and superfluous nuclear membrane and nucleoplasm is lost. The Golgi apparatus detaches from the now completed acrosomal cap and moves posteriorly as the acrosome starts to change its shape. Nine coarse fibres form along the axis of the developing tail, each aligned with an outer microtubule doublet of the flagellum (see Fig. 3.6 for details). In the final phase, the mitochondria migrate to the anterior part of the flagellum, and condense around it as a series of rods forming a spiral (Fig. 3.6). The superfluous cytoplasm appears 'squeezed' down the spermatid and is shed as the spermatozoa are released (6). The mature spermatozoon has remarkably little cytoplasm left.

Fig. 3.6 (*opposite*) (a) Diagram of a primate spermatozoon (50 μm long) showing, on the left, the main structural regions and, on the right, the boundaries between them. The surface membrane structure within each region is highly characteristic, having a unique lipid, saccharide, surface charge and antigenic composition that differs from that in adjacent regions. These differences are probably maintained both by the inter-regional boundary structures and by the underlying molecular attachments to cytoskeletal elements. The differences are important functionally (Chapter 8). (b) Sagittal section of head, neck and top of midpiece. Note the elongated nucleus with highly compact chromatin and the acrosomal sac. The posterior end of the nuclear membrane is the only part to retain nuclear pores and forms the implantation fossa, which is connected to the capitulum by fine filaments. The capitulum, in turn, is connected to the outer dense fibres by two major and five minor segmented columns and is also the site of termination of the two central microtubules of the flagellum. The more distal of the two centrioles degenerated late in spermiogenesis. In some species both centrioles are lost; they are not essential for sperm motility, only for the initial formation of the axoneme during spermiogenesis.

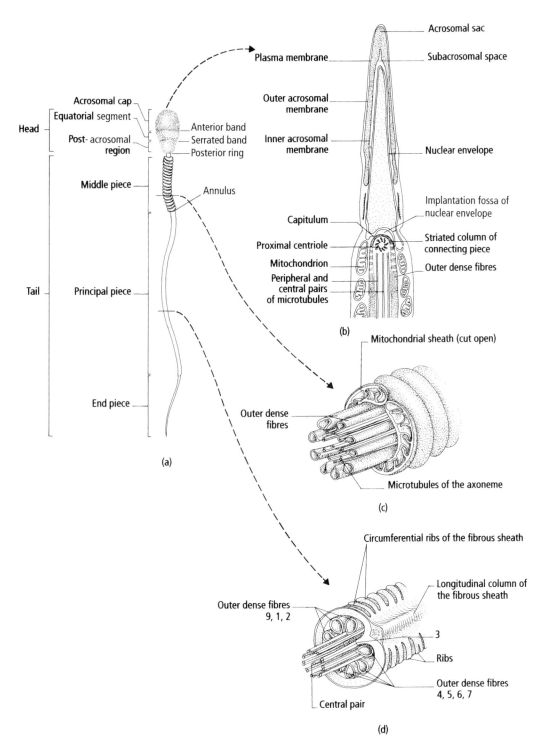

Fig 3.6 (*continued*) (c) Sketch of midpiece (surface membrane removed). Note the sheath of spiral mitochondria, and the axoneme of the tail comprising nine circumferential doublets of microtubules and two central microtubules; peripheral to each outer doublet is a dense fibre. (d) Section and sketch of principal piece (surface membrane removed). The mitochondria is replaced by a fibrous sheath comprising two longitudinal columns interconnected by ribs. The two fibrous sheath columns connect to underlying outer dense fibres 3 and 8. The outer dense fibres terminate towards the end of the principal piece, and the fibrous sheath then attaches directly to outer microtubules 3 and 8 before itself fading away in the end piece.

and, not surprisingly, requires a large number of genes for its successful completion. The processes of mRNA production and translation continue unabated throughout meiosis and into the early stages of spermatid development, and the activity of large numbers of specific genes and their protein products have now been identified. However, only genes present on the autosomes are expressed, both the sex chromosomes being inactivated during meiotic prophase. Later in the process of spermatid formation, the autosomes also cease RNA production and their DNA becomes highly condensed or *heterochromatic*. This chromosomal condensation is achieved through the action of a set of nuclear basic proteins that are unique to the testis, and which are then replaced during spermatid elongation with protamines (small basic proteins comprising 50% arginine). Protamine molecules on adjacent regions of DNA are cross-linked to each, via disulphide bonds on their constituent cysteines. In this way, spermatozoa develop tight units of heterochromatin in which genetic expression is completely absent.

The continuation of protein synthesis into the period of spermatid development raises the possibility of so-called *'haploid gene expression'*. As a result of meiotic crossing over and the independent segregation of chromosomes and chromatids at meiosis, each haploid set of genes is unique, and so the profile of mRNAs that each transcribes might also be unique. Their translation into proteins might then render the spermatids recognizably different in their structure or properties. It might then be possible to separate spermatids and spermatozoa into subpopulations based on their carriage of distinctive genetic alleles. Such a separation might also occur in the female genital tract, thereby exerting 'natural selection' on a population of spermatozoa that was genetically and, via haploid expression, phenotypically heterogeneous. A great deal of work has been done in an attempt to demonstrate haploid expression. It has been given impetus by the thought that X- and Y-bearing spermatozoa might be separated, thereby allowing pre-fertilization sex 'selection'. However, decisive evidence for haploid expression has been very difficult to find. This is not entirely surprising as spermatids exist in a syncytial mass of cytoplasm, giving opportunity for mRNAs and proteins to diffuse into all spermatids regardless of their genotype. Additionally, the premeiotic inactivation of the X and Y chromosomes (see above) makes selection for sex by this approach very unlikely. Recently, however, the separation of X- and Y-bearing spermatozoa has been claimed not on the basis of the *differential expression* of the sex chromosomes, but as a result of their *different total DNA contents* (see Fig. 1.2). Thus, the percentage difference in DNA content of X- and Y-bearing

human spermatozoa is 2.9% (boar 3%; bull 3.8%; stallion 4.1%; ram 4.2%). The separation of fertile spermatozoa by flow cytometry can achieve enrichment rates of over 75% in large farm animals. However, the prospects for a 100% successful separation of human spermatozoa for use therapeutically by this approach are low.

ORGANIZATION OF SPERMATOGENESIS

Each mature spermatozoon is one sibling in a large clonal family, derived from one parental spermatogonium type A. The family is large because of the number of premeiotic mitoses, and the spermatozoa are only siblings and not 'identical twins' because meiotic chiasmata formation ensures that each is genetically unique despite having a common ancestral parent. Within each testis tubule, many such clonal families develop side by side, and there are 30 or so tubules within each rat testis. How is the development of these families organized temporally and spatially?

How long does spermatogenesis take?

One way to measure the length of time it takes to complete parts of the spermatogenic process is to 'mark' cells at different points during the process, and then to measure the rate of progress of the labelled cells through its completion. For example, if radioactive thymidine is supplied to the resting primary spermatocytes as they engage in the final round of DNA synthesis prior to entry into meiosis, the cell nuclei will be labelled and their progress through meiosis, spermiogenesis and spermiation can be followed. In this way, the amount of time required for each spermatogenic step can be measured. In Fig. 3.7 (see foldout, facing page 48), the times required for each step in the rat are represented visually in blocks, the length of each being a measure of relative time. The absolute time for the whole process, from entry into first mitosis to release of spermatozoa, is recorded for several species in Table 3.1 (column 1).

Species vary in the total time required for spermatogenesis as well as for each of its individual component steps. However, one dramatic observation that has come from this sort of study is that within a species the rate of progression of cells through spermatogenesis is remarkably constant. Thus, the spermatogonia type A within any testis of a given species seem to advance through spermatogenesis at the same rate, and take the same total time for completion of the process. Hormones, or other externally applied agents, do not seem to speed up or slow down the spermatogenic process. They may affect

Table 3.1 Kinetics of spermatogenesis

Species	Time for completion of spermatogenesis (days)	Duration of cycle of the seminiferous epithelium (days)
Man	64	16
Bull	54	13.5
Ram	49	12.25
Boar	34	8.5
Rat	48	12

whether or not the process occurs at all, but not the rate at which it occurs. This remarkable constancy suggests a high level of intrinsic organization.

The spermatogenic cycle

So far we have considered the process of spermatogenesis from the viewpoint of a single spermatogonium type A generating a clone of spermatozoa at a constant and characteristic rate. Once this process has been commenced at a particular point in any tubule, new spermatogonial stem cells at the same point do not commence the generation of their own clones until several days have elapsed. Remarkably, it has been found that this interval between successive entries into spermatogenesis is also

constant and is also characteristic for each species (Table 3.1, column 1). Somehow, the stem cell population measures, or is told, the length of this time interval.

In the case of the rat, this cyclic initiation of spermatogenesis occurs about every 12 days. As this 'spermatogenic cycle' is one quarter of the 48–49 days required for completion of mature spermatozoal production, so it follows that four successive spermatogenic processes must be occurring at the same time (Fig. 3.7). The advanced cells, in those spermatogenic clones that were initiated earliest, are displaced progressively by subsequent clones from a peripheral towards a luminal position in the tubule. Thus, a transverse section through the tubule will reveal spermatogenic cells at four distinctive stages in the progression towards spermatozoa, each cell type representing a stage in separate, successive cycles (Fig. 3.7).

As the interval between successive entries of spermatogonial stem cells into spermatogenesis is constant, and as the rate of progress of cells through spermatogenesis is constant, it must follow that the cells in successive cycles will always develop in parallel. Therefore, the sets of cell associations in any radial cross-section through a segment of tubule taken at different times will always be characteristic (Figs 3.7 & 3.8). For example, as the cycle interval is 12 days, and as it also takes 12 days for the six mitotic divisions, entry into meiosis will always be occurring just as a new cycle is initiated by the first division of a spermatogonium type A (Fig. 3.8, column 8). Similarly, it takes 24 days (i.e. two cycles) for the pre-

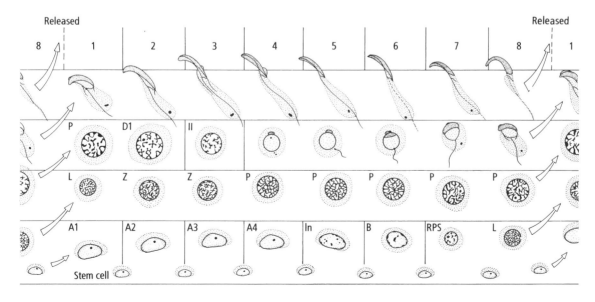

Fig. 3.8 The sections indicated in Fig. 3.7 are summarized here. Read from left to right. Bars between cells indicate cell divisions; otherwise the changes are not quantal but occur by progressive differentiation. Abbreviations as for Fig. 3.7. RPS: resting primary spermatocyte.

meiotic spermatocyte to complete meiosis and the early phase of spermatid modelling. So not only will entry into mitosis and entry into meiosis coincide, but so will the beginning of spermatid elongation (Fig. 3.8, column 8). These events will also coincide with release of spermatozoa at spermiation, as it takes a further 12 days for the completion of spermatid elongation.

Up to this point, we have considered the organization of one spermatogonium type A and its descendants. However, there are thousands of spermatogonia type A in any one testis at any one time: what sort of organization among them is there?

Cycle of the seminiferous epithelium

If all of the stem cells in the pubertal testis were to contribute their first spermatogonia type A and enter mitotic activity at exactly the same time, then, as the time to complete spermatogenesis is constant, all the individual developing clones would release their spermatozoa at the same time. Moreover, as the stem cells fed out spermatogonia A into subsequent spermatogenic sequences at intervals of 12 days (in the rat), periodic pulses of spermatozoa would occur every 12 days, which could result in an episodic pattern of male fertility. This problem could be circumvented if the pubertal spermatogonia type A initiated mitotic activity not synchronously but randomly. Then, although different clones would all develop at both the same rate and the same cyclic interval, their relative times of entry into spermatogenesis would be staggered, thus eliminating the pulsatile release of spermatozoa and smoothing it into a continuous flow. The testis functions in a manner somewhat between these two extremes, although the end result, as we will see, is continuous sperm production.

Close examination of the testes of most mammals, excluding man, shows that in a cross-section through a tubule the same set of cell associations is observed, regardless of the point on the circumference studied (Fig. 3.9a–c). Many discrete clones are present around the basement membrane in any cross-section of tubule, and the fact that each of the clones is at the same stage of development with the same set of cell associations means that all the clones in that section of tubule must be synchronized in absolute time. It is as though, at puberty, a message passed circumferentially around a segment of tubule, stimulating the stem cell population in that segment to initiate spermatogonial type A production. Once activated synchronously, the constancy of both the spermatogenic cycle and the spermatogenic rate would keep all clones locked together. The fact that all the clones within one segment of tubule are synchronized is fortuitous, experimentally and clinically, because it enables us more easily to visualize the spermatogenic cycles initiated by individual spermatogonia type A, grouped as they are in a tubule cross-section. This spatial coordination of adjacent spermatogenic cycles gives rise to the 'cycle of the seminiferous epithelium', as the whole epithelial cross-section goes through cyclic changes in patterns of cell association (Fig. 3.8).

The human testis is somewhat atypical, as a cross-section through an individual tubule reveals a degree of spatial organization that is more limited to 'wedges' (Fig. 3.9d). It is as if the activator message at puberty did not get all the way round a cross-section of tubule, and so the coordinated clonal development of different spermatogonia type A was initiated over a smaller area. This does not mean, of course, that the control of either the spermatogenic cycle or the rate of spermatogenesis in man differs fundamentally from control mechanisms in other species. It means merely that the spatial coordination between individual clones is not so great.

Spermatogenic wave

One final feature of the organization of spermatogenesis needs to be addressed: the *spermatogenic wave*. If a rat seminiferous tubule is dissected and laid out longitudinally, and cross-sections are taken at intervals along it and classified according to the set of cell associations in it, a pattern, similar to that in Fig. 3.10, will then often result. Quite extraordinarily, it seems that adjacent tubule segments, each containing synchronized spermatogenic clones, were activated serially at puberty. Thus, in Fig. 3.10, the most advanced segment (7) is at the centre; moving along the tubule in either direction leads to sets of cell associations characteristic of progressively earlier stages of the cycle of the seminiferous epithelium. It is as though the central segment was activated first at puberty, and then a hypothetical 'activator message' spread along the tubule in both directions, progressively initiating mitosis and, thereby, the first cycle of clonal growth. The resulting appearance in the adult testis, as shown in Fig. 3.10, is called the spermatogenic wave.

It is important not to confuse the wave with the cycle of the seminiferous epithelium, although both phenomena appear very similar. Imagine that, whereas the sequence of cell associations forming the *wave* could be recorded by travelling along the tubule with a movie-camera running, the same sequence of cell associations would only be captured in the *cycle* by setting up the

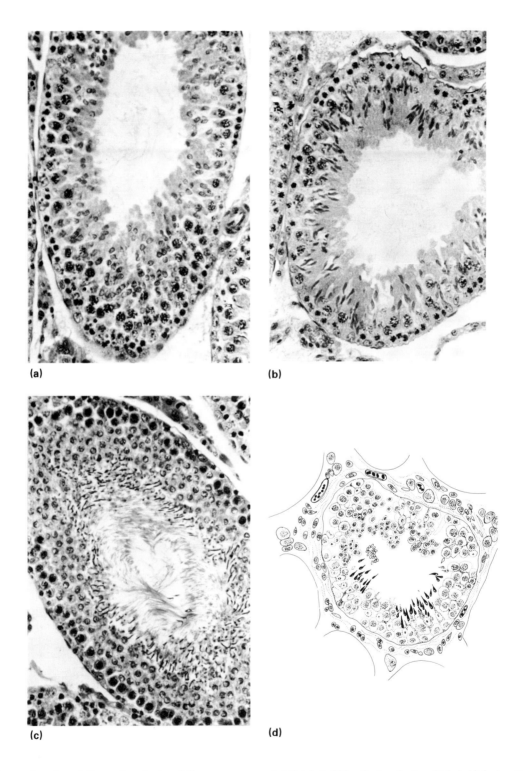

(a)

(b)

(c)

(d)

Fig. 3.9 (a–c) Cross-sections through three adjacent rat seminiferous tubules. Note that, within each tubule, the sets of cell associations along all radial axes are the same. However, each tubule has a different set of cell associations from its neighbour. Thus, tubule (a) has the type 1 set (Figs 3.7 & 3.8), tubule (b) has the type 3 set, and tubule (c) the type 8 set. Tubule (d) is a sketch of a human testis tubule. Note that different sets of cell associations are evident along different radial axes through the same tubule.

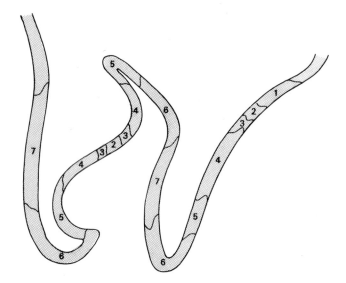

Fig. 3.10 Dissected seminiferous tubule from a rat testis. Note that whole segments of the tubule are at the same stage (numbered) of the cycle of the seminiferous epithelium, and that adjacent segments tend to be either just advanced or just retarded.

movie-camera on time-lapse at a fixed point in the tubule. Thus, the *wave occurs in space, while the cycle occurs in time.*

An organizing role for the Sertoli cell?

These observations on spermatogenesis imply a remarkable degree of temporal and spatial organization among the spermatogenic cells, the basis for which is not understood. It is tempting to invoke a role for the Sertoli cell. Thus, the linkage of adjacent Sertoli cells to each other by the zonular tight and gap junctions that make up the blood–testis barrier (see 'The compartments of the testis', Fig. 3.1), provides a potential route for communication and synchronization along and around the tubule. Moreover, each Sertoli cell spans the tubule from peritubular basement membrane to lumen, thereby providing a potential radial conduit for communication, through which all its associated spermatogenic cells could be locked into the same rate of developmental progression. This latter possibility is made more attractive by the observation that the Sertoli cell engages in intimate associations with the cells of the spermatogenic lineage. These associations are of three types: (1) pachytene spermatocytes communicate with, and receive material from, the Sertoli cell via gap junctional complexes; (2) most spermatocytes and spermatids form so-called *ectoplasmic specializations* with Sertoli cells (Fig. 3.1), and these are largely thought to be concerned with anchoring the cells and remodelling them during

spermiogenesis; (3) elongating spermatids and Sertoli cells form heavily indented *tubulobulbar complexes* through which the Sertoli cell is thought to remove material during cytoplasmic condensation (Fig. 3.1). Finally, the Sertoli cell itself shows characteristic changes in morphology and biochemistry in concert with the cycle of the seminiferous epithelium. For example, the volume, lipid content, nuclear morphology and number and distribution of secondary lysosomes vary cyclically, as do the synthesis and output of a number of testicular proteins, such as ABP, SGP1&2, transferrin and plasminogen activator. Interestingly, the output of the latter protein is high at around the time of spermiation and the passage of preleptotene spermatocytes into the adluminal compartment, suggesting a potential use for its proteolytic activities. However, although the evidence linking Sertoli cell activity to the spermatogenic cycle is compelling, whether this association *reflects* or *directs* the spermatogenic cycle remains to be determined.

Summary

The production of spermatozoa is a complex and highly organized process, which is now well described, even if its coordinating control remains imperfectly understood. Whether or not the Sertoli cell plays a role in control of spermatogenic rates, cycles and waves remains to be established. However, it is established that the Sertoli cell does play a critical role in mediating the actions of hormones and other agents on spermatogenesis. We will now consider endocrine activity within the testis and its relationship to the process of spermatogenesis.

TESTICULAR ENDOCRINE ACTIVITY AND THE CONTROL OF SPERMATOGENESIS

Hormone production by the testis

The most important hormones produced by the testis are the androgens, which, as discussed in Chapter 1, play a major and essential role both during development and in reproductive and sexual function in the mature male (see Chapter 7). In addition, the testis produces oestrogens, and in some species such as the horse, in large amounts. The growth factors, inhibin, activin and Müllerian inhibiting hormone (MIH), are also testicular products, as is the peptide hormone, oxytocin. Where does each of these come from and what does each do locally and systemically?

Steroids of the testis

As discussed in Chapter 2, the androgens comprise a class of steroid with distinct structural and functional features. The principal testicular androgen is testosterone. It is synthesized from acetate and cholesterol (Fig. 3.11) by the Leydig cells of the interstitial tissue. Thus, Leydig cells isolated in culture produce testosterone. The enzyme, 3β-hydroxysteroid dehydrogenase, involved in A5 to A4 conversions (Fig. 3.11) can be localized to Leydig cells, and in pathological conditions and seasonal breeders, changes in testosterone output corre-

late with changes in the morphology of the smooth endoplasmic reticulum in Leydig cells.

In man, 4–10 mg of testosterone are secreted daily. The hormone rapidly enters both the blood (in the ram, 80 ng/ml of testicular venous blood) and lymph (50 ng/ml), but although the concentrations in each are similar, the major quantitative output is to the blood, with its greater flow (17 ml blood/min and 0.2 ml lymph/min). However, the lymphatic flow is important as the lymph drainage pathways carry testosterone adjacent to the testicular excurrent ducts and the male accessory sex glands,

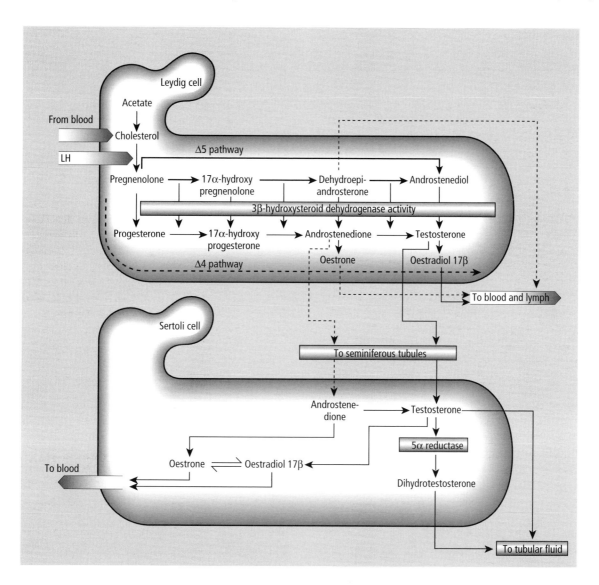

Fig. 3.11 Summary of the steroidogenic pathway in the human testis. The principal (Δ5) path for testosterone synthesis in the Leydig cells of the human is indicated by heavy lines, but the Δ4 pathway is also used and may be more important in other species. In addition to testosterone, some of the intermediates in the pathway are released into the blood—androstenedione at 10% and

DHA at about 6% of testosterone levels in man. Some testosterone and androstenedione enter Sertoli cells. Here they may bind to androgen receptors directly or after conversion to the more potent dihydrotestosterone. Androgens may also be converted to oestrogens. In humans prepubertally, this occurs predominantly in the Sertoli cell, but postpubertally in the Leydig cells.

which are stimulated by it. Not all the testosterone goes into the blood and lymph. Testosterone is relatively lipid soluble (see Chapter 2), and passes through the cellular barriers between the testicular compartments to enter the seminiferous tubules. Here, much of it is converted to the more active androgen, dihydrotestosterone, by 5α-reductase activity in the Sertoli cells. The androgens are then bound by two molecules. Within the Sertoli cells, an androgen receptor binds testosterone, making this cell a target for androgen stimulation (see below). Within the tubular fluid of the adluminal compartment, the androgen-binding protein (ABP) secreted by the Sertoli cells binds testosterone, and carries the hormone in the testicular fluid that flows out into the excurrent ducts, where it stimulates their activity (see Chapter 7).

Testosterone can also be converted to oestrogens in the Sertoli cells, although recent evidence suggests that this pathway operates mainly in fetal life in humans. As the testis matures, most oestrogen is derived directly from Leydig cell activity. It is not clear what the role, if any, of the oestrogen might be. In adult man, the output is relatively small, but in the boar and stallion, it matches the androgen output.

Growth factors and peptides of the testis

The growth factors, inhibin and activin (see Table 2.6 & Fig. 2.5), are both produced by the Sertoli cells. About 25% leaves the testis via lymphatic flow, with most of the remainder passing into the fluid of the seminiferous tubule, from which they are absorbed during passage through the epidiymis. Thus, these growth factors eventually reach the systemic circulation, bind to target cells in the pituitary (see Chapter 5), and thereby function as endocrine hormones. However, they also have paracrine and autocrine roles within the testis, as receptors binding inhibin have been located on Leydig cells, and those binding activin have been identified on both Sertoli cells and cells of the spermatogenic lineage (see 'Hormonal dependence of spermatogenesis' below). Output of MIH from the adult testis is low, its primary role being in embryonic, fetal and neonatal life (see Chapter 1).

Oxytocin (see Chapter 2, 'Small peptides') is produced in the Leydig cells and has been shown to stimulate seminiferous tubule motility, via an action on the peritubular myoid cells. Its passage from the testis in the lymph may give it a further paracrine function: the stimulation of epididymal motility.

Hormonal dependence of spermatogenesis

The production of androgens and spermatozoa are interrelated functionally. Thus, at puberty, androgens rise and spermatogenesis commences. In some adult mammals, androgen and sperm production do not occur throughout the year (e.g. the roe-deer, ram, vole, marine mammals and possibly some primates). Rather, the behaviour and morphology of the males show seasonal variations, which reflect the changing levels of androgen output. In these *seasonal breeders*, the spermatogenic output also varies in parallel with the changing endocrine pattern. The capacity of the Sertoli cells to bind, and indeed to generate via 5α-reductase activity, potent androgens indicates that they are a target for androgen action. They also have a central role in spermatogenesis. If androgens are neutralized, by blocking their output, spermatogenesis proceeds only as far as the very early preleptotene stages of meiosis. The male becomes *aspermatogenic*. Restoration of androgens, restores spermatozoal output. This causal association between the presence of androgens and the process of spermatogenesis ensures that mature spermatozoa are always delivered into an extragonadal environment suitably androgenized for their efficient transfer to the female genital tract. Given the importance of androgens for spermatogenesis and male function, how is their output regulated?

It has been known for many years that removal of the pituitary gland (*hypophysectomy*) causes the testes to shrink, sperm output to decline, and spermatogenesis to arrest at the primary spermatocyte stage. The Leydig cells become involuted, testosterone output falls, and the testosterone-dependent male genitalia hypotrophy. If testosterone is given at the time of hypophysectomy, spermatogenesis continues, albeit at a slightly reduced level, and the secondary sex characters show little sign of regression, although the Leydig cells do still involute. These experiments established that the secretion of testosterone by the Leydig cells was under the control of the pituitary.

What is the nature of this control? In experiments on the rat, it was shown that if, after hypophysectomy, *luteinizing hormone* (LH; see Table 2.4 for details) is administered instead of testosterone, then not only are secondary sex characters and spermatogenesis maintained, but the Leydig cells do *not* involute and testosterone output is maintained. Further confirmation of the role of LH comes from the administration to an intact adult male of an antiserum to bind free LH. The level of plasma testosterone falls and regression of the androgen-dependent secondary sex characters follows. The results of these two experiments suggest strongly that pituitary-derived LH stimulates the Leydig cells to produce testosterone. Subsequently, LH has been shown to bind specifically to high-affinity LH receptors on the surface of the Leydig cells. As a result, intracellular cAMP levels rise within 60 seconds, and testosterone output rises within 20–30 minutes. In contrast, LH neither binds to, nor stimulates, isolated tubules. LH does not act alone on the Leydig cells; two other hormones also influence its activity. Both prolactin, a second anterior pituitary hormone (Table 2.5), and inhibin bind to receptors and facilitate the stimulatory action of LH. Neither hormone alone, however, stimulates testosterone production.

It seems clear, then, that LH stimulates Leydig cells to produce testosterone, which passes into the tubules, binds to androgen receptors within the Sertoli cells, and thereby supports spermatogenesis. However, it is important to stress that although testosterone or LH administered immediately after hypophysectomy can prevent aspermatogenesis in rats, some reduction in testis size and a 20% reduction in sperm output does occur. In many other species, including primates, the decline in sperm output after hypophysectomy is even more severe, despite administration of high doses of testosterone or LH. Moreover, if *aspermatogenesis* is allowed to develop after hypophysectomy of any species, sperm production cannot be restored with LH or testosterone alone. For complete restoration and maintenance of spermatogenesis, LH stimulation of Leydig cells is not enough. The same requirement applies to the restoration of spermatogenesis in seasonally fertile animals. Another pituitary hormone, the follicle stimulating hormone (FSH), is required in each of these conditions (Table 2.4).

FSH binds to receptors detected on the basal cell membrane of Sertoli cells to stimulate: adenyl cyclase activity; RNA and protein synthesis; mobilization of energy resources; production of testicular fluid; and the output of Sertoli cell proteins, such as ABP and inhibin. Critically, production of intracellular androgen-receptor proteins is also stimulated by FSH, which renders the Sertoli cell responsive to the action of androgens. Androgens in turn stimulate the appearance of FSH receptors. Thus, FSH and testosterone act synergistically on the Sertoli cell to allow spermatogenesis to go to completion. Here is direct evidence for a permissive role, for the Sertoli cell in the events of spermatogenesis. There is a final piece of evidence that supports this conclusion. Vitamin A has long been recognized as essential for spermatogenesis; its deficiency is associated with subfertility, and in its complete absence spermatogenesis arrests at preleptotene stages. Receptors for vitamin A are present in Sertoli cells.

SUMMARY

In this chapter we have considered the two major products of the testis and their relationship. The main endocrine secretion of testosterone by the Leydig cells is dependent primarily upon pituitary LH. Some of the testosterone enters the seminiferous tubules, where it acts on the Sertoli cells, together with FSH, in order to help maintain the cellular product of the testis: the spermatozoa. Without testosterone, spermatogenesis ceases. Although both FSH and testosterone are essential for full spermatogenesis, their actions appear to be purely permissive. They determine whether or not spermatogenesis will take place, but do not regulate the rate at which cells develop along the spermatogenic lineage, the frequency with which the stem cells provide spermatogonia type A, or the spatial coordination between adjacent clones of spermatogenic cells. These processes appear to be regulated internally in some way, probably involving the activity of Sertoli cells.

FURTHER READING

Cupps PT (Ed) (1991) *Reproduction in Domestic Animals,* 4th edition. Academic Press, London.

Erickson RP (1993) Molecular genetics of mammalian spermatogenesis. In: *Genes in Mammalian Reproduction* (Ed. Gwatkin RBL), pp. 1–26. Wiley-Liss, New York.

Fawcett DW & Bedford JM (1979) *The Spermatozoon.* Urban & Schwarzenberg, Munich.

Griswold MD, Morales C & Sylvester S (1988) Molecular biology of the Sertoli cell. *Oxford Reviews in Reproductive Biology* **1,** 124–161.

Heckert L & Griswold MD (1993) Expression of the FSH receptor in the testis. *Recent Progress in Hormone Research* **4,** 61–77.

Johnson LA (1994) Isolation of X- and Y-bearing sperm for sex preselection. *Oxford Reviews in Reproductive Biology* **1,** 304–326.

Maddocks S & Setchell BP (1988) The physiology of the endocrine testis. *Oxford Reviews in Reproductive Biology* **1,** 53–123.

Parvenin M, Vihdo KK & Tappari J (1986) Cell interactions during the seminiferous epithelial cycle. *International Reviews in Cytology* **1,** 115–151.

Roosen-Runge EC (1962) *Biological Reviews* **3,** 343–377.

Setchell BP (1978) *The Mammalian Testis.* Paul Elek, New York.

C H A P T E R 4

Ovarian Function

In Chapter 1, we described how the fetal ovary formed from an indifferent genital ridge. It was observed that the ovary differentiated later than the testis and that its endocrine activity was not required during fetal life. The female phenotype develops spontaneously in the absence of gonadal hormones, and constitutes almost a 'neutral pattern' of sexual differentiation when compared to that of the male. However, ovarian endocrine activity must occur postnatally if full sexual maturation is to occur at puberty. In addition, from puberty onwards, the ovarian germ cells must generate haploid gametes for fertilization by spermatozoa. As in the testis, the production of gametes by the ovary is coordinated with its endocrine activity. However, adult ovarian function shows a major difference from testicular function: relatively few oocytes are released, and their release is not in a continuous stream like spermatozoa, but occurs *episodically* at *ovulation*. The release of the two major steroid secretions of the ovary, the *oestrogens* and *progestagens*, reflects this episodic release of oocytes. Thus, the period prior to ovulation is characterized by *oestrogen dominance*, and the period following ovulation is characterized by *progestagen dominance*. Once this sequence of oestrogen–

ovulation–progestagen has been completed, it is repeated. Therefore we speak of a *cycle of ovarian activity*.

The cyclic release of steroids imposes a corresponding cyclicity on the whole body and, in most species, on the behaviour of the adult female. These cycles are called the *oestrous cycle* in animals and the *menstrual cycle* in higher primates. The reason for this cyclicity of female reproductive activity lies in the fact that the genital tract of the female mammal, unlike that of the male, must serve two distinct reproductive functions, each with different demands. It must act to transport gametes to the site of fertilization, and it also provides the site of implantation of the conceptus and its subsequent development. Each cycle of the female reflects these two roles. During the first, oestrogenic part of the cycle, the ovary prepares the female for receipt of the spermatozoon and fertilization of the oocyte; during the second, progestagenic part of the cycle, the ovary prepares the female to receive and nurture the conceptus should successful fertilization have occurred. Sandwiched between these two endocrine activities of the ovary, the oocyte is released at ovulation. In this chapter, we will consider the sequence of changes *within the ovary itself* by which a coordinated and cyclic pattern of production of the oocyte and ovarian steroids is achieved. In later chapters, we will consider how this ovarian cyclicity leads to the oestrous and menstrual cycles.

THE ADULT OVARY

The adult ovary is organized on a pattern comparable to the testis (Fig. 4.1) with *stromal tissue*, containing the *primordial follicles* (homologous to tubules), and glandular tissue, the so-called *interstitial glands* (homologous to Leydig cells). The primordial follicle, comprised of flattened mesenchymal cells condensed around a primordial germ cell (Figs 4.2a & 4.4a), constitutes the fundamental functional unit of the ovary. In the first part of this chapter, we will pursue the formation, growth and fate of a single follicle in some detail. In the second part of the chapter, we will relate the activity of a single follicle to the activity of the ovary as a whole.

The pattern of gamete production in the female, like that in the male, shows processes of cell proliferation by

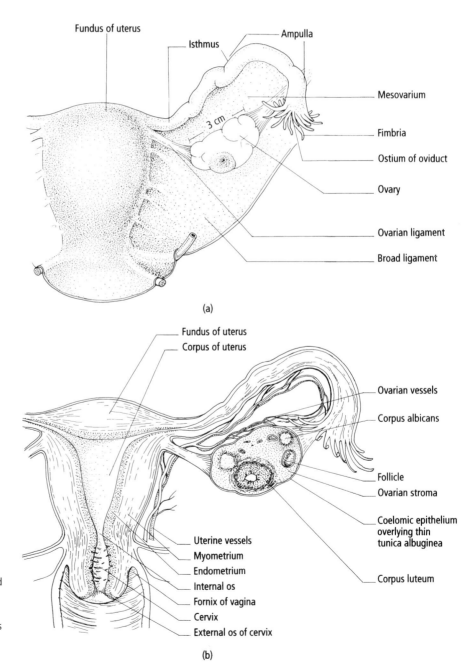

Fig. 4.1 Posterior views of human uterus and one oviduct and ovary: (a) intact; (b) sectioned. The ovaries have been pulled upwards and laterally, and would normally have their long axes almost vertical. Note that all structures are covered in peritoneum except the surface of the ovary and oviducal ostium. The ovary has a stromal matrix of cells, smooth-muscle fibres and connective tissue containing follicles, corpora lutea and corpora albicans, and interstitial glands. Anteriorly at the hilus the ovarian vessels and nerves enter the medullary stroma via the mesovarium.

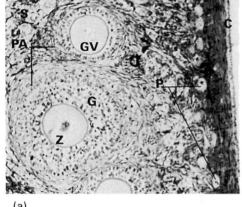

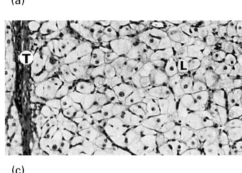

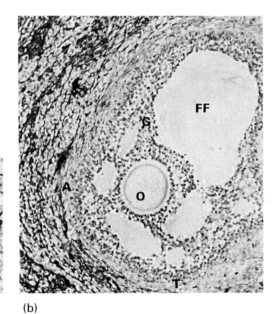

(a)

(c)   (b)

Fig. 4.2 (a) Primordial follicles (P) in the stroma (S), lying adjacent to the coelomic epithelium (C), grow in size to give preantral follicles (PA) with enlarged oocytes containing a 'nucleus' or germinal vesicle (GV) surrounded by a zona pellucida (Z) and proliferated granulosa cells (G). (b) Further growth leads to antral follicles (A) containing fully grown oocyte (O) and follicular fluid (FF). The thecal tissues (T) differentiate. (c) After a preovulatory growth spurt by the follicle and its ovulation, granulosa cells luteinize (L) within a fibrous thecal capsule (T).

mitosis, genetic reshuffling and reduction by *meiosis*, and packaging of the chromosomes during *oocyte maturation*. In the female, however, the need for proliferation by mitosis is not as great because only one or a few oocytes are shed during each cycle, unlike the massive sperm output of the testis. The female therefore uses a different approach to the mitotic phase of gametogenesis. The primordial germ cells that entered the gonad continue their mitotic proliferation well after ovarian morphology is established. At this stage they are known as *oogonia* (compared to spermatogonia in the proliferative phase of the male). However, unlike the situation in the male, the mitotic phase of the oogonia *terminates finally* before birth (human, cow, sheep, goat, mouse), or shortly thereafter (rat, pig, cat, rabbit, hamster), by which time *all* oogonia have entered into their first meiotic division, thereby becoming *primary oocytes* (compared to primary spermatocytes in the male).

This termination of mitosis and early entry into meiosis is evidently evoked by a *meiosis initiation factor* derived from cells of the ingrowing mesonephric tissue, as the removal of this tissue prevents meiosis. The consequence of this early termination of mitosis is that, by the time of birth, *a woman has all the oocytes within her ovary that she will ever have* (Fig. 4.3). If these oocytes are lost, for example by exposure to X-irradiation, they cannot be replaced from stem cells and the woman will be infertile.

This situation is distinctly different from that in the male in which the mitotic proliferation of spermatogonial stem cells continues throughout adult reproductive life.

During their progress through the first meiotic prophase, the oocytes organize the condensation of surrounding ovarian mesenchymal cells, probably mostly derived from invading mesonephric cords, to form the *granulosa cells* of the primordial follicles. The follicular cells secrete a basement membrane, *membrana propria*, around the outside of this cellular unit (Fig. 4.4a). With the formation of the primordial follicles, the oocytes within arrest their progress through first meiotic prophase at *diplotene*, with the chromosomes within a nucleus generally known as the *germinal vesicle* (see Fig. 1.1 for details of meiosis). The oocyte halted at this point in meiosis is said to be at the *dictyate stage* (also called *dictyotene*).

The primordial follicle may stay in this arrested state for up to 50 years in women, with the oocyte metabolically ticking over and waiting for a signal to resume development. The reason for storing oocytes in this extraordinary protracted meiotic prophase is unknown. Although a few follicles may resume development sporadically and incompletely during fetal and neonatal life, regular recruitment of primordial follicles into a pool of growing follicles occurs first at puberty. Thereafter, a few follicles recommence growth every day, so that a continuous trickle of developing follicles is formed. When a primordial follicle commences growth it passes through three stages of development on route to ovulation: first it becomes a *primary* or *preantral* follicle, then a *secondary* or *antral* follicle (also called *vesicular* or *Graffian* follicle), and finally a *preovulatory* follicle in the run up to ovulation itself. The time taken to traverse each of these stages is not known precisely for any species, but the figures in Table 4.1 show that the preantral phase is the longest and the preovulatory phase is the shortest. We will now trace the development of one such primordial follicle.

Follicular growth and development

Primordial to antral stages

The earliest preantral phase of follicular growth is characterized by an increase in the diameter of the primordial follicle from 20 μm to between 200 and 400 μm, depending on the species (Fig. 4.4a,b). A major part of this growth occurs in the primary oocyte, which increases its diameter to its final size of 60–120 μm. This oocyte growth is not accompanied by the reactivation of meiosis. Indeed, the dictyate chromosomes are actively synthesizing large amounts of ribosomal and mRNA, the latter being used to generate stores of proteins that are essential for later stages of oocyte maturation and for the first day or so of development of the fertilized oocyte (see Chapter 9). It therefore constitutes part of the packaging process of gametogenesis.

Over this period, the oocyte reaches its final size. Early on in the growth phase, it secretes glycoproteins, which condense around it to form a translucent acellular layer called the *zona pellucida*. The zona separates the oocyte from the surrounding granulosa cells, which divide to become several layers thick (Fig. 4.4b). However, contact with the oocyte is maintained via cytoplasmic processes, which penetrate the zona and form gap junctions at the oocyte surface. Gap junctions also form in increasing numbers between adjacent granulosa cells, thus providing the basis for an extensive network of intercellular communication. Through this network, low molecular weight, biosynthetic substrates, such as amino acids and nucleotides, are passed to the growing oocyte for incorporation into macromolecules. This nutritional network is important, as the granulosa layer is completely *avascular*: no blood vessels penetrate the membrana propria.

In addition to oocyte growth and granulosa cell proliferation, the preantral follicle also increases in size through the condensation of ovarian stromal cells on the membrana propria. This loose matrix of spindle-shaped cells is called the *theca* of the follicle (Fig. 4.4b). With

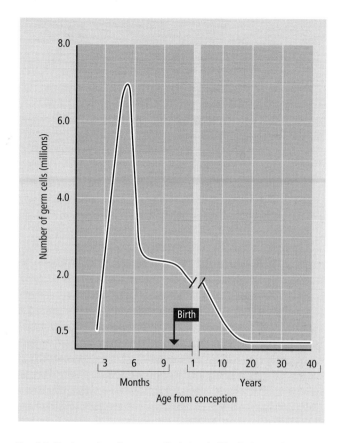

Fig. 4.3 Numbers of ovarian germ cells during the life of a human female. After an initial period of migration, mitotic proliferation commences at around 25–30 days, and continues until around 270 days (birth). The first meiotic prophases can be detected at around 50–60 days, and the first diplotene-stage chromosomes at around 100 days. All germ cells are at the dictyate stage by birth. Atresia of oocytes in the meiotic prophase is first overt by about 100 days, and continues throughout fetal and neonatal life. The period from entry of first germ cells into meiosis to full attainment of the dictyate stage by all germ cells varies between species; examples, in days post-conception (date of birth in brackets), are: sow 40–150 (114); ewe 52–110 (150); cow 80–170 (280); rat 17.5–27 (22).

further development and proliferation, the thecal cells can be distinguished as two distinct layers (Fig. 4.4c): an inner glandular, highly vascular *theca interna*, surrounded by a fibrous capsule, the *theca externa*.

The granulosa cells also continue to proliferate, resulting in a further increase in follicular size. As they proliferate, a viscous fluid starts to appear between them (Figs 4.2b & 4.4c). This *follicular fluid* is comprised partly of mucopolysaccharides secreted by the granulosa cells and partly of a serum transudate. The drops of fluid coalesce to form a single *follicular antrum* (Fig. 4.4d). The appearance of the follicular antrum marks the beginning of the *antral phase* of development. Increase in follicular size from now on depends mainly

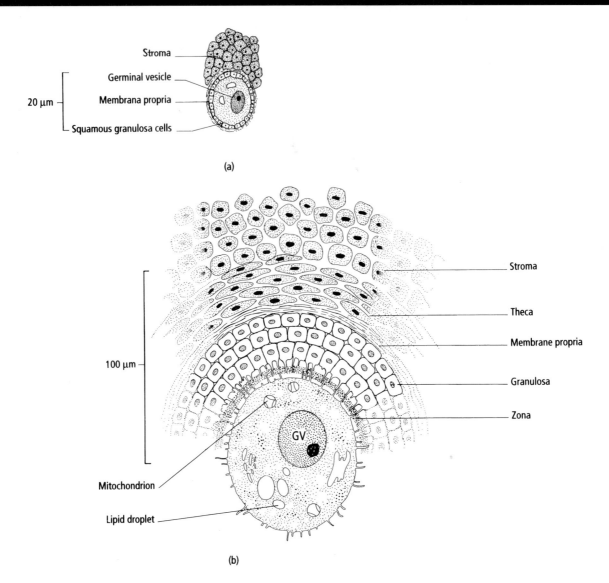

(a)

(b)

Fig. 4.4 The morphology of follicular development. (a) Primordial follicle embedded in stromal cells: note outer membrana propria, within which squamous granulosa cells enclose a primary oocyte at the dictyate stage with a germinal vesicle (equivalent to the nucleus). (b) Some stromal cells have condensed on this preantral primary follicle to form a thecal layer; the granulosa cells have divided and become cuboidal, the oocyte has grown and secreted the zona pellucida. Contact between the granulosa cells and the primary oocyte is maintained by cytoplasmic processes through the zona, which interdigitate with the many microvilli present on the oocyte. Within the oocyte: mitochondria increase in number and are small and spherical with columnar cristae; the smooth endoplasmic reticulum breaks up into numerous small vesicles; the Golgi complex breaks into small vesicular units often associated with lipid droplets. The chromosomes are still active synthetically.

on an increase in the size of the follicular antrum and the volume of follicular fluid, although granulosa cells do continue to proliferate.

Although the oocyte does not increase in size over this antral period, it is not inactive: synthesis of RNA and turnover of protein continues. As the follicular antrum grows, the oocyte, surrounded by a dense mass of granulosa cells called the *cumulus oophorus*, becomes suspended in fluid. It is connected to the rim of peripheral granulosa cells only by a thin 'stalk' of cells (Fig. 4.4d).

This maturing antral follicle is now ready to enter its preovulatory phase and to approach ovulation. However, before this remarkable process is described, we will first examine the underlying control of growth from primordial to expanded antral follicle.

How is the growth of the follicle regulated?

We do not understand why each day a few primordial follicles start to develop as preantral follicles, nor how

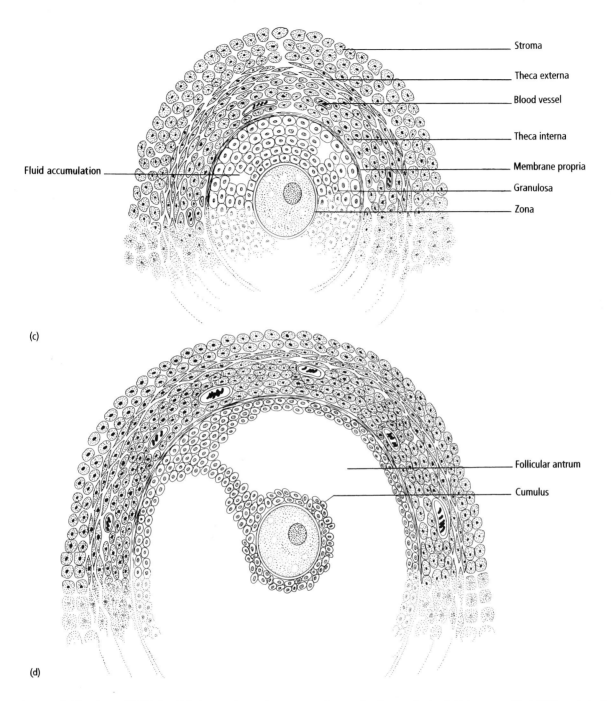

Fig. 4.4 *(Continued)* (c) Early antral follicle in which the granulosa and thecal cells have proliferated. The thecal cells now comprise two layers, an outer fibrous theca externa and an inner theca interna, rich in blood vessels and large, foamy cells with abundant smooth endoplasmic reticulum. Within the granulosa layer coalescing fluid drops are evident. (d) Expanded antral follicle, with a fully developed follicular antrum, leaving the oocyte surrounded by a distinct and denser layer of granulosa cells, the cumulus oophorus. The oocyte itself shows only a small increase in size.

those that do are selected. It does seem clear that the initiation and early progress through preantral follicular development occurs independently of any direct *extra-ovarian* controls, although the paracrine action of cytokines, such as the epidermal growth factor (EGF) within the ovary, may well be involved. However, there comes a point when further follicular development *does* require external support, and, as was found for spermatogenesis in the male, this external support is provided by the pituitary.

Table 4.1 Duration of phases of follicular development in the non-pregnant animal

Species	Preantral phase (days)	Antral phase (hours)	Preovulatory phase (hours)	Luteal phase (days)
Mouse	14–30	4	11	2
Human	85*	8–12	37	12–15
Sheep	NK	4–5	22	14–15
Cow	NK	c.10	40	18–19
Pig	NK	c.10	41	15–17
Horse	NK	c.10	40	15–16

* Also includes very early antral development (see Table 4.2).
NK, not known

Removal of the pituitary (hypophysectomy) prevents the complete antral development of follicles. The precise stage at which follicles arrest in the absence of a pituitary depends on the species: in the rat, arrest occurs at late preantral to early antral stages, but in the human, it occurs when antral follicles are slightly more advanced, with a diameter of 2 mm. The granulosa cells in arrested follicles show reduced protein synthetic activity, accumulation of lipid droplets and pyknotic nuclei. Death of the oocyte follows. The follicle is invaded by leucocytes and macrophages, and becomes fibrous scar tissue. This process is called *atresia*, and the follicles are said to be *atretic*.

Atresia is prevented only by the presence of pituitary luteinizing hormone (LH) and follicle stimulating hormone (FSH). These hormones bind to follicular *FSH and LH receptors* that first appear on cells in the late preantral and early antral follicles. In hypophysectomized animals, early antral follicles can be saved from atresia simply by administering gonadotrophins. The effect of the gonadotrophins is to stimulate further antral growth. What do the gonadotrophins do and where in the follicle do they do it?

When the distribution of receptors in early antral follicles is analysed, it is found that only the *cells of the theca interna bind LH* whereas only the *granulosa cells bind FSH*. Moreover, the effects of hormone binding at each of these sites produces very different consequences, of which the most important is the nature of the sex steroids synthesized by each cell population.

The antral follicles produce and release increasing amounts of steroids as they grow under the influence of the gonadotrophins. The main oestrogens produced are oestradiol 17β and oestrone. Antral follicles also account for 30–70% of the circulating androgens, mainly androstenedione and testosterone, found in women (the remainder coming from the adrenal). Each class of these families of sex steroid is produced in different parts of the follicle. Thus, the antral follicle can be dissected surgically, so that its granulosa cells become separated from the cells of the theca interna. When grown separately *in vitro*, the *thecal cells are found to synthesize androgens* from acetate and cholesterol. This conversion is greatly stimulated by LH (Fig. 4.5). Only very limited oestrogen synthesis by these cells is possible, particularly in the early stages of antral growth. The granulosa cells, in contrast, are incapable of forming androgens. If, however, the granulosa cells are supplied with exogenous androgens, they possess enzymes that will readily aromatize them to oestrogens. *This aromatization is stimulated by FSH* (Fig. 4.5).

Thus, the androgens produced by developing follicles are derived exclusively from thecal cells, whereas the oestrogens can arise via two routes. One involves *cell cooperation* in which *thecal androgens* are *aromatized* by the *granulosa cells*. The other route is by *de novo* synthesis from acetate in thecal cells. The balance between these two potential sources of oestrogen varies with different species, but it seems probable that *all* of the oestrogen within the follicular fluid and *most* of the oestrogen released from the follicle to the blood results from cell cooperation.

Both the production of steroids and the increase in size of antral follicles are intimately interlinked. It has become clear recently that the steroids, in addition to their release systemically via secretion into the blood, also have a *local intrafollicular role*. Oestrogens, progestagens and androgens can all be detected in follicular fluid. The *androgens*, as well as serving as substrates for conversion to oestrogens, also *stimulate aromatase activity*. The *oestrogens* can bind to receptors in the granulosa cells, which are then *stimulated to proliferate* and also to synthesize yet more oestrogen receptors. As the granulosa cells are the major site of conversion of androgens to oestrogens, a system of *positive feedback* is operating in which *oestrogen stimulates further oestrogen output*. This process therefore culminates towards the end of antral expansion in a *surge of circulating oestrogens* from the *most advanced follicles*. Daily monitoring of urinary oestrogen levels therefore provides a good guide to the state of maturity of the most mature follicles (Table 4.2).

Steroids are not the only agents of *paracrine* activity within the follicle, as the production and activity of a number of cytokines are also stimulated by the gonadotrophins. These act to modulate the actions of steroids and gonadotrophins, and some of them are listed together with their postulated actions in Table 4.3.

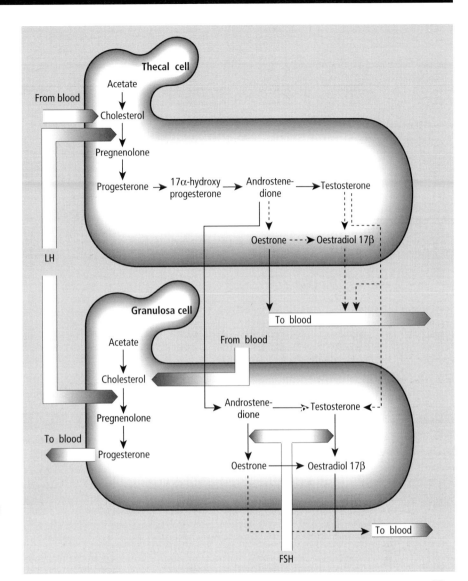

Fig. 4.5 Scheme outlining the principal steroidogenic pathways in follicular cells. Bold arrows represent the main activities of antral follicles. Light arrows represent activities developing in preovulatory follicles and persisting into the luteal phase. In the corpora lutea, the oestrogenic synthetic capacity of thecal cells only persists in those species in which the thecal cells become incorporated into the corpus luteum. Where alternative pathways exist, the minor pathway is indicated by a dashed line.

In particular, note that *activin suppresses androgen output by the theca* but *stimulates the granulosa cells to develop aromatizing capacity*, while *inhibin stimulates androgen output* and *moderates aromatizing activity*. As activin is present early in the antral phase but inhibin only later, these cytokines may act to regulate the balance between androgen output and conversion, so avoiding excess androgen production prematurely.

Oestrogens, in conjunction with FSH, have a crucial further role within the expanded antral follicle. Together these hormones stimulate the *appearance of LH-binding sites* on the *outer layers of granulosa cells*, which hitherto lacked them. These LH-binding sites are critical for successful entry of the expanded antral follicle into the preovulatory phase of follicular growth.

Ovulation

Just as late preantral and early antral follicles trickling through the first, hormone-independent phase of follicular growth will become atretic unless exposed to tonic levels of FSH and LH levels, so the expanding antral follicles will also die unless a *brief surge of high levels of LH* coincides with the appearance of LH receptors on the outer granulosa cells. If an LH surge occurs when both the granulosa and thecal cells can bind LH, then entry into the preovulatory phase of growth occurs. If it does not, the expanded antral follicle dies.

The surge of LH affects these advanced follicles in two ways: first, it causes terminal growth changes in both the

Morphology	Day of menstrual cycle	Diameter mm§	FSH/LH receptors present	(Oestrogen) pmoles/L in peripheral blood
Preantral	Throughout	<0.5	-	NA
Very early antral	Throughout	Up to 2	-	NA
Early antral	1–6	2–7	+	<120
Expanding antral*	6–10	7–10	+	100–200
Expanded antral	10–12	10–20	+	200–400
Preovulatory	13–14	20–25	+	800 and >†

* In naturally cycling women a single dominant follicle emerges at this point and only it grows thereafter.
† 10^3 to 10^4 higher oestrogen concentrations within the follicular fluid itself.
§ Recent advances in ultrasound technology now make it possible to monitor these final stages of follicular growth in the conscious subject, and thereby to ascertain how near the follicles are to ovulation. See fig. 14.9.
NA, Not applicable.

Table 4.2 Human follicular development

follicle cells and the oocyte, which results in the oocyte's expulsion from the follicle at *ovulation*; second, it changes the whole endocrinology of the follicle, which becomes a *corpus luteum* at ovulation.

Preovulatory growth

Within 3–12 hours, depending on the species, of the beginning of a surge of LH, dramatic changes occur in the oocyte. The nuclear membrane surrounding the dictyate chromosomes breaks down, and the arrested meiotic prophase is ended. The chromosomes progress through the remainder of the first meiotic division (see Fig. 1.1 for details of meiosis), culminating in an extraordinary cell division in which *half* the chromosomes, but almost *all* the cytoplasm, goes to one cell: the *secondary oocyte* (Fig. 4.6). The remaining chromosomes are discarded in a small bag of cytoplasm called the *first polar body*, which dies subsequently. This unequal division of cytoplasm conserves, for the oocyte, the bulk of the materials synthesized during earlier phases.

The chromosomes in the secondary oocyte immediately enter the second meiotic division and come to lie on the second metaphase spindle. Then, suddenly, meiosis arrests yet again, and the oocyte is ovulated in this *arrested metaphase state*. We do not know the biological significance of this second meiotic arrest, but it seems to be caused by the presence of a protein complex called *cytostatic factor*. One component of this complex is a specific protein called *c-mos*. Just how c-mos functions to

cause arrest is the subject of much research, and is considered in more detail in Chapter 8.

The termination of the dictyate stage and the progress of *meiotic maturation* through to *second metaphase arrest* and ovulation is accompanied by *cytoplasmic maturation* of the oocyte (Fig. 4.6). The intimate contact between the oocyte and the granulosa cells of the cumulus is broken by withdrawal of the cytoplasmic processes. The Golgi apparatus of the oocyte synthesizes lysosomal-like granules, which migrate towards the surface of the oocyte to assume a subcortical position (*cortical granules*). Protein synthetic activity continues at the same rate, but new and distinctive proteins are synthesized. This activity prepares the oocyte for fertilization (see Chapter 8). If an oocyte is shed from its follicle prematurely, or removed surgically prior to completion of these maturational events, then its fertilizability is much reduced. For this reason, in clinical *in vitro* fertilization programmes, human oocytes aspirated from preovulatory follicles may be cultured for a few hours before addition of spermatozoa. This procedure improves the chances of a complete maturation by the oocyte.

Meiotic and cytoplasmic maturation of the oocyte is stimulated by the surge of LH, yet it is clear that LH cannot, and does not, bind to the oocyte itself. Therefore its effect must be mediated via the cells of the follicle, where it is thought to act by suppressing cAMP levels.

In addition to acting on follicle cells in order to generate signals to the oocyte, LH also affects directly the growth and endocrinological activity of the follicle cells themselves. A final and considerable increase in follicu-

Cytokine	Source	Actions
Epidermal growth factor (EGF)	Preantral granulosa cells (GCs) and thecal cells	Supresses LH receptors and oestrogen synthesis in GCs
Insulin-like growth factors 1 and 2 (IGF1,2)	Mitotic GCs at antral stage (interstitial cells in primates?); FSH/LH modulate their activities	Promote FSH-induced mitosis/differentiation of GCs, and androgen synthesis by thecal cells
Transforming growth factor β (TGF-β)	GCs	GC proliferation and FSH-induced differentiation
Fibroblast growth factors (FGFs)	Early immature follicles, thecal cells and corpus luteum	Block oestrogen output, stimulate progesterone output, breakdown of basement membrane
Activin	GCs in early antral phase (stimulated by FSH and androgen) and luteal cells (stimulated by LH/HCG)	Stimulates FSH-induced terminal differentiation of GCs and arrests atresia; attenuates LH-induced androgen rise in theca
Inhibin	GCs in later antral phase, and corpus luteum	Moderates activin-induced stimulation of GCs; enhances LH-induced androgen rise in theca and progesterone rise in corpus luteum

Table 4.3 Cytokines and follicular development

lar size occurs (reaching 25 mm in diameter and more in the human; Table 4.2), almost exclusively due to a rapid expansion of the volume of follicular fluid. This expansion is accompanied by a loosening of the intercellular matrix between the more cortical layers of granulosa cells and an increase in total blood flow to the follicle.

Preovulatory endocrine changes

The preovulatory growth in follicular size is matched by changes in the pattern of steroid secretion. Within 2 hours or so of the beginning of the LH surge, there is a transient rise in the output of follicular oestrogens and androgens, followed by a decline. This rise coincides with distinctive changes in the thecal layer, which appears transiently stimulated and hyperaemic. The outer cells of the granulosa layer also show a marked change in their properties a few hours after the LH peak.

First, they no longer convert androgen to oestrogen, but instead *synthesize progesterone*. Second, LH *stimulates this synthesis of progesterone* via the newly acquired LH receptors. Third, the cells have lost, or reduced, their capacity to bind oestrogen and FSH. This acquisition of the capacity to respond to LH by synthesizing progesterone results in a release of progesterone from the follicle, which becomes significant in the human several hours prior to ovulation, although in most species only just before or immediately after ovulation.

Although this preovulatory phase of follicular growth is the shortest (see Table 4.1), it is also the most dramatic. It culminates in the remarkable process of ovulation.

The process of ovulation

By the end of the preovulatory phase of follicular growth, the rapid expansion of follicular fluid has

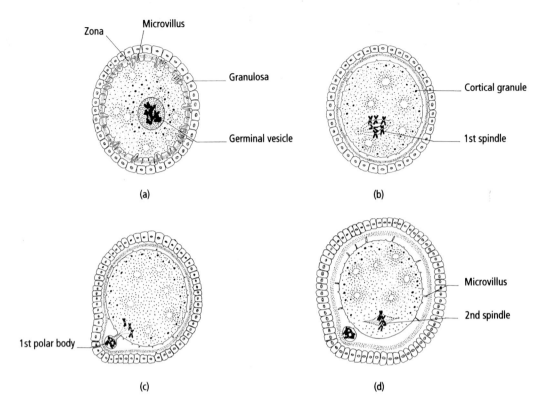

Fig. 4.6 Reactivation of meiosis in the preovulatory oocyte. A few hours after LH stimulation of the follicle (a → b), the germinal vesicle breaks down, and the chromosomes complete prophase and arrange themselves on the first meiotic spindle. Meanwhile, cytoplasmic contact between oocyte and granulosa cells ceases, and cortical granules made by the Golgi apparatus migrate to the surface. Subsequently, the first meiotic division is completed with expulsion of the first polar body (b → c). The chromosomes immediately enter the second meiotic division but stop at second metaphase. The cytoplasm around the eccentrically placed spindle is devoid of cortical granules and the overlying membrane lacks microvilli (c → d). The oocyte is ovulated in this arrested state (except in dogs and foxes in which the oocyte is ovulated at metaphase 1, and the first polar body is extruded after ovulation).

resulted in a relatively thin peripheral rim of granulosa cells and regressing thecal cells, to which the oocyte, with its associated cumulus cells, is attached only by a tenuous and thinning stalk of granulosa cells. The increasing size of the follicle and its position in the cortex of the ovarian stroma causes it to bulge out from the ovarian surface leaving only a thin layer of epithelial cells between the follicular wall and the peritoneal cavity. At one point in this exposed periphery, the wall becomes even thinner and avascular; the cells in this area dissociate and then appear to degenerate, and the wall balloons outwards. The follicle then ruptures at this point, the *stigma*, causing the fluid to flow out on to the surface of the ovary carrying with it the oocyte and its surrounding mass of cumulus cells.

In many species, including humans, the ovarian surface is directly exposed to the peritoneal cavity, but in some (e.g. the sheep, horse and rat) a peritoneal capsule or bursa encloses the ovary to varying degrees and acts to retain the oocyte cumulus mass(es) close to the ovary.

There they are collected by cilia on the *fimbria* of the oviduct, which sweep the cumulus mass into the *oviducal ostium* (Fig. 4.1). The residual parts of the follicle within the ovary collapse into the space left by the fluid, the oocyte and the cumulus cells, and within this cavity a clot forms. Thus, the postovulatory follicle is comprised of a fibrin core, surrounded by several collapsed layers of granulosa cells, enclosed within a *fibrous outer thecal capsule*.

The precise biochemistry of the ovulatory events is unclear. Proteolytic enzymes are important. The levels of *plasminogen activator* (a protease which also cleaves procollagenase to generate active collagenase) secreted by the granulosa cells reach a peak at the time of ovulation, while those of its inhibitor reach a nadir. Another protease, *renin*, is present in thecal cells, and the follicular activity of a trypsin-like enzyme is elevated. These enzymes may collaboratively digest the follicular wall at the stigma, and their activity seems to be stimulated by the LH surge.

The corpus luteum

Formation and function

The collapsed follicle now transforms into a *corpus luteum* (Fig. 4.2c). Within the follicular antrum the fibrin core undergoes fibrosis over a period of several days; the membrana propria between the granulosa and thecal layers breaks down and blood vessels invade. Both luteinizing granulosa cells and cells from the theca interna contribute to the corpus luteum, although many thecal cells also disperse to the stromal tissue. The granulosa cells cease dividing and hypertrophy to form large lutein cells, rich in mitochondria, smooth endoplasmic reticulum, lipid droplets, Golgi bodies and, in many species, a carotenoid pigment, lutein, which may give the corpora lutea a yellowish or orange tinge. This transformation is referred to as *luteinization* and is associated with a steadily increasing secretion of progestagens from the corpus luteum. The thecal cells form smaller lutein cells, produce progesterone and androgens, and seem to be richer in LH receptors. After LH or CG stimulation, these smaller cells can serve as a stem cell population for the more endocrinologically active, large lutein cells in some species.

In most species, the principal progestagen secreted from the large lutein cells is progesterone, but secretion of significant quantities of 17α-hydroxyprogesterone in primates and of 20α-hydroxyprogesterone in the rat and hamster also occurs. In a few species, notably the great apes and humans, and to a lesser extent the pig, the corpus luteum also secretes oestrogens, particularly oestradiol 17β. Its source also seems to be the large luteal cells, using as substrate androgens derived from the small luteal cells. In most species (e.g. the monkey, sheep, cow, rabbit, rat and horse), however, the corpus luteum secretes only trivial amounts of oestrogen.

The corpus luteum also secretes two other hormones. Inhibin is secreted in high amounts in higher primates. It acts to promote production of progesterone (see also Chapter 5). The second hormone is oxytocin (Fig. 2.7), which comes from the large lutein cells and the importance of which will become evident (see 'Luteolysis' below).

Maintenance

The endocrine support of the corpus luteum, like its composition and secretory pattern, shows considerable variation between species. The conversion of a follicle to a corpus luteum requires that high surge levels of LH provoke ovulation. This gonadotrophin is then also required, albeit at much lower levels, for the maintenance of the corpus luteum. However, in some species, prolactin is also an important component of the so-called *luteotrophic complex*. In those species in which prolactin is luteotrophic (Table 4.4), prolactin receptors can be detected on granulosa cells from the preovulatory stage onwards.

Luteolysis

The life of the corpus luteum in the non-pregnant female varies from 2 to 14 days (Table 4.1 & Chapter 10). Luteal regression or *luteolysis* involves a collapse of the lutein cells, ischaemia and progressive cell death with a consequent fall in the output of progestagens. The whitish scar tissue remaining, the *corpus albicans*, is absorbed into the

Species	LH*	Prolactin	Oestrogen	FSH†
Human	++	+?	Luteolytic?	-
Cow	++	+?	Luteolytic?	-
Sheep	++	+?	-	-
Pig	+	+?	+	-
Rabbit	+	?	+++	+
Rat/mouse	+	+++	-	-
Dog	+	+	?	?
Hamster	+	+	?	+

* LH is essential to stimulate the initiation of luteinization in all species. Here its role subsequently is assessed. It may be directly luteotrophic, but in species in which follicles are growing through the luteal phase, the LH may also stimulate some oestrogen synthesis locally (see The ovarian cycles of humans in relation to those of the cow, pig, sheep and horse in text).

† FSH may only be luteotrophic indirectly by stimulating oestrogen.

NK, Not known.

Table 4.4 Hormones luteotrophic for the non-pregnant corpus luteum of different species

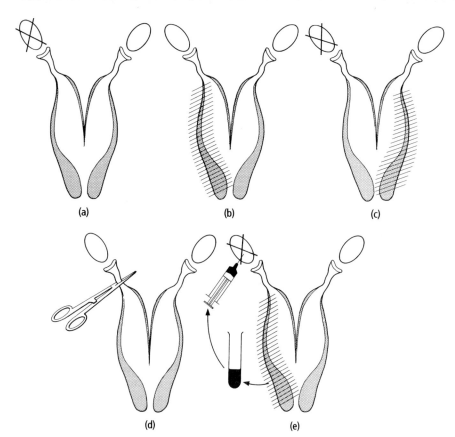

Fig. 4.7 Non-pregnant sheep uterus and ovaries. (a) A single corpus luteum present in the left ovary regresses (X). (b) Removal of the ipsilateral uterine horn (hatched) prevents regression. (c) Removal of the contralateral horn does not prevent regression. (d) Clamping the blood supply between horn and ovary prevents regression. (e) If the endometrium of the removed ipsilateral horn is homogenized and reinjected into the ovarian artery, the corpus luteum regresses (compare b with e).

stromal tissue of the ovary over a period that varies from weeks to months, depending on the species. Luteolysis can be caused by withdrawal or inadequacy of the luteotrophic complex. However, in many species, it is not primarily a failure of luteotrophic support, but active production of a *luteolytic factor* that brings about normal luteal regression.

In many mammals studied, *with the notable exception of primates*, it is found that luteal life can be prolonged by *hysterectomy* (removing the uterus). However, if the endometrium of the excised uterus is homogenized and injected, then luteolysis does occur (Fig. 4.7e). Luteal prolongation can also be achieved by ligating the tissues, including the blood vessels, between the uterus and the ovary (Fig. 4.7d). These results led to the suggestion that a humoral factor passes from the endometrium to the ovary, and causes luteolysis. Two observations on the ewe suggested that this humoral factor was highly labile. First, if instead of removing the whole uterus, which in the ewe has two separate horns, only one horn was removed, then only the corpus luteum on the opposite side regressed (Fig. 4.7b,c). Second, if the whole uterus was transplanted elsewhere in the body, luteolysis was prevented unless the ovaries had also been transplanted with the uterus as a unit.

The identity of the endometrial substance responsible has now been established for the sheep, cow, guinea-pig and horse as *prostaglandin* $F_{2\alpha}$ ($PGF_{2\alpha}$; see Fig. 2.4). $PGF_{2\alpha}$ is first secreted about 10–15 days after corpus luteum formation. It passes from the endometrium into the uterine vein in a series of pulses at roughly 6-hour intervals. From the vein, it passes by a local counter-current transfer to the ovarian artery, and thence to the ipsilateral ovary. Corpus luteum regression follows shortly thereafter. If this $PGF_{2\alpha}$ is neutralized experimentally by specific antibodies, then luteolysis is prevented. Conversely, injections of exogenous $PGF_{2\alpha}$ 1 or 2 days in advance of its endogenous output leads to advanced luteolysis; indeed, $PGF_{2\alpha}$ will cause regression of luteinized cells *in vitro*.

The production of $PGF_{2\alpha}$ is stimulated by the oxytocin secreted from the corpus luteum, and if oxytocin is neutralized, then luteolysis is delayed. One effect of $PGF_{2\alpha}$ is, in turn, to stimulate more oxytocin release from the corpus luteum, thereby effecting a positive feedback loop. However, oxytocin is present in the corpus luteum well before 10–15 days of luteal life, so why does it not stimulate $PGF_{2\alpha}$ secretion earlier? The answer seems to be that oxytocin receptors in the endometrium, which are required for its action, are absent until this time. What causes their appearance is uncertain.

The control of luteolysis in *primates*, unlike that for other species, does *not* involve uterine prostaglandins. The levels of prostaglandins secreted do rise in the late luteal phase, but neither hysterectomy nor antibodies to prostaglandins prolong luteal life. Moreover, injections of prostaglandins are without effect on the corpus luteum, unless very high doses are used when a transient drop in progesterone output occurs. What then causes luteolysis in the primate? One possibility is that the very low levels of LH during the luteal phase of primates (see 'The ovarian cycles of humans in relation to those of the cow, pig, sheep and horse' below) are insufficiently luteotrophic and the corpus luteum simply regresses slowly.

HOW MANY FOLLICLES OVULATE?

The number of follicles ovulating in any one cycle is characteristic for each species and ranges from one to several hundred. There appear to be two points during follicular maturation at which this number can be regulated, and in both cases it is the balance between survival and atresia that is critical. Thus, in humans about 15–20 early antral follicles are rescued from atresia and *recruited* for development at the start of each menstrual cycle. FSH is the crucial hormone for follicular recruitment, although of course LH is necessary for fully functional follicles. Thus, the number of follicles recruited can be increased if endogenous FSH levels are augmented by exogenous FSH or reduced if FSH levels are diminished too low. This observation must mean that a delicate interplay between the levels of circulating gonadotrophins and the follicular levels of FSH receptors determines the number of follicles recruited from the early antral pool. *If appropriate hormone levels and acquisition of sufficient receptors coincide, then follicular development continues. If hormone levels are inappropriate when the receptors develop, then follicular atresia ensues.*

Of the 15–20 follicles recruited in the human, usually only one emerges as *dominant* during further follicular growth. It is only this one dominant follicle that subsequently undergoes preovulatory growth, develops LH receptors on its granulosa cells, and subsequently, therefore, is able to ovulate in response to a surge in LH. We do not know why one follicle emerges, but it is possible that cytokines are again involved. FSH stimulates the production of insulin-like growth factor 1 (IGF-1) from granulosa cells and this IGF-1 appears to be involved in the stimulation of aromatase activity, and thus oestrogen production. If IGF-1 activity is neutralized, the effectiveness of FSH in stimulating oestrogen output is reduced. FSH is also involved in the regulation of ovarian production of the endogenous IGF-binding proteins described in Chapter 2 (see 'Binding proteins'). These can be produced by the theca, but FSH *suppresses* their synthesis and so prevents them from sequestering the IGF-1 from its receptors. It is possible, therefore, that among growing follicles, IGF-binding proteins are produced preferentially by those follicles with the *least good* supply of FSH and/or the *fewest FSH receptors*. The capture of IGF-1 by its binding protein would further impair the effectiveness of FSH and so promote a downward spiral of the follicle to atresia. In contrast, those follicles best able to respond to FSH would spiral upwards to expansion and ovulation. However, other factors may well be involved in the selection of the dominant follicle.

FOLLICULAR DEVELOPMENT AND THE OVARIAN CYCLE

In the previous sections, we have described the development of an *individual follicle* through to either ovulation and luteinization, or to atresia *en route*. But each of the two ovaries has many primordial follicles, and we must now consider the relationships between the various follicles developing at different times, and in both ovaries, in order to obtain an overall picture of ovarian function and cyclicity.

The ovarian cycle

One complete *ovarian cycle* is the interval between successive ovulations, where each ovulation is preceded by a period of oestrogen dominance. As the oestrogens are derived from the follicles, the period prior to ovulation is often called *the follicular phase of the cycle*. Correspondingly, the postovulatory period is often called the *luteal phase*, because progesterone is derived from the corpus luteum. The duration of the ovarian cycle and its constituent follicular and luteal phases in various species are summarized in Table 4.5. It is immediately clear that there are major differences between species in both the absolute length of the ovarian cycle and in the relative duration of its follicular and luteal components. In fact, these apparent major differences mask a fundamentally similar organization, and result only from minor but significant modifications to a basic pattern. First, we will discuss the human ovarian cycle, as it is the easiest to understand. We will then relate that pattern to those of other species.

Table 4.5 Ovarian cyclicity in different species

Species	Length of cycle (days)	Follicular phase (days)	Luteal phase (days)
Human	24–32	10–14	12–15
Cow	20–21	2–3	18–19
Pig	19–21	5–6	15–17
Sheep	16–17	1–2	14–15
Horse	20–22	5–6	15–16
Mouse and rat* (+ infertile male)	13–14	2	11–12
Rabbit* (+ infertile male)	14–15	1–2	13
Mouse and rat	4–5	2	2–3
Rabbit	1–2	1–2	0

* See text for discussion.

The ovarian cycles of humans in relation to those of the cow, pig, sheep and horse

Figure 4.8a shows a basic outline of two sequential human ovarian cycles. The pattern of measured blood steroids and gonadotrophins is indicated below, and the activities of the four follicular stages are indicated above. A continuous trickle of developing early antral follicles occurs throughout the cycle; growth of these follicles does not require gonadotrophic support, and the follicles do not secrete significant levels of steroids, and thus do not affect blood levels of steroids. These early antral follicles are doomed to atresia, however, unless rescued by FSH and LH. During the luteal phase, LH and FSH levels are negligible and so atresia occurs. In contrast, during the first part of an ovarian cycle, LH and FSH levels are sufficient to support the expansion of new antral follicles. One of these rescued preantral follicles will survive to become the dominant antral follicle, secreting high levels of oestrogens at maturity 8–12 days later: witness the rising blood oestrogen levels. This advanced follicle is then converted to a preovulatory follicle by the transient high levels of LH that may be measured in the blood at this time. The other antral follicles become atretic. The successful ovulatory follicle forms a corpus luteum, which secretes progesterone and oestrogen, until luteolysis 14 or so days later. A new cycle then begins as tonic gonadotrophin levels are elevated.

Two important features emerge from this description of the human ovarian cycle: first, we need to understand what controls the fluctuating levels of gonadotrophin output (this will be discussed in detail in the next chapter); second, the *complete antral expansion phase, ovulation and the complete luteal phase occupy one complete ovarian cycle*. This second feature distinguishes humans

from the cow, pig, horse and sheep. In these species, as indicated for the pig in Fig. 4.8b, significant antral expansion can occur while the luteal phase of the previous crop of follicles is still proceeding, and this growth is possible because *FSH and LH in these species do not fall to negligible levels during the luteal phase* (see Chapter 5 for explanation of why this should be). Thus, in these species the complete antral expansion phase, ovulation and the complete luteal phase *occupy longer than one complete ovarian cycle* (follow and compare the lines with solid arrows in Fig. 4.8a,b). It is almost as though in the cow etc., the human cycle has been 'telescoped' by 'pushing' the follicular part of one cycle backwards into the luteal half of the previous cycle.

The ovarian cycles of most other species can be understood quite simply in terms of the above discussion. We will examine those of the rat, mouse and rabbit because they exhibit a certain distinctive feature of great importance in understanding the ovarian cycles of humans and the large farm animals.

The ovarian cycles of the rat and mouse

The ovarian cycles of the rat and mouse are basically of the 'telescoped' sort seen in large farm animals. They show, however, a curious and distinctive feature: the cycle differs in length, depending upon whether or not the female mates. If the female has an infertile mating at the time of ovulation, for example with a vasectomized male, her luteal phase is 11–12 days in duration (often called *pseudopregnancy*), and the ovarian cycle resembles that of the pig. However, if she fails to mate at the time of ovulation the luteal phase is only 2–3 days long. In the latter case the corpora lutea only become transiently functional in producing progestagens, secreting a small amount of progesterone, but mainly 20α-hydroxyprogesterone.

The explanation for this curious phenomenon lies in the *mechanical stimulus to the cervix* provided naturally by the penis at coitus. The presence of such stimulation is relayed via sensory nerves from the cervix to the central nervous system, and activates the release of *prolactin* from the pituitary. This hormone, as was seen earlier (Table 4.4), is an essential part of the luteotrophic complex in the rat and mouse, and without it luteal life is abbreviated from the 'normal' extended pattern characteristic of the large farm animals to only 2 or 3 days. The rat and mouse derive increased reproductive efficiency from this evolutionary modification, as without the abbreviating device they would only be fertile every 13 or 14 days instead of every 4 or 5. As their pregnancy only lasts 20–21 days, this is a highly significant economy. Evolutionary pressures for such a

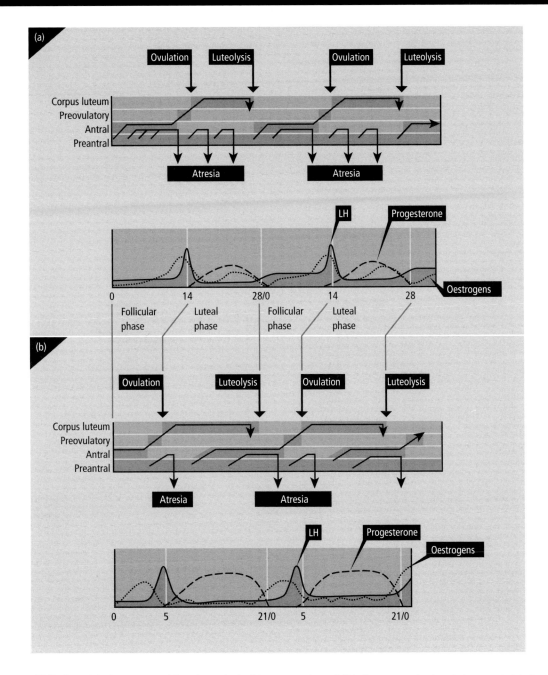

Fig. 4.8 Summary of follicular activity in two sequential ovarian cycles in: (a) the human; and (b) the pig. The presence of the different follicular stages is indicated. Relative blood hormone levels are as indicated. The history of an ovulatory follicle from preantral to luteolysis or atresia is indicated by the continuous lines marked with closed arrows.

truncation of the luteal phase would not apply to the pig, cow, sheep and horse where pregnancy is a very extended event compared with the luteal phase.

This neat neural device which shortens the luteal phase introduces an important new concept into our discussion. It illustrates how the central nervous system (CNS) can influence ovarian function. The following two chapters will discuss this topic in detail. We will firstly look, however, at one other type of modification to the ovarian cycle that re-emphasizes the role that the CNS can play.

The ovarian cycle of the rabbit

A female rabbit caged alone shows little evidence of a cycle: blood levels of oestrogens are high; progestagens are low; she is always on heat and ready to mate, but ovulation cannot be detected. Yet her ovary contains waves of expanding antral follicles. It is as though she is in a continuous follicular phase. If she is mated with a vasectomized buck, or if her cervix is stimulated mechanically, she ovulates 10–12 hours later and has a luteal phase, or 'pseudopregnancy' of 12 or so days. If the vasectomized buck is left in with her, she will show a 14-day cycle with a 2 + 12 day follicular–luteal pattern, much like the porcine pattern.

As in the rat, cervical stimulation in the rabbit is the source of a sensory input to the CNS, which, in this case, induces a surge of LH, high levels of which rescue any expanded antral follicles from atresia and ovulate them. In essence, the rabbit has abbreviated her cycle even more than the rat or mouse by eliminating the luteal phase completely. This phenomenon is known as *induced ovulation*, and is seen in a number of species (e.g. the rabbit, cat, ferret, camel and llama). Our discussion on different ovarian cycles is summarized in Fig. 4.9.

INTERSTITIAL GLANDS

We have discussed at length the activities of the maturing follicles composed of primordial follicles and accreted stromal cells: the thecal layers. However, within the stroma, and between the developing follicles, lie the interstitial glands. Do they have a role to play in ovarian function and cyclicity?

The interstitial glands show considerable variation among different species, and little is known of their activity. Interstitial glands are composed of aggregates of steroidogenic-like cells, which contain extensive smooth endoplasmic reticulum and lipid droplets. In the rabbit, in which these cells are well developed, they synthesize progesterone and 20α-hydroxyprogesterone, and are sensitive to the LH activity observed after coitus. In the rat, the interstitial glands are also stimulated by the surge of LH to form progestagens. Interstitial cells have also been implicated in the synthesis of androgens in the human, rat and rabbit ovaries, and it is possible that this tissue serves as an additional source of androgens for both secretion and aromatization in the follicles.

SUMMARY AND CONCLUSIONS

Follicular growth and differentiation through to a functional corpus luteum is a complex process that is only completed successfully by under 0.1% of follicles. Most become atretic at some point during the process. The resumption of development by primordial follicles is hormone independent. During this phase the oocyte undergoes its major growth, and the follicle acquires receptors for FSH and oestrogen (granulosa cells) and for

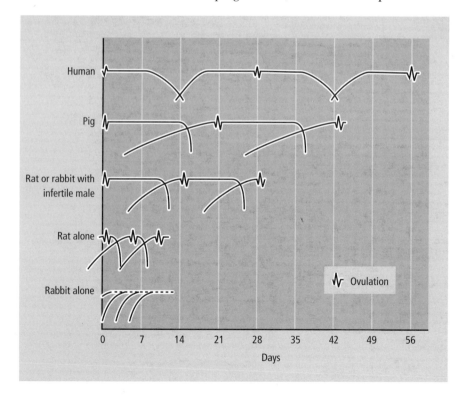

Fig. 4.9 Comparison of ovarian cycles of the human, the pig, the rat or rabbit caged with an infertile male, and the rat or rabbit caged alone. Day 0: first ovulation. Each continuous line represents a complete and successful sequence of growth of one crop of antral follicle(s) through preovulatory follicle(s) to corpora lutea and ultimately luteolysis.

LH (thecal cells). During the much shorter phase of antral expansion, growth of follicular cells occurs with a rising output of oestrogens and androgens. Adequate levels of LH and FSH are required for this phase to be completed successfully, and the higher the levels of FSH, the more follicles are likely to survive. Towards the end of this phase, oestrogen output increases from the dominant antral follicles, and their outer granulosa cells acquire LH receptors. During the short preovulatory phase, a large surge of LH gives a sudden stimulus via the LH receptors on both granulosa and thecal cells. This surge of LH first stimulates, and then stops, the endocrine activity of the thecal cells, and also switches off the aromatizing activity of granulosa cells and diverts them to the production of progestagens. This dramatic switch in endocrine activity is accompanied by the renewed meiotic and cytoplasmic maturation of the oocyte and culminates in ovulation. During this phase the granulosa cells of many species acquire prolactin receptors. The final luteal phase of follicular development is characterized by a rising output of progestagens, on which may be superimposed, in primates and pigs, a rise in oestrogen output. A luteotrophic complex of some, or all, of the three hormones, prolactin, LH and oestrogen, supports the luteal phase. During this phase, oxytocin output increases and receptors for it develop in the endometrium towards the end of the luteal phase. The oxytocin stimulates prostaglandin production, at least in non-primates. The whole sequence terminates with luteolysis, due either to the production of the uterine luteolytic hormone $PGF_{2\alpha}$ or due to failure of luteotrophic support.

The ovary is like the testis in producing both gametes and steroids. However, whereas in the male the gametes and the androgens are produced continuously and concurrently, in the female the ovarian output is cyclic and shows two distinct phases separated by the ovulatory release of an oocyte. The follicular phase, prior to ovulation, is dominated by a rising output of oestrogen and some androgens, whereas in the luteal phase, after ovulation, progestagens predominate. In many species the follicular phase is brief, and the major part of follicular growth occurs during the luteal phase of the previous cycle. In humans, however, the follicular phase is more extended.

We have considered the cyclicity of reproduction in the female from the viewpoint of the ovary. However, the cyclic output of steroids imparts cyclicity to the physiology and behaviour of the whole female as we will see in more detail in Chapter 7. This external manifestation of the internal ovarian cycle can be recognized in many mammals by a behavioural characteristic. Female mammals of most species are only receptive to males (*on heat*), and therefore willing to mate, around the time of ovulation. This period of heat is conditioned by the hormonal milieu of the preovulatory phase, and can be accompanied by marked changes in behaviour pat-

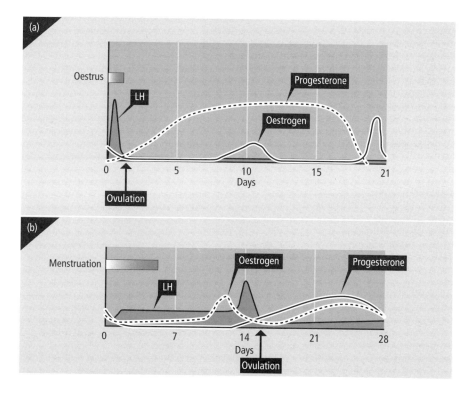

Fig. 4.10 The ovarian cycles in (a) the cow; and (b) the human, expressed as oestrous and menstrual cycles. The first day of the cycle coincides with the beginning of either oestrus or menstruation (plasma concentrations not to scale).

terns. The frisky state observed in some species, particularly in horses, was given the name *oestrus* (or 'attacked by gadflies'), and thus the internal ovarian cycle is manifested externally in the so-called *oestrous cycle*. Day 1 of each oestrous cycle is generally considered to be *the day of first appearance of oestrous behaviour*.

Higher primates show little evidence of oestrus and thus it is inappropriate to speak of their oestrous cycle. However, in higher primates, another external manifestation of ovarian cyclicity is observed: the shedding of bloody endometrial tissue via the vagina at the end of the luteal phase. This hormonally conditioned event is called *menstruation* (i.e. monthly event), and is used as the external basis for the measurement of the *menstrual cycle* in which the *first day of menstruation* is considered Day 1 of the cycle, as may be observed in Fig. 4.10. Although both the oestrous and menstrual cycles reflect ovarian cyclicity, the starting days for each cycle occur at different points in the underlying ovarian cycle.

We have seen in this chapter that the levels of gonadotrophins critically affect the successful completion of the ovarian (and thus of the oestrous/menstrual) cycle, and that the circulating levels of gonadotrophins vary naturally during the cycle. Moreover, we have observed that the CNS can affect the levels of circulating gonadotrophins. In the next chapter we will examine how gonadotrophin levels are regulated, and how the CNS exerts its effects.

FURTHER READING

Alilo HW & Dowd JP (1991) The control of corpus luteum function in domestic ruminants. *Oxford Reviews in Reproductive Biology* **13,** 203–237.

Baird DT & Smith KB (1993) Inhibin and related peptides in the regulation of reproduction. *Oxford Reviews in Reproductive Biology* **15,** 191–232.

Bromsel-Helmreich O (1985) Ultrasound and the preovulatory human follicle. *Oxford Reviews of Reproductive Biology* **7,** 1–72.

Cupps PT (Ed.) (1991) *Reproduction in Domestic Animals,* 4th edition. Academic Press, London.

Findlay JK (1994) Peripheral and local regulators of folliculogenesis. *Reproduction, Fertility and Development* **6,** 127–139.

Hillier SC (1985) Sex steroid metabolism and follicular development in the ovary. *Oxford Reviews of Reproductive Biology* **7,** 168–222.

Hirschfield AN (1991) Development of follicles in the mammalian ovary. *International Reviews in Cytology* **124,** 43–101.

Hsueh AJW & Schomberg DW (1993) *Ovarian Cell Interactions.* Springer Verlag, Berlin.

McNeilly AS, Crow W, Brooks J & Evans G (1992) Luteinising hormone pulses, follicle-stimulating hormone and control of follicular selection in sheep. *Journal of Reproduction and Fertility* (Suppl.) **45,** 5–19.

Richardson MC (1986) Hormonal control of ovarian luteal cells. *Oxford Reviews of Reproductive Biology* **8,** 321–378.

Urban AC & Veldhuis JD (1992) Endocrine control of steroidogenesis in granulosa cells. *Oxford Reviews in Reproductive Biology* **14,** 225–262.

Webley GE & Hearn JP (1994) Embryo–maternal interactions during the establishment of pregnancy in primates. *Oxford Reviews in Reproductive Biology* **16,** 1–32.

The Regulation of Gonadal Function

In Chapters 3 and 4, we saw how the follicle stimulating hormone (FSH), the luteinizing hormone (LH) and prolactin regulate the cellular and endocrine functions of the ovary and testis. In this chapter, we will discuss the mechanisms by which secretion of these pituitary hormones themselves is regulated. Pituitary hormone secretion is influenced by the central nervous system (CNS), in particular the *hypothalamus*. So we must also consider CNS function, and, in particular, how the CNS mediates hormonal and environmental influences on reproduction.

THE HYPOTHALAMIC–PITUITARY AXIS

The pituitary

The pituitary gland lies in the *hypophyseal fossa* of the *sphenoid bone*, overlapped by a circular fold of dura mater, the *diaphragma sellae*, which has a small central opening through which the pituitary stalk, or *infundibulum*, passes (Fig. 5.1). The gland has an extremely rich

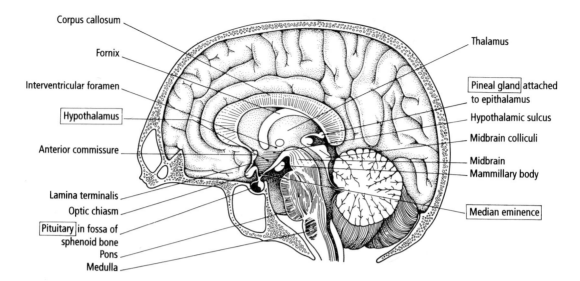

Fig. 5.1 A sagittal section of the human brain with the pituitary and pineal glands attached. Note the comparatively small size of the hypothalamus, which is shaded, and its rather compressed dimensions ventrally. The pineal is attached by its stalk to the epithalamus (habenula region) and lies above the midbrain colliculi. The interventricular foramen is the communication between lateral and third ventricles, the latter being a midline, slit-like structure, which has been opened up by the midline cut exposing this medial surface of the brain. The thalamus (above) and the hypothalamus (below) form one wall (the right in this view) of the third ventricle, which is best viewed in Fig. 5.3.

blood supply derived from the internal carotid artery via its superior and inferior hypophyseal branches. Venous drainage is by short vessels, which emerge over the surface of the gland and enter neighbouring dural venous sinuses.

There are three lobes of the pituitary gland. The anterior lobe (*adenohypophysis*) is derived embryologically from a small diverticulum (*Rathke's pouch*) pinched off from the dorsal pharynx. It is closely apposed to a distinct and smaller posterior lobe (*neurohypophysis*), which is derived embryologically from neurectoderm. This origin is reflected in its connection by a stalk of nervous tissue, the infundibulum, to the overlying hypothalamus in the region of the *median eminence* (Figs 5.1 & 5.2). An *intermediate lobe* of the pituitary (a small division of the adenohypophysis) lies between the two.

The anterior lobe of the pituitary contains a variety of cell types, among them the *gonadotrophs* (basophilic cells containing granules of FSH or LH) and *lactotrophs* (acidophilic cells containing prolactin). In addition to FSH, LH and prolactin, the anterior pituitary secretes *growth hormone* (GH or *somatotrophin*) from *somatotrophs*, *adrenocorticotrophic hormone* (ACTH or *corticotrophin*) from *corticotrophs*, and *thyroid stimulating hormone* (TSH or *thyrotrophin*) from *thyrotrophs*. The posterior lobe of the pituitary secretes the two nonapeptide hormones: *arginine vasopressin* (AVP or *antidiuretic hormone*: ADH) and *oxytocin*. Both lobes of the pituitary are connected anatomically and functionally to the overlying hypothalamus.

The hypothalamus

The hypothalamus is a relatively small area at the base of the brain. It is part of the *diencephalon* and lies between the midbrain (caudally) and the forebrain (rostrally; Fig. 5.1). The boundaries of the hypothalamus are conventionally described as the *hypothalamic sulcus* separating it from the *thalamus* (above), the *lamina terminalis* (in front) and a vertical plane immediately behind the *mammillary bodies* (behind; Fig. 5.1). The hypothalamus is split symmetrically into left and right halves by the *third ventricle*, containing cerebrospinal fluid and lying in the midline so that the hypothalamus forms its floor and lateral walls (Figs 5.1 & 5.3). Despite its small size, the hypothalamus is an extremely complicated structure with many diverse functions, which include the regulation of sexual and ingestive behaviours, the control of body temperature and the integration of autonomic activity. Each function is associated with one or more small areas of the hypothalamus consisting of aggregations of neurons, called *hypothalamic nuclei* (Figs 5.2 & 5.3). The functions of the hypothalamus associated with reproduction involve the *supraoptic, paraventricular, arcuate, ventromedial* and *suprachiasmatic* nuclei, and also two less easily defined

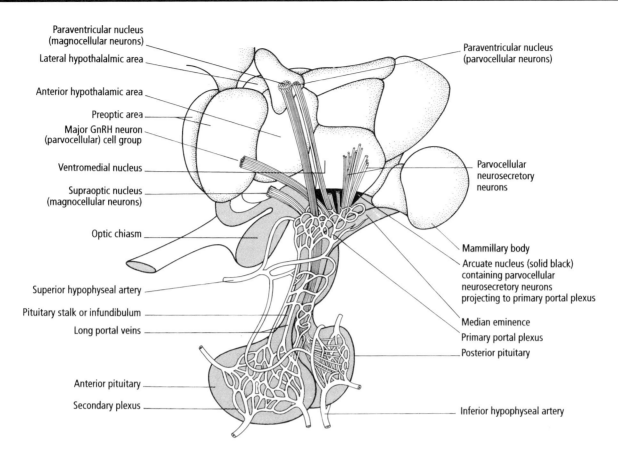

Paraventricular nucleus
(magnocellular neurons)

Lateral hypothalalmic area

Anterior hypothalamic area

Preoptic area

Major GnRH neuron
(parvocellular) cell group

Ventromedial nucleus

Supraoptic nucleus
(magnocellular neurons)

Optic chiasm

Superior hypophyseal artery

Pituitary stalk or infundibulum

Long portal veins

Anterior pituitary

Secondary plexus

Paraventricular nucleus
(parvocellular neurons)

Parvocellular
neurosecretory
neurons

Mammillary body

Arcuate nucleus (solid black)
containing parvocellular
neurosecretory neurons
projecting to primary portal plexus

Median eminence

Primary portal plexus

Posterior pituitary

Inferior hypophyseal artery

Fig. 5.2 A highly schematic and enlarged view of the human hypothalamus. Note the portal capillary system derived from the superior hypophyseal artery and running between the median eminence/arcuate region of the hypothalamus and the anterior lobe of the pituitary. The ventromedial, paraventricular and arcuate nuclei are well defined and contain neurons that terminate in close association with the portal capillaries. The anterior hypothalamic and preoptic areas should be viewed as a continuum in functional and anatomical terms. The major group of GnRH-containing neurons is indicated in the preoptic area.

areas, the *medial anterior hypothalamic* and *medial preoptic areas*. These nuclei have either direct neural or indirect vascular connections with the pituitary gland.

The hypothalamus also has rich interconnections with many parts of the brain; in particular with the *autonomic areas* and *reticular core* of the *brainstem* (especially monoaminergic cell groups) and also areas of the *limbic forebrain*, such as the *amygdala, hippocampus* and *septum*. Of major importance is the photic input from the *retina* to the *suprachiasmatic nuclei*, as this is the route by which the hypothalamus is made aware of the cycles of light and dark that can, in turn, affect reproductive function.

Connections between hypothalamus and pituitary

The posterior pituitary

The axons of the magnocellular neurons in the supraoptic and paraventricular nuclei project directly to the posterior lobe of the pituitary via the hypothalamo-hypophyseal tract (Fig. 5.2). The hormones of the posterior pituitary (vasopressin and oxytocin) are synthesized in the neuronal cell bodies of these two hypothalamic nuclei, each packaged by specifically associating with a binding protein (*neurophysin*), and transported along the axons by a process of axoplasmic flow to be stored in the posterior lobe of the pituitary (Fig. 2.7). From here, release of the hormone into the bloodstream occurs. This system of neurons in the hypothalamus is called the *magnocellular* (i.e. large-celled) neurosecretory system (Fig. 5.4a).

The anterior pituitary

In contrast to the direct neural connections linking the hypothalamus and posterior pituitary, the hypothalamus communicates with the anterior lobe by a vascular route (Fig. 5.2). A variety of neurohormones is synthesized in hypothalamic neurons, designated the *parvocellular* (i.e. small-celled) neurosecretory system. The axons of these neurons terminate in the pericapillary space of the

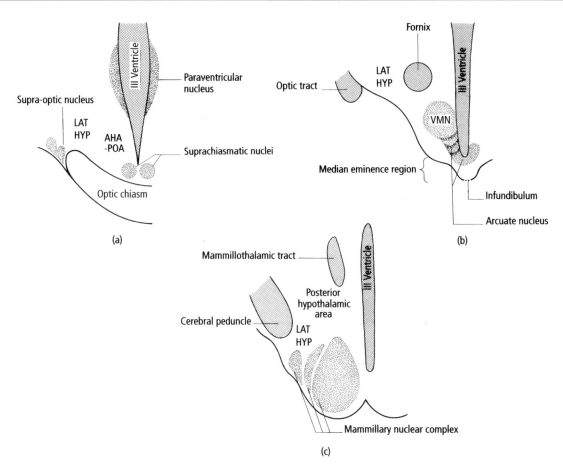

Fig. 5.3 Three coronal sections at different anterior–posterior levels of the human hypothalamus (consult Fig. 5.2 to construct planes of sections). (a) Through the optic chiasm, note the third ventricle in the midline flanked by the paraventricular (magnocellular) nuclei, which, together with the supraoptic nuclei, synthesize oxytocin and vasopressin. The latter are then transported along the hypothalamo-hypophyseal tract (axons of neurons in these nuclei) to the posterior pituitary. The region of the suprachiasmatic nuclei and the anterior hypothalamic–preoptic area (AHA–POA) are shown. (b) Through the infundibulum and showing the relationship between the arcuate and ventromedial nuclei (VMN). The capillary loops of the portal plexus are found in this median eminence region. (c) Through the level of the mammillary bodies and showing the mammillary nuclear complex. The area labelled LAT HYP in all three sections is the lateral hypothalamus, and is composed of many ascending (largely aminergic neurons from the brainstem) and descending (from the rostral limbic and olfactory areas) nerve fibres. This pathway represents a major input/output system for the more medially placed hypothalamic nuclei.

primary portal plexus of vessels in the lateral and medial palisade zones of the external layer of the median eminence. These capillaries are derived from the superior and inferior hypophyseal arteries and pass from this area of the hypothalamus down to the anterior pituitary. By this route, the neurohormones manufactured by hypothalamic neurons reach and act upon the gonadotrophs, thyrotrophs, corticotrophs, somatotrophs and lactotrophs in order to regulate the synthesis and release of their various hormones. Several hypothalamic areas contain parvocellular neurons, including the paraventricular nucleus, periventricular and medial preoptic area, and the arcuate and ventromedial nuclei (Fig. 5.4b).

Hypothalamic control of gonadotrophin secretion

The glycoprotein hormones LH and FSH are secreted by gonadotrophs in the anterior pituitary. Immunocytochemical studies have revealed that each hormone is generally elaborated in a different cell type, but occasionally both may be found in the same cell. The synthesis and secretion of FSH and LH depend on gonadotrophin releasing hormone or GnRH (a decapeptide; see Chapter 2, Fig. 2.6). Immunocytochemical techniques have located GnRH (and its precursor peptide) in a major group of neurons within a continuum that includes the caudal medial septal nucleus, and the periventricular medial preoptic and adjacent anterior

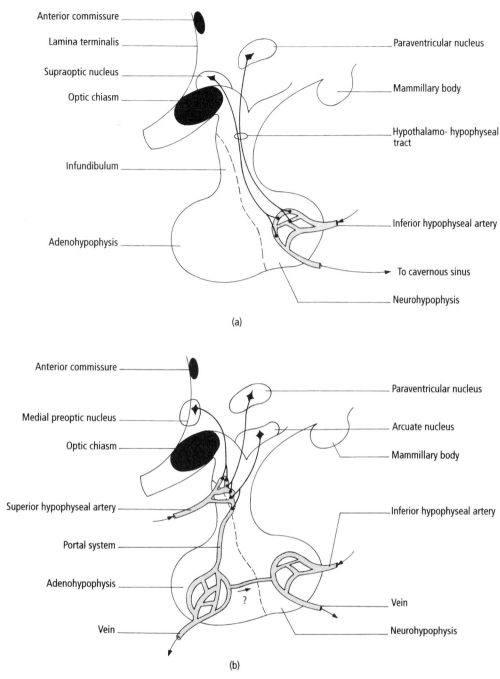

Fig. 5.4 (a) Schematic representation of the magnocellular neurosecretory system. Neurons in the supraoptic and paraventricular nuclei send their axons via the hypothalamo-hypophyseal tract, and through the infundibulum to the neurohypophysis, where terminals lie in association with capillary walls, the site of neurosecretion. (b) Schematic representation of the parvocellular neurosecretory system. Neurons in the medial preoptic area (e.g. GnRH-containing), anterior periventricular nucleus, medial parvocellular paraventricular nucleus (e.g. vasoactive intestinal polypeptide- and CRF-containing), and arcuate nucleus (e.g. dopamine-containing) send their axons down to the portal vessels in the external layer (palisade zone) of the median eminence, where neurosecretion occurs.

hypothalamic areas (Fig. 5.5a). GnRH-containing neurons have also been demonstrated in the arcuate nucleus of primate species, but not reliably in the brains of non-primate species. Nerve terminals containing GnRH are particularly common in association with portal capillaries of the lateral palisade zone (Fig. 5.5a), which is thus the primary site of GnRH neurosecretion.

GnRH is assumed to be the most important final common mediator of all influences on reproduction conveyed through the CNS. Any abnormality in GnRH syn-

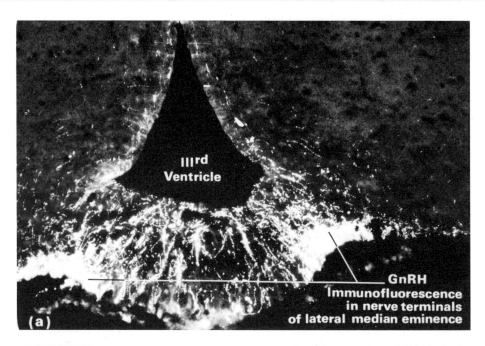

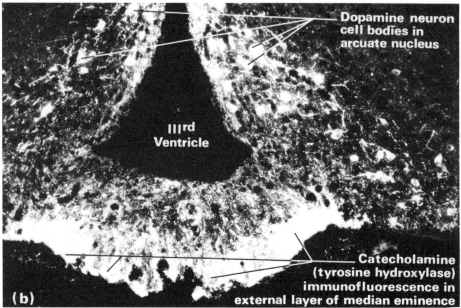

Fig. 5.5 (a) GnRH detected by immunofluorescence in the median eminence (middle region). Note the very dense network of GnRH neuron terminals in the lateral region (lateral palisade zone), but no fluorescent cell bodies are visible. (b) Tyrosine hydroxylase immunofluorescence in an adjacent section through the median eminence. The dense terminal fluorescence in the external layer largely represents dopamine, but noradrenaline-containing terminals will also be labelled by this procedure. Large dopamine-containing cell bodies are clearly visible in the arcuate nucleus.

thesis, storage, release or action will result in partial or complete failure of gonadal function. Destruction of GnRH-producing neurons in the hypothalamus, or immunization against the peptide, prevents gonado-trophin function and results in gonadal atrophy. In the former case, this can be reversed by appropriate treat-ment with intravenous synthetic GnRH. Concurrent sampling of portal blood to measure GnRH secretion and of peripheral blood to measure gonadotrophins, has established the relationship between GnRH and gonadotrophin secretion. Both are secreted in a pulsatile manner, approximately one pulse being measured about every hour or so and therefore called *circhoral pulses*. Each peripheral LH peak coincides with a GnRH pulse (Fig. 5.6), although every GnRH pulse is not necessarily followed by an LH pulse. This '*circhoral clock*' (sometimes called the GnRH pulse generator) controlling pulsatile GnRH secretion seems to reside in the hypothalamus. Thus, in rhesus monkeys in which the mediobasal hypothalamus has been isolated from the rest of the brain by surgical knife cuts, the pulsatile secretion of LH and FSH is preserved. Moreover, electrophysiological recording in the mediobasal hypothalamus has revealed a marked increase in multiunit neuronal activity in synchrony with each peripheral LH pulse. However, whether the pulse generator is comprised of GnRH neurons themselves or is the product of another system that integrates the activity of GnRH cells is uncertain.

The pulsatile mode of GnRH secretion is crucially important for gonadotrophin secretion. Thus, after removal of endogenous GnRH by destruction of the mediobasal hypothalamus, LH and FSH secretion can only be restored by use of an intravenous infusion pump programmed to deliver exogenous GnRH in pulses at a frequency approximating that seen naturally. Continuously infused GnRH does not restore regular menstrual cycles in females (Fig. 5.7). This critical requirement for pulsatile GnRH secretion can be understood in terms of the regulation by GnRH of its receptors on the gonadotrophs. Thus, the response to a first

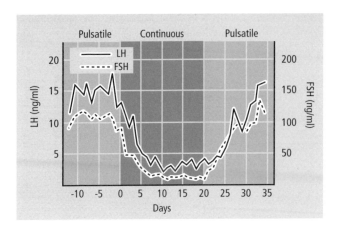

Fig. 5.7 Suppression of plasma LH and FSH concentrations after initiating (Day 0) a continuous GnRH infusion (1 μg/minute) in an ovariectomized, hypothalamically-lesioned rhesus monkey in whom pulsatile gonadotrophin secretion had been re-established by pulsatile GnRH infusions (1 μg/minute for 6 minutes once per hour). The inhibition of gonadotrophin secretion was reversed by reinstating the pulsatile pattern of GnRH administration on Day 20.

GnRH pulse is an initial release of stored LH and FSH occurring within minutes and lasting for 30–60 minutes. Associated with this release is the movement of secretory granules into a zone beneath the plasmalemma and a decrease in their size, perhaps indicating maturation of their contents. In consequence of this mobilization of granules (the 'self-priming' effect), a second exposure to GnRH results in a much larger 'primed' release of LH. Over a period of hours to days with exposure to pulsatile GnRH, gonadotrophin biosynthesis is also stimulated. After binding, some GnRH-receptor complexes remain on the surface while others are internalized via coated pits to lysosomal structures where degradation or other processing of the peptide can occur. Continuous exposure of gonadotrophs to GnRH, or infusion of a long-acting GnRH analogue, results in maintained occupancy of the receptors and is followed eventually by receptor down regulation (see Chapter 2) and a reduction in pituitary LH and FSH content and secretion.

Summary

Hypothalamic GnRH neurons regulate the synthesis and secretion of FSH and LH by the anterior pituitary. The GnRH is released as a series of pulses into the portal vessels; it reaches and binds to receptors on the gonadotrophs, and drives gonadotrophin secretion in a similar, pulsatile manner. Alterations in the output of GnRH, LH and FSH may be achieved, therefore, by increasing or decreasing the amplitude or frequency of

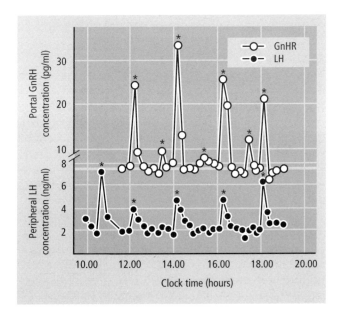

Fig. 5.6 Concentrations of GnRH in portal plasma (○) and LH in jugular venous plasma (●) of four ovariectomized ewes. Asterisks indicate secretory episodes (pulses) of GnRH and LH.

these pulses of GnRH or by modulating the response of the gonadotrophs to the pulses. As we shall now see, each of these mechanisms is employed.

REGULATION OF GONADOTROPHIN SECRETION IN FEMALES

During the menstrual cycle, the output of gonadotrophins is regulated mainly by the secretory products of the ovary. In order to understand the dynamic relationship between the ovary and gonadotrophin output two phenomena must be examined: (1) *a depressant effect* on gonadotrophin output induced by elevation of the plasma concentrations of oestrogens, progestagens and inhibin; and (2) an increase, or *surge*, of LH and FSH secretion induced principally by oestradiol. Throughout our discussion we will use data from the primate as a model, referring to other species where corroborative evidence or species differences exist.

Oestradiol and feedback control of FSH and LH secretion

Plasma concentrations of circulating FSH and LH increase markedly after ovariectomy or the menopause (Fig. 5.8b). This rise is largely attributable to removal of oestradiol, as infusion of this hormone results in the rapid decline of FSH and LH levels. The important characteristics of this oestrogen action are: (a) only low circulating levels of oestradiol are required to exert a marked effect; and (b) the effect is very rapid in onset, detectable within 1 hour and maximal by 4–6 hours. As oestradiol is acting to suppress gonadotrophin levels, the process is termed *negative feedback*.

In contrast, if plasma concentrations of oestradiol increase greatly, for example 200–400% above those seen in the early follicular phase of the cycle, and remain at this high level for 48 hours or so, then LH and FSH secretion is enhanced not suppressed (Fig. 5.9). Under these conditions we speak of a *surge* of LH and FSH. The term *positive feedback* is often used to describe the relationship whereby high levels of oestradiol increase the secretion of gonadotrophins, and thereby of oestradiol.

Thus, oestradiol has a dual function in regulating gonadotrophin secretion. At low circulating levels, it exerts rapidly expressed, negative feedback control over FSH and LH. At high, maintained circulating levels, positive feedback becomes the dominant force and an LH and FSH surge is induced.

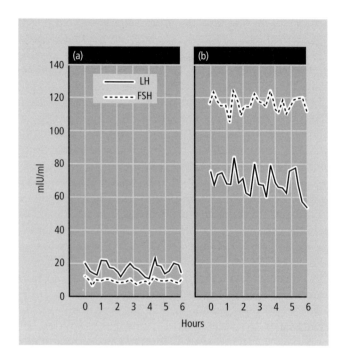

Fig. 5.8 Circulating FSH and LH levels in: (a) a woman at Day 7 in the (early) follicular phase; and (b) a postmenopausal woman. The difference between the two represents the negative feedback effect of oestradiol. Note: (1) the reversal of LH and FSH levels between (a) and (b), which may reflect preferential inhibition of FSH secretion in (a) due to inhibin (see text); (2) the changes in pulsatility of FSH and LH between (a) and (b).

Progesterone and feedback control of FSH and LH secretion

The high plasma concentration of progesterone, such as is seen in the luteal phase (4–8 ng/ml in humans), has two effects. First, the negative feedback effects of oestradiol are enhanced, FSH and LH secretion being held down to a very low level. Second, the positive feedback effect of oestradiol is blocked. Thus, injection of oestradiol into women during the progesterone-dominated phase of the menstrual cycle (luteal phase) is not followed by an LH surge.

Inhibin and feedback control of FSH secretion

In the last chapter, we identified a number of auto- and paracrine factors that appear to influence follicular and luteal development and function (see Table 4.3). Two of those factors, the activins and inhibins (Fig. 2.5), also act *endocrinologically* to influence FSH secretion. Thus, intravenous administration of inhibin into ewes: reduces

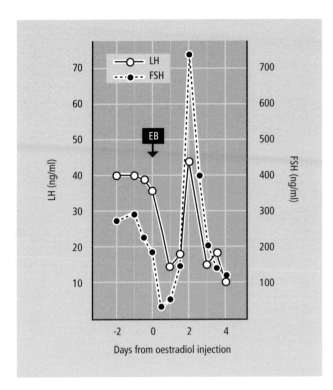

Fig. 5.9 The negative and positive feedback effects of a large injection of oestradiol benzoate (EB) in a female rhesus monkey. Note the early action of the oestrogen is to decrease FSH and LH levels (negative feedback, maximal at 6 hours or so). But if oestradiol levels are of sufficient magnitude and duration, a surge of the gonadotrophins occurs after 36–48 hours (positive feedback). Note that one effect naturally follows the other and each is critically dependent on time–dose relationships with the steroid.

pituitary and plasma FSH concentrations; prevents the rise in FSH, which usually follows ovariectomy; and blocks the FSH, but not the LH, response to a GnRH agonist (Fig. 5.10). Infusion of antibodies to inhibin during the late antral phase causes an increase in FSH, but not LH, concentrations in plasma (Fig. 5.11). Clearly, these experimental data strongly indicate a role for inhibin in the negative feedback regulation of FSH secretion. More recently, activins have been shown to have marked FSH releasing properties, although the significance of this property for the physiological control of the cycle is not presently clear.

Role of feedback in the menstrual cycle

Let us now re-examine the blood levels of steroids, inhibin and gonadotrophins during a normal human menstrual cycle (Fig. 5.12), and interpret them in the light of positive and negative feedback.

The follicular phase of the cycle

By convention, the menstrual cycle begins with the first day of menstruation. Just prior to this event, luteal oestrogen, progesterone and inhibin levels fall, negative feedback inhibition is relaxed, and FSH levels rise, followed, after a day or so, by LH. These rises permit antral growth to proceed, resulting, by the mid-follicular phase, in a rising output of oestrogens, androgens and inhibin. In consequence, FSH levels fall and LH levels flatten or rise only slowly. The selection of a dominant follicle leads to a further rise in oestrogen (together with their biosynthetically associated androgens), culminating in an oestrogen (and androgen) surge. While this output of oestrogen reflects the development of only the *most advanced* follicle(s), the inhibin output provides an indication of the activity of *all* the developing antral follicles. Thus, while the inhibin provides a measure of the *number* of follicles recruited, oestrogen output usually reflects the maturity of the follicle destined to ovulate. During the preovulatory phase, the surge in oestrogens triggers a rapid rise in LH and FSH levels via a positive feedback effect, and ovulation follows. As a result, inhibin, androgen and thus oestrogen outputs fall, as progesterone levels rise. The LH and FSH levels now fall equally precipitously because, at least in part, they lack a continuing positive feedback stimulus. The first half of the cycle is complete.

The crucial role of the oestrogen surge in triggering the LH surge has been shown in female rhesus monkeys by actively immunizing them against oestradiol. Neither an LH surge nor ovulation occur. However, if the synthetic oestrogen stilboestrol (which is not neutralized by the antiserum) is given to the immunized animal, the neutralization of endogenous oestradiol is bypassed and the synthetic oestrogen reinstates an LH surge. Thus, the pattern of FSH, LH and steroid secretion during the first half of the cycle is explicable largely in terms of the positive and negative feedback effect of oestrogens (primarily oestradiol) and inhibin on gonadotrophin secretion. The length of the follicular phase appears to be determined by the rate at which the main source of oestrogen, the principal preantral follicle, matures, and, as it is a major arbiter of cycle length, it has been described as the *ovarian* or *'pelvic' clock*.

The luteal phase of the cycle

The luteal phase of the cycle is characterized by rising plasma progesterone and 17α-hydroxyprogesterone concentrations, which peak around 8 days after the LH surge. In higher primates (but not other species), the luteinized cells of the corpus luteum also make large amounts of oestrogen and inhibin (Fig. 5.12). In all species, the progesterone depresses levels of

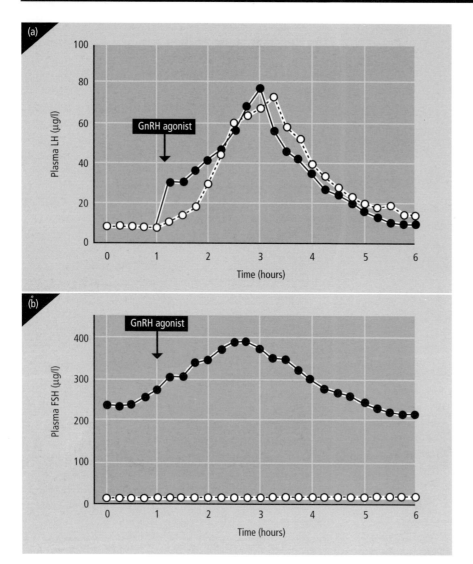

Fig. 5.10 Changes in plasma LH (a) and FSH (b) concentration following an injection of a GnRH agonist (→), in control ovariectomized ewes (●) and ovariectomized ewes treated with inhibin (○). Note the complete absence of an FSH response to GnRH in the inhibin-treated ewes in (b).

gonadotrophins (negative feedback), a depression most marked in higher primates in which inhibin is also active (compare levels of gonadotrophins in pig and human in Fig. 4.8). Growth of antral follicles is therefore suppressed and so androgens are also at a low level. Although in primates oestrogens from the corpus luteum may rise to levels which previously induced positive feedback, they fail to induce an LH surge. This failure reflects the effects of the uniquely high levels of progesterone found during the luteal phase of the cycle that inhibit positive feedback by oestradiol (see 'Progesterone and feedback control of FSH and LH secretion' above). At the end of the luteal phase, if conception has not occurred, oestrogens, progesterone and inhibin decline at luteolysis, the negative feedback effect of these hormones is relaxed, and LH and FSH levels rise, again permitting the rescue of preantral follicles and the initiation of another cycle.

Sites of positive and negative feedback

Having described the interrelationships between ovarian hormones and gonadotrophins during the menstrual cycle, we must now consider the sites at which these hormones exert their regulatory influences. There are two obvious candidates. First, the anterior pituitary: the hormones might regulate FSH and LH secretion by a (direct) action on the gonadotrophs, decreasing (negative feedback) or increasing (positive feedback) their sensitiv-

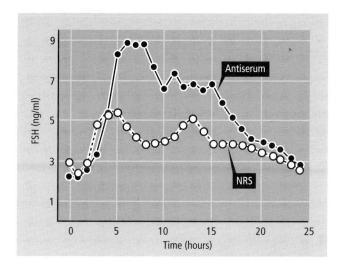

Fig. 5.11 The effect of intravenous infusion of normal rabbit serum (NRS) (○) or inhibin antiserum (●) on plasma FSH levels during the late antral phase. Note the marked rise in plasma FSH concentrations in the female rats immunized passively against inhibin.

ity to GnRH pulses of invariant frequency and magnitude arriving from the hypothalamus. There are abundant receptors for oestrogens, progestagens and inhibin in the anterior pituitary, emphasizing the potential importance of this site. Second, the hypothalamus: the ovarian hormones might change the GnRH signal either directly by effecting the GnRH neurons in the hypothalamus, or indirectly by changing the activity of other neural systems that exert a modulatory influence on GnRH release in the median eminence. Large regional concentrations of receptors for oestradiol and progesterone exist in a continuum between the medial preoptic and anterior hypothalamic areas and also in the ventromedial hypothalamus (particularly the arcuate nucleus and median eminence). In contrast, there is no direct evidence to support a hypothalamic site of action of inhibin. Of course, these two potential sites of feedback are not mutually exclusive, and a third possibility is, therefore, that ovarian hormones alter *both* the GnRH signal *and* the response of the anterior pituitary to it.

The anterior pituitary

The most convincing demonstration that the pituitary can respond to both the negative and positive feedback effects of oestradiol comes from experiments on ovariectomized rhesus monkeys. Large lesions of the mediobasal hypothalamus, which destroy the arcuate and ventromedial nuclei and a large part of the median eminence (see Fig. 5.2), result in abolition of GnRH

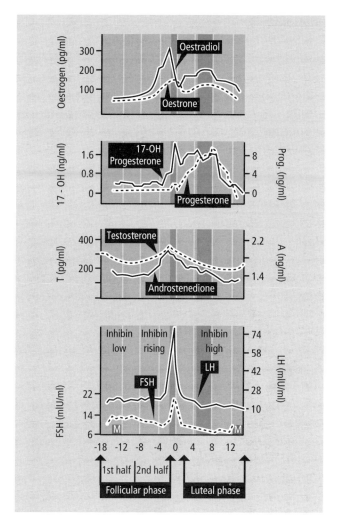

Fig. 5.12 Serum hormone levels during the human menstrual cycle. Note the units of measurements; FSH and LH are expressed in milli-international units/ml; testosterone and oestradiol in pg/ml; 17α-hydroxyprogesterone, progesterone and androstenedione in ng/ml. Note that in primates inhibin is high in the luteal stage, contributing to the greater suppression of FSH during the luteal stage seen in primates. M: menstruation.

output, and a decrease in serum FSH and LH to undetectable levels. Using a programmable intravenous infusion pump, hourly pulses of GnRH restore pulsatile LH and FSH secretion. The subsequent injection of oestradiol, so as to reach surge levels, results first in a fall, and then in a dramatic rise (surge) in serum FSH and LH levels. In the presence, therefore, of invariant GnRH pulses of constant amplitude and frequency, controlled by the infusion pump, oestradiol can exert both its negative and positive feedback effects on gonadotrophin secretion. Clearly, this action can only have been mediated via the anterior pituitary. Clinical studies on postmenopausal or hypogonadal women point to the same conclusion.

Studies on the pituitary sensitivity of women undergoing normal cycles provide further support for this conclusion. These experiments were performed by measuring the change in plasma levels of LH and FSH induced by pulses of GnRH administered on different days of the menstrual cycle (similar results have been obtained for the rat oestrous cycle). The results in Fig. 5.13 show that secretion of FSH and LH by the pitu-

itary, in response to constant GnRH pulses, increases during the follicular phase, reflecting an increased sensitivity of the secretory response with rising levels of plasma oestradiol. However, notice that the responsiveness of the pituitary to GnRH *remains* very high in the luteal phase of the cycle, a time when FSH and LH levels are normally at their lowest, and when it is impossible to induce an LH surge with an oestrogen surge (see above). This must mean that oestradiol and/or progesterone, which are both high at this time, exert an important com-

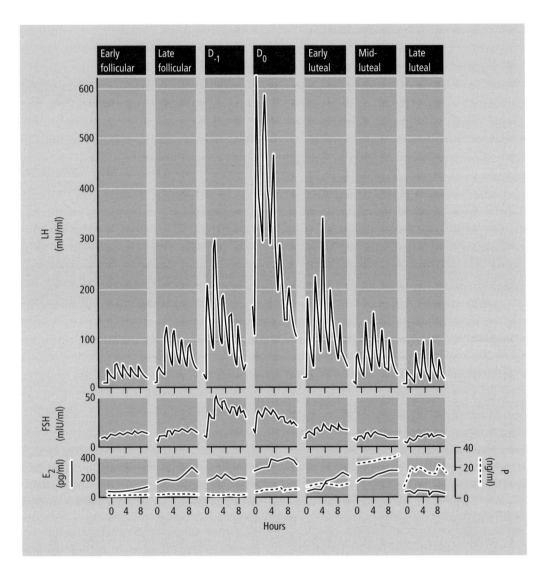

Fig. 5.13 The release of LH and FSH in response to 'pulse injections' of GnRH (10 μg at 5 × 2-hour intervals) during various phases of the menstrual cycle. The magnitude of the LH and FSH responses can be measured against the naturally changing circulating levels of oestradiol and progesterone. Note the correlation between oestrogen levels and increasing pituitary responsiveness to the injected GnRH. Note particularly that this remains high in the early and mid-luteal phases, at times when further LH/FSH surges cannot be elicited by oestradiol (see text). D_{-1} and D_0: the day before, and the day of, the spontaneous LH surge.

ponent of their negative feedback action on FSH and LH secretion somewhere other than the pituitary. This site is the hypothalamus as we will see in the next section. The demonstration that inhibin is able to reduce the FSH secretory response to GnRH (Fig. 5.10) strongly indicates that its effects are mediated by actions on the anterior pituitary.

Taken together, these findings imply that modulation of pituitary sensitivity to GnRH pulses can and does occur, but is not in itself sufficient to explain the changes observed in a normal cycle. Steroid-dependent alterations in the amplitude and frequency of the GnRH signal also play an important role.

The hypothalamus

The pattern of LH and FSH pulses varies during the menstrual cycle. Thus, during the follicular phase, LH is secreted in a series of high-frequency, low-amplitude pulses occurring approximately once every hour. By contrast, the luteal phase of the cycle is characterized by a pattern of high-amplitude, low-frequency, irregular LH pulses, often with long intervals between them of up to 6 hours (Fig. 5.14). As these gonadotrophin pulses indicate underlying GnRH secretory episodes, it is likely that changes in steroid hormone output influence GnRH output directly. The experimental manipulation of the

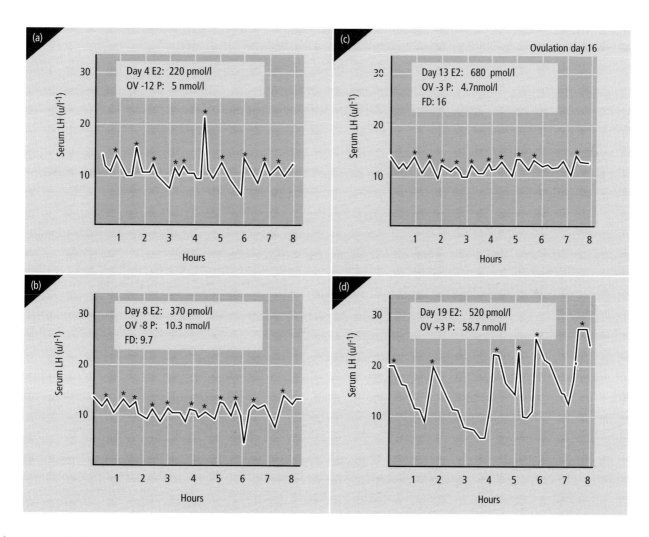

Fig. 5.14 LH pulsatility during a human ovulatory cycle (ovulation occurred on Day 16) in 10-minute serum samples. (a) Early follicular phase. (b) Mid-follicular phase. (c) Late follicular phase. (d) Early luteal phase. Note the unique high amplitude, low-frequency pulses characteristic of the luteal phase when progesterone plasma concentrations are high. * LH peaks. E2: 17β-oestradiol; FD: follicular diameter (mm); OV: +/− number of days after/before ovulation. P: progesterone.

steroid environment confirms this suspicion. Thus, *progesterone acts primarily to reduce pulse frequency* while *oestrogen acts to reduce pulse amplitude.*

More direct information on the modulation of GnRH secretory activity by steroids requires measurement of the peptide in portal blood. Obviously this is impossible in humans and, even in experimental animals, presents considerable methodological difficulties. However, it has been achieved in experiments on the rat, sheep and rhesus monkey, and the data show clearly that *GnRH secretory activity is subject to modulation by steroid feedback* during the menstrual and oestrous cycles, so confirming the indirect evidence described above. Thus, in rats, GnRH secretion is increased on the afternoon immediately before the LH surge. Similarly, the LH surge induced by exogenous oestradiol administration in rhesus monkeys and ewes is associated with elevated

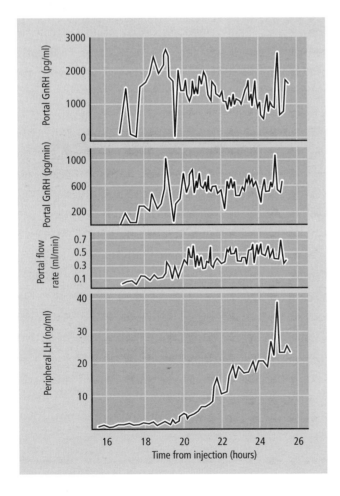

Fig. 5.15 GnRH concentration in portal blood, GnRH secretion rate and, portal blood-flow rate and peripheral plasma LH levels in an ovariectomized ewe given an injection of 50 μg oestradiol monobenzoate to induce LH surge. Note the increased frequency of GnRH and LH pulses during the LH surge.

concentrations of GnRH in portal blood (Fig. 5.15), indicating a hypothalamic site of positive feedback. In contrast, direct evidence for a hypothalamic site of negative feedback by oestradiol is less clear, as it is difficult to measure with confidence reductions in the already low levels of GnRH.

Sites of hypothalamic feedback

Where in the hypothalamus do steroids act? The problem of defining sites of steroid feedback has been approached using the classical techniques of neurobiology, namely the intracerebral implantation of sex steroids and other substances, or localized lesions of the suspected sites of steroid action. There are considerable species differences in the results of these studies, which may reflect the variable extent to which environmental factors influence reproduction (see 'Summary' at end of chapter). The apparent variability in the location of GnRH-containing neurons within the hypothalamus of different species may also be a factor. Thus, in rats, the major group of GnRH neurons is located in the medial preoptic area. Knife cuts placed behind the optic chiasm sever the axons of these neurons on their way to the median eminence, prevent oestrous cyclicity and may also disrupt gonadal function in males. However, in female rhesus monkeys, similar cuts do not disrupt cyclicity, and this may be due, at least in part, to the presence of a second, major group of GnRH neurons in the arcuate nucleus unaffected by the lesion and able, therefore, to sustain gonadotrophin secretion.

Experiments in which oestradiol has been placed into the hypothalamus have consistently implicated the arcuate nucleus as a site of negative feedback influence on GnRH secretion. Thus, in rats, oestradiol infused into the arcuate nucleus suppressed gonadotrophin secretion without detectable amounts of the steroid reaching the anterior pituitary. Similar experiments in female rhesus monkeys showed that oestradiol, in amounts 1000-fold less than those administered systemically, exerted marked negative feedback actions when infused into the arcuate, premammillary and ventromedial nuclei (Fig. 5.16). These actions of oestradiol seem not to involve GnRH-containing neurons directly, but appear to be mediated indirectly through other neural systems with convergent effects on GnRH neurosecretion.

Attempts to establish a site of positive feedback effects of oestradiol have had variable success. However, in rats, implantation of oestradiol in the anterior hypothalamic–preoptic area continuum has been shown to induce an LH surge without evidence of diffusion of the steroid into the anterior pituitary. There are few data from the primate on this issue of positive feedback actions of oestradiol exerted in the brain.

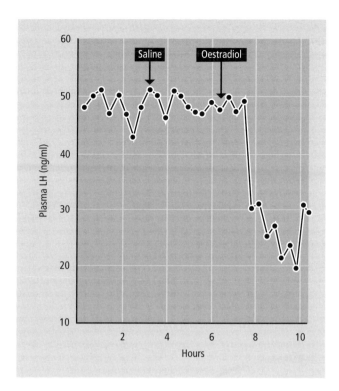

Fig. 5.16 The negative feedback effect of oestradiol (800 pg) on plasma LH when infused into the hypothalamic arcuate nucleus of ovariectomized rhesus monkeys. Note the lack of response to earlier, control injections of saline and the decline in LH concentrations within a short time after oestradiol injection.

The negative feedback actions of progesterone during the luteal phase of the cycle appear to operate in the arcuate region, where oestradiol-induced progesterone receptors are found in considerable number.

Taken together, these data indicate that while oestradiol and progesterone can act directly on the anterior pituitary to exert feedback effects on gonadotrophin secretion, the hypothalamus is also an important site for these effects via the modulation of GnRH secretion. Inhibin appears to exert its effects on FSH secretion primarily in the anterior pituitary.

How do steroids achieve their feedback effects?

What are the *cellular* mechanisms by which negative and positive feedback are exercised? What does oestradiol do to gonadotrophs to increase or decrease their sensitivity to GnRH? How does oestradiol induce a GnRH surge? How does progesterone decrease the frequency of GnRH pulses during the luteal phase?

Within the anterior pituitary, oestradiol appears to exert its positive feedback effects by inducing and maintaining GnRH receptors and by sensitizing the self-priming process whereby GnRH induces its own receptors. Small amplitude GnRH pulses, which do not

by themselves cause an LH pulse (Fig. 5.6), may prime a full LH response to the next adequate GnRH pulse. Figure 5.17 shows that the presence of oestradiol enhances this interaction between GnRH and its receptor, perhaps thereby contributing to the magnitude of the oestradiol-induced LH surge. There is less information on the ways in which oestradiol causes a *decrease* in gonadotrophin secretion. It is not known whether the steroid contributes to a decrease in GnRH receptors or an uncoupling of the receptors from subsequent biochemical events in the gonadotrophs. Nor is there detailed information on the way that inhibin exerts its selective effects on FSH secretion.

How does oestradiol induce a GnRH surge in the hypothalamus during positive feedback? The preoptic and anterior hypothalamic areas are rich in oestradiol receptors, and the GnRH content of neurons in this area changes in response to oestradiol. However, use of double-staining techniques has shown that oestradiol binding sites are not found in neurons that contain GnRH, indicating that steroid effects on GnRH secretion must be mediated indirectly via other neural systems, which are oestrogen targets and which converge

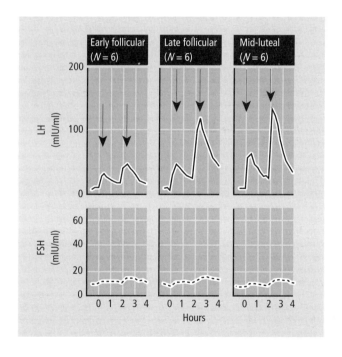

Fig. 5.17 The 'self-priming' effect of GnRH and its dependence on steroid hormone levels. The figure shows the enhanced response to a second injection of GnRH during three stages of the menstrual cycle (arrows indicate GnRH injection). Note the large increase between the early and late follicular phases, which correlates well with the rise in oestradiol secretion, and, furthermore, how this effect remains pronounced in the mid-luteal phase when oestradiol, as well as progesterone, secretion is high.

onto GnRH neuronal cell bodies or terminals. Among the hypothalamic neural systems that have been implicated in the regulation of GnRH secretion are those containing dopamine, the amino acid, γ-aminobutyric acid (GABA), and the opioid peptide, β-endorphin (Fig. 5.18).

Within the arcuate nucleus is a group of neurons known as the *tuberoinfundibular dopamine, or TIDA, neurons* (Figs 5.3 & 5.5b). These neurons take up radio-labelled oestradiol and project to the region of the portal capillaries where they lie in close association with GnRH terminals (Figs 5.5a,b & 5.18). The evidence for a functional relationship between these dopamine and GnRH terminals relies on pharmacological and correlative neurochemical data. Thus, treatment with dopamine or its agonists decreases plasma levels of LH, presumably by decreasing GnRH release. However, the physiological importance of dopamine in the regulation of GnRH secretion remains uncertain.

Some 70–80% of GABA-containing neurons in the preoptic and mediobasal hypothalamic areas accumulate oestradiol, and systemic oestradiol treatment (causing a surge of LH) is associated with a marked decrease in GABA release in the mediobasal hypothalamus. It has been suggested, therefore, that some of the effects of oestradiol on LH secretion are mediated by local, oestrogen-sensitive GABA neurons whose axons converge onto GnRH-containing neurons (Fig. 5.18). Using the technique of *in vivo* microdialysis, in which extracellular transmitter levels in small regions of the brain can be measured, it has been shown that fluctuations in hypothalamic GABA release are highly correlated with LH pulses in the peripheral circulation.

The opioid peptide, β-endorphin, is found in a major group of neurons in the arcuate nucleus which richly innervate the medial preoptic area containing the majority of GnRH-containing neurons. Hypothalamic β-endorphin concentrations fluctuate during the cycle, with the highest

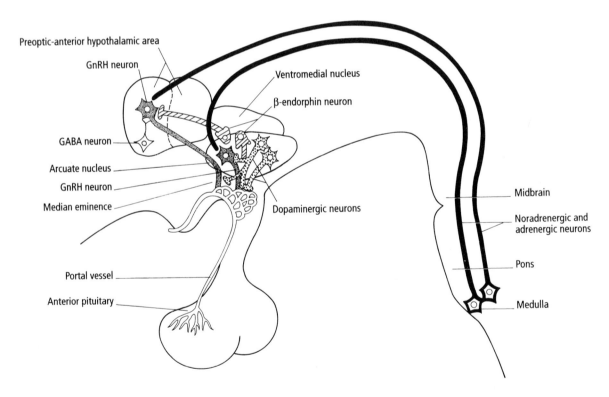

Fig. 5.18 Schematic diagram to show some of the postulated neurochemical interactions that may control GnRH secretion. GnRH neurons lie mainly in the medial preoptic area, but also in the arcuate nucleus in primate species. They project to the portal vessels in the median eminence, especially to the lateral palisade zone. Dopamine neurons in the arcuate nucleus may modulate GnRH release by effects in the external layer of the median eminence. Neurons within the hypothalamus that contain β-endorphin also modulate anterior pituitary secretion, perhaps by modulating GnRH neuron activity in the medial preoptic area or also by altering dopamine neuron activity and hence GnRH (and dopamine) neurosecretion. Noradrenergic and adrenergic neurons in the medulla project to the medial anterior hypothalamus and preoptic area and have been seen by pharmacological techniques to enhance GnRH secretion. GABA-containing neurons have been shown to accumulate oestradiol and may exert local control over GnRH neuron activity.

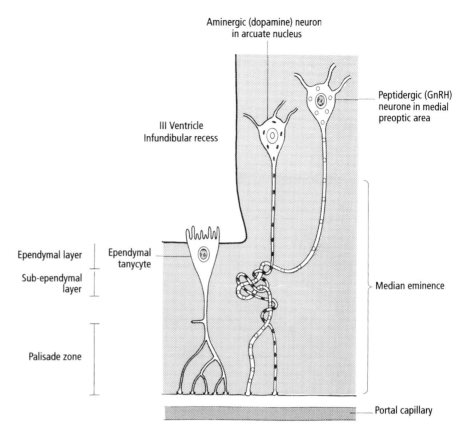

Aminergic (dopamine) neuron
in arcuate nucleus

Peptidergic (GnRH)
neurone in medial
preoptic area

III Ventricle
Infundibular recess

Ependymal layer

Sub-ependymal
layer

Ependymal
tanycyte

Median eminence

Palisade zone

Portal capillary

Fig. 5.19 A schematic illustration of the anatomical relationships of the ependymal tanycytes to the capillaries of the portal plexus and the juxtaposition of their 'terminals' to the terminals of monaminergic (e.g. dopamine) and peptidergic (e.g. GnRH) neurons. It has been suggested that the tanycytes could convey biologically active substances from the CSF in the ventricle to the anterior pituitary via the portal capillaries. That they *can* do this seems clear, but the physiological importance of this route remains to be defined.

levels in the luteal phase and the lowest levels in the follicular phase. When given intraventricularly, β-endorphin suppresses pulsatile, and preovulatory surge, gonadotrophin release, whereas naloxone, an antagonist of β-endorphin at predominantly μ-opiate receptors, accelerates LH pulses and markedly elevates serum LH levels. Thus, the β-endorphin-containing neurons in the arcuate nucleus may mediate the negative feedback effects of gonadal steroids, particularly progesterone. It remains a puzzle as to how these various neural mechanisms, which can affect GnRH secretion and are sensitive to steroid hormones, interact under physiological circumstances.

Finally, some interest has focused on ependymal tanycytes, which form a direct link between the base of the third ventricle and the pituitary portal plexus (Fig. 5.19) via which biologically active substances in the cerebrospinal fluid (amines, peptides and steroids) might reach the anterior pituitary. There is little doubt that tanycytes are able to perform this function, but it is far from clear why this system might be required in addition to all the others discussed above, particularly as the transport mechanism in the tanycytes appears to be non-selective.

REGULATION OF GONADOTROPHIN SECRETION IN MALES

The neuroendocrine mechanisms that govern testicular function are fundamentally similar to those that regulate ovarian activity. Hypothalamic GnRH, acting via a pulsatile secretion into the portal system on the pituitary of the male, is responsible for the secretion of gonadotrophins, which regulate the endocrine and spermatogenic activities of the testis (see Chapter 3). Gonadotrophin secretion in the male is also under negative feedback control by the testis via at least two hormonal products, which are differentially concerned with the regulation of LH and FSH. The major difference between the sexes in the control of gonadal activity is that gamete and attendant steroid hormone production in the male occur continuously after puberty and not cyclically, as is the case in females. The lack of cyclicity is reflected in the absence of a positive feedback influence of testicular products on gonadotrophin release, as there is no abrupt change in gamete or hormone production to underlie it.

Testosterone and the regulation of the pituitary–Leydig cell axis

In Chapter 3, we saw that testosterone secretion by the testis is the result of LH stimulation of the Leydig cells. It is now widely accepted that testosterone is, in turn, the principal hormone responsible for regulating LH secretion. Immunization against, and thus neutralization of, testosterone in rhesus monkeys results in increased circulating levels of LH. Conversely, administration of testosterone causes an abrupt decline in LH levels in castrate males of all species (Fig. 5.20). The negative feedback effect of testosterone is achieved largely by decreasing the frequency of episodic LH peaks via an effect on the hypothalamus, but there is also some change in pulse amplitude, reflecting a changing responsiveness of the pituitary to GnRH. Androgen receptors are found in abundance in both the hypothalamus and pituitary, and implantation of testosterone in the mediobasal, periarcuate hypothalamus of castrate male rats causes a significant fall in circulating LH levels. At least in rodents, 5α-dihydrotestosterone has some effect on LH secretion, whether given systemically or implanted in the hypothalamus. Testosterone also inhibits FSH secretion, but its effects are less than on LH. A more complete suppression of FSH comes from the combined action of androgens and the second testicular hormone, inhibin.

Inhibin and the regulation of the pituitary–seminiferous tubule axis

The role of inhibin in the regulation of FSH secretion in males is more controversial than in females, as its levels in testicular lymph, rete testis fluid and semen is some 100-fold lower than the level in follicular fluid. However, an increased output of testicular inhibin from the Sertoli cells does appear to be related directly to the successful completion of spermiogenesis. Thus, failure to complete spermatogenesis in man is correlated with depressed inhibin levels and elevated serum FSH levels. Conversely, stimulation of spermatogenesis in hypospermatogenic men is accompanied by a rising output of spermatozoa and inhibin and by declining serum FSH levels. As FSH stimulates Sertoli cells, and as these cells are strongly implicated in the support of spermatogenesis (see Chapter 3) as well as being the site of inhibin production, such a feedback loop makes biological sense.

IS THE HYPOTHALAMO-PITUITARY–GONADAL AXIS SEXUALLY DIMORPHIC?

We have seen that females exhibit both a negative feedback regulation by sex steroids and a positive feedback response to oestradiol. Does this represent a fundamental difference between males and females, or can males also show a gonadotrophin surge if given an oestradiol injection? The answer depends greatly on the species studied and on the effects of fetal or neonatal hormones on the physiology of the developing brain.

If ovaries are transplanted into normal male rats castrated in adulthood, they fail to show any cyclic changes. In contrast, ovaries transplanted into recipient males *castrated at birth* undergo cyclic ovulation. Clearly, the presence of the testis at birth prevents subsequent support of ovarian cyclicity, and testicular androgens are responsible for this effect. Thus, female rats injected with testosterone during the first few days after birth do not show oestrous cycles in adulthood. Their ovaries contain follicles that secrete oestrogens (they are said to be in constant oestrus), but as ovulation does not occur there are no corpora lutea.

The neonatal androgen causes acyclicity by suppression or modification of the oestradiol positive feedback mechanism. Thus, if male or female rats are castrated in adulthood and are subsequently injected with oestradiol, only the females show a surge of gonadotrophins

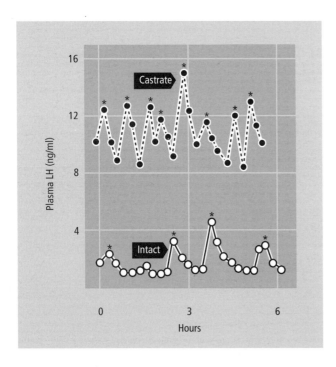

Fig. 5.20 The negative feedback effect of testosterone on plasma LH levels in male red deer. *Upper part:* levels in the castrate male. *Lower part:* levels in the intact male. Note the increased frequency of pulses (*) after castration.

(Fig. 5.21). If the same experiment is undertaken with female rats given testosterone during the first few days of life, no surge is observed. Does neonatal testosterone act on the ovary, pituitary or hypothalamus to suppress the positive feedback response? The ovaries of androgenized females are quite capable of secreting surge levels of oestrogen if transplanted into normal females, so their functional capacity does not seem to be grossly impaired. Similarly, pituitaries from males or androgen-exposed females can function cyclically if transplanted adjacent to a normal female's hypothalamus. The 'masculinizing' effect of neonatal testosterone would therefore appear to be exerted on the hypothalamus. Precisely where in the hypothalamus is not firmly established.

It has been widely assumed that this 'masculinization' of the brain occurs in all species. Indeed, the phenomenon has been demonstrated in many rodents, sheep and some carnivores, although the 'critical period' of sensitivity to the effects of androgens varies considerably. For

example, guinea-pigs have a gestation period of 68 days, compared to 21 days in the rat, and are born in a state of relative maturity. The critical period during which androgens exert their effects on the brain is pre- and not postnatal. Similar considerations apply to the large domestic animals and carnivores.

In contrast, experiments on primates indicate that 'masculinization' of the hypothalamus does not occur in the same way, and that the capacity for positive feedback exists in normal male monkeys and men. Thus, in castrated male monkeys (as well as in hypogonadal and castrated men) an administered oestrogen surge reliably induces a gonadotrophin surge (Fig. 5.22). Indeed, ovaries transplanted into castrated male monkeys undergo apparently normal monthly ovulatory cycles, a marked contrast to the results of similar experiments in rats. However, the positive feedback action of oestradiol cannot be elicited in *intact* male monkeys for reasons that are not clear. Female rhesus monkeys exposed to high levels of testosterone during fetal life are also able to show menstrual cycles as adults, although puberty occurs slightly later than usual. Indeed, in some of these monkeys the external genitalia are so 'masculinized' that menstruation occurs through a penis-like phallus! Similarly, human females exposed to high levels of

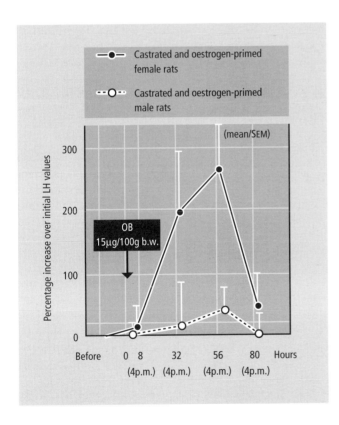

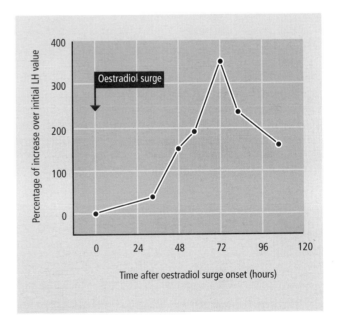

Fig. 5.21 The positive feedback effect of oestradiol in castrate male and female adult rats. The LH response, measured as a percentage change from resting values, to a single large injection of oestradiol benzoate (OB) is shown to be present only in the female and not in the male. This is taken to be evidence of sexual differentiation of the hypothalamus, neonatal androgens in the male preventing the ability to respond to an oestrogen surge with an LH surge in the adult.

Fig. 5.22 Positive feedback effects of oestradiol in a castrate male talapoin monkey. Unlike the results in Fig. 5.21, the normal male monkey is able to respond to an oestradiol surge with an LH surge similar to that seen in the female. This suggests that exposure of the primate's brain to testosterone *in utero* does not 'masculinize' the hypothalamus as has been demonstrated in non-primate mammals.

androgens during gestation, for example as the result of the adrenogenital syndrome, also have menstrual cycles as adults, often after a somewhat delayed puberty.

Clearly, 'masculinization' of the hypothalamic mechanism underlying positive feedback does not occur in normal male primates, including men, and, thus, by these criteria there are no apparent hypothalamic endocrine consequences in female primates, including women, exposed to high levels of androgens *in utero*. We must conclude, therefore, that sexual differentiation of the brain is not a global phenomenon so far as the mechanisms regulating ovarian and testicular function in the adult are concerned. This conclusion is reminiscent of the less overt effects that neonatal androgens appeared to have on sexual behaviour in adult primates, as compared to their rather dramatic effects on sexual behaviour in rats and other non-primate species (see Chapter 1).

THE REPRODUCTIVE FUNCTIONS OF PROLACTIN

Prolactin is made in the pituitary lactotrophs, which are distributed evenly throughout the anterior pituitary. It is stored in secretory granules and released in a pulsatile manner, which probably reflects the pulsatile release of controlling hypothalamic hormones.

Hypothalamic control of prolactin secretion

Unlike other pituitary hormones, prolactin is secreted in large amounts when the vascular links between the pituitary and hypothalamus are *disconnected*. This observation has led to the search for a hypothalamic factor(s) that inhibits release (*prolactin inhibitory factor or PIF*) and therefore acts as the primary controller of prolactin secretion. However, it is now clear that prolactin releasing factors also exist and that prolactin secretion is the result of complex interactions between these various influences.

Prolactin inhibitory factors

Probably the most important PIF is the catecholamine, dopamine (Fig. 2.9), which is found in neurons of the arcuate nucleus, the axons of which project to the portal capillaries in the medial and lateral palisade zones of the external layer of the median eminence (Fig. 5.5). Dopamine secreted into the portal blood from the terminals of this TIDA system is carried to the lactotrophs, which contain dopamine receptors. The amine and its agonists (e.g. *bromocriptine*) suppress prolactin secretion (Fig. 5.23) while dopamine antagonists (e.g. *haloperidol, metoclopramide* and *domperidone*) increase prolactin secre-

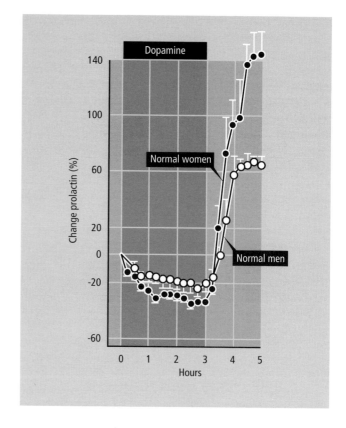

Fig. 5.23 The effects of dopamine infusion on prolactin levels in normal men and women. Note the rapid onset of effect of dopamine and the rebound increase in prolactin levels on stopping the infusion. In women with hyperprolactinaemia, bromocriptine, a dopamine-receptor agonist, will cause serum prolactin concentrations of 80–100 ng/ml (normally they are less than 18 ng/ml) to fall to below 10 ng/ml within 5 or 6 hours.

tion by direct actions on the dopamine receptors of the lactotroph. Having bound to the receptors, dopamine is internalized and acts, at least in part, to increase lysosomal degradation of prolactin within the secretory granules, thereby making less of the hormone available for release. While dopamine is undoubtedly an important PIF, two other 'hormones', GABA (see 'How do steroids achieve their feedback effects?' above) and GnRH-associated peptide (GAP) (Fig. 2.6), are both found in portal blood and can also inhibit prolactin release, but no clear physiological role has yet been assigned to them *in vivo*.

Prolactin releasing factors

Several hormones can stimulate prolactin secretion, although a physiological role for all of them has not been established. The *vasoactive intestinal polypeptide* (VIP; see Chapter 2, 'Small peptides') is an extremely potent releaser of prolactin, and receptors for the peptide have been demonstrated on the lactotrophs. Furthermore, a

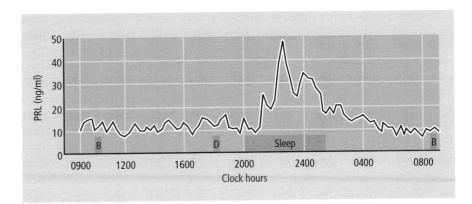

Fig. 5.24 Daily variation of prolactin secretion. Note the onset of sleep is associated with the rise in serum prolactin concentrations, which then begin to fall towards the period of wakening. B: breakfast; D: dinner.

decrease in dopamine secretion from the median eminence is associated with an increased prolactin secretory response to VIP. Although only very low levels of VIP are present in neurons in the brains of normal males or non-lactating females, VIP is measurable in portal blood. In suckling female rats, VIP-containing neurons are readily seen in the parvocellular paraventricular nucleus, and a rich terminal plexus becomes visible in the external layer of the median eminence. Moreover, increased amounts of VIP mRNA are also detectable in the paraventricular neurons at this time. These observations suggest that VIP may be particularly important as a releasing factor during periods of high prolactin secretion, for example during lactation (see Chapter 13). The observation that VIP may be elaborated by lactotrophs has led to the suggestion that it may also be involved in an autocrine regulation of prolactin secretion. Although hypothalamic thyrotrophin releasing hormone (TRH) stimulates prolactin secretion, and receptors for it have been localized to lactotrophs, its physiological role in the regulation of prolactin secretion has not been established.

Oestrogens also induce hyperprolactinaemia, probably by decreasing the sensitivity of lactotrophs to dopamine and perhaps by increasing the number of TRH receptors. Chronic exposure of lactotrophs to oestrogen results in increased pituitary DNA and RNA synthesis with consequent hyperplasia and increased prolactin secretion. The prolactin secretory response to oestrogens is, therefore, not acute and the steroid should not be regarded as a releasing factor in the way that VIP is.

In addition to these humoral controls, prolactin secretion is also subject to daily variation, the highest plasma concentrations occurring during the nocturnal sleep period. Reversal of the sleep–waking cycle results in reversal of the daily rhythm of prolactin secretion, demonstrating that it is sleep-entrained rather than light-entrained (Fig. 5.24). In a normal sleep–waking cycle,

prolactin release begins to increase 1–1.5 hours after sleep onset and is achieved by progressive increases in pulse amplitude. Plasma concentrations are elevated during the remaining hours of sleep and fall in the early morning, shortly before awakening. Lowest concentrations are found between about 10.00 a.m. and 12.00 noon. Interestingly, the bursts of prolactin secretion seem to occur during slow wave, or 'non-rapid eye movement' sleep (non-REM), while REM (or paradoxical) sleep is associated with the smallest episodic prolactin pulses.

An oestrous rhythm of prolactin secretion in animals, such as the rat, has also been demonstrated, with a prolactin surge occurring coincident with the LH surge. However, in women, it seems generally to be agreed that no clear menstrual rhythm in serum prolactin levels exists and nor does prolactin secretion alter significantly after the menopause. Changes in prolactin secretion during pregnancy and lactation will be discussed in Chapters 10 and 13. Prolactin is also seen to be released by acute stressors, both in animals and humans. The functions of prolactin as a 'stress hormone' when released in this way are unclear.

Feedback regulation of prolactin secretion

Dopamine released from TIDA neurons into the portal capillaries is generally regarded as being the main controller of prolactin secretion. What regulates the activity of TIDA neurons? It appears that the answer to this question is prolactin itself. Increases in circulating prolactin levels result in an increase in dopamine turnover within TIDA neuron terminals of the median eminence and a reduction, therefore, in prolactin secretion. The increase in dopamine turnover is related to an increase in tyrosine hydroxylase activity, the enzyme that is rate-limiting in the intraneuronal synthesis of dopamine. This so-called 'short-loop' feedback control of hypothalamic TIDA neuron activity, and hence dopamine release, by circulating prolactin is illustrated in Fig. 5.25.

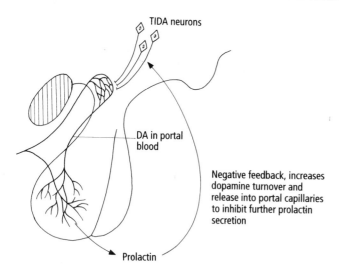

Fig. 5.25 Schematic summary of the proposed negative feedback relationship between prolactin and dopamine (PIF). Prolactin is believed to accelerate dopamine (DA) turnover in the arcuate nucleus neurons (tuberoinfundibular dopamine or TIDA neurons), and the catecholamine is then released into the portal capillaries and thereby reaches the lactotrophs. Hyperprolactinaemia could be caused by a failure either of PIF activity at the dopamine receptor level in the anterior pituitary, or a reduction of TIDA neuron activity in the hypothalamus.

Functions of prolactin

Prolactin generally seems to function as an ancillary hormone, promoting the activities of other hormones. Its extramammary functions in mammals are too numerous even to list, but include such diverse examples as regulation of kidney and adrenocorticotrophic activity (these tissues having higher prolactin-binding activity even than mammary tissues), in addition to synergistic actions with ovarian and adrenal steroids and gonadotrophins. During the luteal phase in the rat, sheep and goat, where prolactin is an essential part of the luteotrophic complex, it acts by increasing the amount of oestradiol receptor in the corpus luteum. Few of these actions have been established in women. There is some evidence that prolactin may participate in the regulation of steroidogenesis in the follicle, particularly the inhibition of progesterone secretion in the early stages of follicular growth and its enhancement in the luteal phase. In addition, prolactin appears to be able to modulate the number of ovarian receptors for LH and so affect steroidogenesis indirectly. However, the exact importance of these findings to our understanding of anterior pituitary regulation of ovarian function remains to be established.

In males, most of the information concerning prolactin and testicular function comes from experiments on rodents. It has been shown that, while exerting little effect on its own, prolactin may increase the number of LH receptors and potentiate the steroidogenic effect of LH on Leydig cells; testicular prolactin receptors seem to be confined to the interstitial tissue of the testis. Similarly, prolactin increases the uptake of androgen and increases 5α-reductase activity in the prostate, while testosterone maintains the number of prolactin receptors in the prostate. Prolactin also potentiates the effects of testosterone on the seminal vesicle. In man, the functional importance of prolactin in testicular and reproductive tract activity is unclear.

Hyperprolactinaemia

Pathological elevation of serum prolactin levels in men is associated with impaired fertility (see Chapter 14), decreased circulating levels of testosterone and loss of libido. In women, the syndrome is characterized by amenorrhoea, and hence infertility (see Chapter 14), with or without galactorrhoea (abnormal milk secretion, see Chapter 13) and loss of libido. Clearly, prolactin, in such circumstances, exerts profound effects on reproductive function. The causes of hyperprolactinaemia are multiple and varied. They may of course be 'physiological', for example in pregnancy and during the first few months of breast feeding. They can be iatrogenic (e.g. dopamine receptor-blocking neuroleptic drugs used widely in psychiatry increase prolactin secretion, as do oestrogens, including oral contraceptives) or be the result of some underlying pathology (e.g. some pituitary tumours are prolactin-secreting; so-called prolactinomas, such as adenomas or microadenomas).

The high prolactin levels are largely due to absence of the fall in prolactin secretion normally seen in the morning on awakening, the high-amplitude prolactin pulses being seen abnormally during the daytime. The syndrome is more common in women than in men. The endocrine profile of women with hyperprolactinaemia is characterized by an absence of pulsatile LH secretion, a reduced pituitary LH response to injected GnRH, a failure of positive feedback, and hence chronic anovulation and amenorrhoea. Contrary to earlier opinion, it is now clear that the ovarian responsiveness to FSH and LH is not necessarily impaired and does not underlie the amenorrhoea; indeed, normal cyclicity can be maintained in the presence of high prolactin levels if exogenous gonadotrophins are administered. Exactly how elevated prolactin levels cause these changes is unclear. The decrease in pituitary sensitivity to GnRH might reflect an indirect effect of prolactin via decreased ovarian oestradiol secretion, rather than an effect exerted directly on the pituitary.

Treatment of hyperprolactinaemia has become relatively effective and simple. Dopamine agonists, such as bromocriptine, are now used to lower serum prolactin concentrations immediately, and daily treatment results in the return of ovulation and cyclicity in the vast majority of women within 2 months. In the case of prolactin-secreting tumours, dopamine agonists have an anti-mitotic action and, in addition to lowering plasma prolactin concentrations, reduce tumour size. However, in the majority of cases, cessation of treatment is followed by a resumption of prolactinoma activity and surgical removal is often necessary. The loss of libido in men and some women with hyperprolactinaemia is unexplained. The suggestion that this represents a direct action of prolactin in the brain remains to be demonstrated.

ENVIRONMENTAL INFLUENCES ON REPRODUCTION

The foregoing discussion has established a major role for the CNS in provision of a GnRH supply and as a target for steroid modulation. The CNS also mediates the effects of environmental factors, such as coital stimuli, olfactory stimuli and light, on the regulation of reproductive activity in many species. In addition, factors such as those arising from social interaction, including anxiety or other forms of emotional distress, which can have profound effects on cyclicity and fertility in men and women, are also mediated by the CNS. Although the effects of these social factors are readily observed in humans, studying causal relationships depends, in large part, upon careful analysis of animals living in social groups, where both behavioural and endocrine variables may be controlled experimentally. We will now examine the way in which these various environmental influences on reproduction are mediated.

Effects of light on fertility

Reproduction is only one of a host of activities in which an individual may engage. To ensure that reproductive activity occurs effectively and with minimum interference from other processes, its appropriate timing is important. There are two levels at which control over timing is evident.

Light and circadian rhythms

The temporal control of reproductive activity in the female is complicated because the production of a viable oocyte is itself a cyclical event that must be matched to

other cyclical events occurring within the life of the animal. In a nocturnally active rodent, for example, potential encounters with mates will be restricted to the hours of darkness. This selection pressure has led to the development of an oestrous cycle that is tightly locked to the best indicator of external time, the daily light–dark cycle. As we saw in Chapter 4, the oestrous cycle of rats is much shorter than the menstrual cycle of primates, but the temporal relationships of oestradiol, LH and FSH secretion are remarkably similar (Fig. 5.26). The really dramatic differences are: (a) the surge of progesterone secretion accompanying the LH surge, which is very important in the cyclical control of sexual behaviour (see Chapter 7 for details); and (b) the LH surge itself is precisely timed to occur between 5 and 7 hours before darkness. This ensures that ovulation occurs 12 hours later during the night when the female is active, because of her nocturnal habits, and is behaviourally receptive (see Chapter 7). The likelihood of conception is therefore

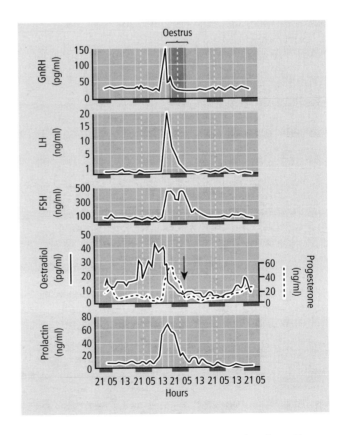

Fig. 5.26 Serum hormone levels during the oestrous cycle in the rat. The black bars represent dark periods (18.00–06.00 hours) centred around midnight, while the arrow denotes the time of ovulation. Also shown are the levels of GnRH measured in portal vein blood, which show a clear surge just prior to the FSH and LH surges. Note that the surge in prolactin is coincident with the gonadotrophin surges.

maximized. In intact animals, the LH surge occurs only every 4 or 5 days because it is dependent on the trigger of rising oestrogen production. However, the neural signal that determines the *time* of the LH surge is present every day. In ovariectomized females having a constantly high level of oestrogen delivered from a subcutaneously implanted capsule, an LH surge occurs every day at precisely the same time. By controlling oestrogen levels, the ovary therefore determines *the day* of ovulation, but a neural timer, controlling a critical period of sensitivity to oestrogen, determines *when*, during that particular day, the LH surge and thus ovulation will occur.

Reversal of the light–dark cycle causes a 12-hour shift in the timing of the critical period of sensitivity to oestrogen, the LH surge and ovulation, demonstrating the essential role of information about light in setting the neural timer. However, even in constant light or dark, daily LH surges will continue, given the appropriate oestrogen environment, indicating that the timing mechanism is a *self-sustaining biological clock* (or oscillator). Under these constant conditions, the oscillator and the rhythm it controls are said to *free-run* with a period of approximately 24 hours, which is therefore termed *circadian* (circa = approximately, diem = day). It is important to realize that the circadian system controls a wide range of other behavioural and endocrine rhythms, both reproductive and non-reproductive, which are held in a very strict, temporal relationship to each other.

The majority of these rhythms are driven by the *suprachiasmatic nuclei* of the hypothalamus. This cluster of neurons, located above the optic chiasm adjacent to the third ventricle (see Fig. 5.3), has the ability to generate a circadian signal even when isolated from the rest of the brain. In life, this approximately 24-hour signal is converted to a precise period of 24 hours by the *entraining* effect of photic stimuli, which reach the suprachiasmatic nuclei via a direct retinal input, the *retino-hypothalamic tract*. Lesions of the suprachiasmatic nuclei disrupt many circadian functions, including the LH surge, and thereby cause a condition of permanent anoestrus. However, the neuroanatomical route linking the circadian signal to the release of GnRH is poorly understood. Furthermore, it is unclear whether there is a circadian rhythm in GnRH secretion, normally peaking in the afternoon, which is amplified by the oestrogen surge, or whether the circadian input leads to a transient increase in the sensitivity of GnRH to the positive feedback effects of oestrogen.

In female primates, patterns of reproductive activity are much more flexible than in rodents, and the social rather than the physical environment has a much more pronounced effect upon reproductive processes. The primate has a menstrual cycle in which ovulation may occur at any time of day; there is not a tightly restricted period of sexual receptivity, or 'heat', and the circadian system makes little contribution to the control of reproductive function. However, another form of temporal programming is apparent in some primates (although not modern human society) and in a wide range of other mammalian groups in which purely circadian influences may not initially be evident.

Light and circannual rhythms

In seasonal environments, where adverse climate and the availability of food will be major determinants of offspring survival and therefore of the reproductive success of the parent, it is adaptive for individuals to ensure that young are born in the equable, productive conditions of spring or early summer. This tight control over birth season, apparent in many domestic and wild species, is achieved by a precise regulation of the *month*(s) of fertility and hence the timing of conception. In species with short gestation times, such as hamsters and birds, winter is a time of infertility with gonadal development suspended until spring. In species with longer gestation times, such as sheep, the anticipation of spring must begin much earlier and seasonal changes in autumn act as a stimulus to reproductive function. This leads to the dramatic spectacle of the *rut* when animals, which have been reproductively quiescent for the entire year, suddenly become sexually active. Males may develop pronounced secondary sexual features, such as antlers, become fertile, aggressive and territorial, and spend their whole time engaged in an intense competition for access to females. Females come into heat and actively show interest in males, accepting their attempts to copulate. In a third group, which includes marsupials, mustelids and seals, the total length of the gestation period can be varied because of delayed implantation and embryonic diapause (see Chapter 9). These processes are sensitive to environmental influences and provide a second level of control over the timing of the birth season. For the reproductive physiologist, these seasonal phenomena offer an important opportunity to investigate the central mechanisms that regulate the fertility of an individual.

In some species, such as deer and ground squirrels, there is good evidence that seasonal cycles are under the control of an endogenous *circannual oscillator*, a biological clock with a period of approximately 1 year. In other species, there is no endogenous rhythmicity and the seasonal rhythms observed in the field are triggered by cyclical stimuli within the environment. Of these, pho-

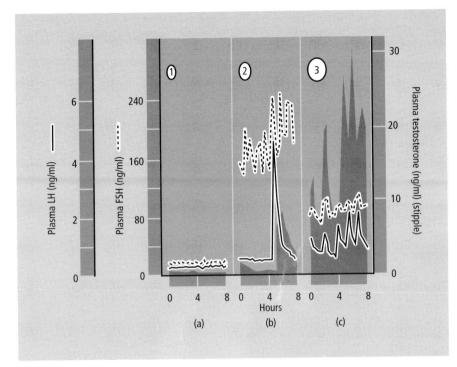

Fig. 5.27 The seasonal sexual cycle of a Soay ram. Changes in the plasma concentration of FSH, LH and testosterone are shown at three times of the year. Testis size is shown inset at each time: (a) in the non-breeding season, when the testes are fully regressed, and LH, FSH and testosterone levels are all low; (b) towards the onset of the breeding season, when the testes are redeveloping and, associated with this process, FSH concentrations are very high; (c) during the mating season, when testosterone levels are very high and this reflects the marked increase in the frequency of pulsatile LH discharge.

toperiod is by far the most important; this is exemplified in the laboratory, where artificial manipulation of day lengths can be used to drive all of the components of the annual reproductive cycle. For example, exposure of Syrian hamsters to less than 12.5 hours of light per day (pseudo-winter) leads to gonadal atrophy and the loss of sexual behaviour. In contrast, these short photoperiods stimulate gonadal activity in species, such as sheep, that normally mate in the autumn. All of these effects are mediated by changes in the frequency of the GnRH pulse generator in the hypothalamus, which then determines the level of secretion of gonadotrophins and steroids (Fig. 5.27).

Photic circannual information obviously has access to the GnRH neurons, but does it use the same pathways that are involved in the circadian control of reproduction? Certainly, the suprachiasmatic nuclei have an important role to play because lesions of these structures completely block photoperiodic sensitivity. However, the pathways involved are not exclusively intrahypothalamic. It is now well recognized that the *pineal gland*, which sits over the dorsal midbrain, attached to the *epithalamus* in the posterior portion of the third ventricle (Fig. 5.1), is the mediator of photoperiodic time measurement. Removal of the gland or interruption of its sympathetic innervation leaves animals insensitive to changing day length. The primary pineal hormone, *melatonin*

(Fig. 2.9), is synthesized and released into the bloodstream only in the hours of darkness, exhibiting a true circadian rhythmicity driven by the suprachiasmatic nuclei. At night, the circadian signal increases sympathetic activation of the gland, resulting in a dramatic rise in the activity of the enzyme, *N-acetyl transferase*, the rate-limiting step in melatonin biosynthesis.

The crucially important feature of the circadian melatonin signal is that it provides a precise representation of the length of the night, so that as days shorten and nights lengthen in autumn, the duration of the nocturnal melatonin peak is increased. Conversely, after the winter solstice, the photoperiod increases and the duration of the melatonin signal falls (Fig. 5.28a,b). The changing shape of the rhythm of circulating melatonin is detected within the hypothalamus and somehow leads to alterations in GnRH secretion. The melatonin signal is such a powerful regulator of neuroendocrine state that in pinealectomized animals the entire reproductive axis can be turned on or off by repeated, nightly administration of programmed infusions of melatonin, which mimic the pattern of its secretion typical of either long or short photoperiods (Fig. 5.28c–f). One important question yet to be answered is how does melatonin work and, in particular, how is it that the same signal leads to opposite neuroendocrine responses in different species?

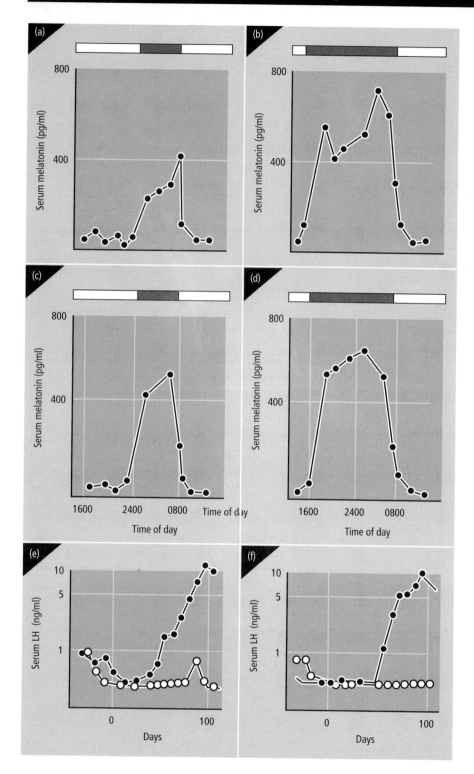

Fig. 5.28 The effects of melatonin on LH secretion in ewes in relation to the light–dark cycle (darkness is indicated by the solid bar over each graph). (a), (b) Serum melatonin profiles of ewes with intact pineals exposed to artificial long (a) and short (b) photoperiods. Note the increased duration of the melatonin signal during the longer night of (b). (c), (d) Serum melatonin profiles of pinealectomized ewes receiving programmed infusions of melatonin designed to mimic the patterns of long and short photoperiods (shown in a and b). (e) Reproductive response (LH secretion) of pineal-intact ewes to artificial long (○—○) or short (●—●) photoperiods (remember that the ewe is an autumn, or 'short-day' breeder). (f) Reproductive response of pinealectomized ewes to programmed infusion of melatonin mimicking long (○—○) or short (●—●) photoperiodic profiles in serum. Clearly, this is virtually identical to the pattern in (e).

Effects of coitus on fertility

In humans and other primates, sheep, rats and many other mammals, ovulation is said to occur 'sponta- neously'. Thus, it depends on an endogenous event timed by the ovary, the oestradiol surge, which results in an ovulatory discharge of LH that may or may not be subject to additional, circadian controls. The neurally mediated variable controlling ovulation, to be discussed next in this chapter, concerns the *'induced' or 'reflex' ovulators*, such as cats, rabbits and ferrets. These animals remain in behavioural oestrus for long periods of time without ovulating until they copulate with a male. Stimulation of the cervix and vagina during coitus evokes the reflex release of an ovulatory surge of LH via afferent, sensory pathways, which gain access to the GnRH release mechanism. Even in these species it seems that the hypothalamo-pituitary axis must be primed with high levels of oestrogen for the neural input to be effective.

Although the data supporting it are less than convincing, there are several reports of reflex ovulation in women. Ovulation has variously been observed to follow coitus very early in the follicular phase (even during menstruation), while termination of abnormally long follicular phases has sometimes been ascribed to coitus-induced, acute LH release. However, the subject has not been studied systematically, and these reports should therefore be viewed sceptically, if with interest.

Similarly, in the rat, pregnancy and pseudopregnancy (i.e. prolongation of luteal life) are also dependent on coitus. In this case, cervical stimulation appears to set up a *diurnal* (during the day) peak in prolactin secretion, in addition to that normally occurring nocturnally. Thus, two daily peaks in prolactin concentration are measurable in the blood and this seems essential for corpus luteum formation, the polypeptide being an essential component of the luteotrophic complex.

Effects of social interaction on fertility

Studies of primates living in social groups have revealed that the social context in which individuals interact can change their endocrine and fertility status. For example, if plasma testosterone levels are measured in male talapoin monkeys in the absence of females, all males, whether single or together as a group, have very similar testosterone levels. However, when oestrogen-treated females are introduced, one male becomes dominant, displays sexual activity with the females, is aggressive to subordinate males and his plasma testosterone levels rise significantly. None of these changes is seen in the subordinate male(s) (Fig. 5.29).

That the behavioural interactions determine the change in plasma testosterone and not vice versa is apparent if all the males are taken out of the group and each replaced alone in turn with the females. In this situation, each male is sexually active and shows a significant rise in plasma testosterone. Addition of the dominant male to a group consisting of the subordinate male alone with females results in the latter male's plasma testosterone declining. These data have been interpreted to suggest that subordination, especially being on the receiving end of aggression, causes the decrease in testosterone levels. Conversely, being aggressive and/or displaying sexual behaviour, as in the dominant male, is associated with increased plasma testosterone. The significance of the elevated testosterone levels in the dominant male, and the mechanism by which it is achieved, are not entirely clear. As we will see in Chapter 7, testosterone levels have but a tenuous relationship to sexual motivation in gonadally intact males and even subordinate males have sufficient levels of the hormone to maintain sexual behaviour. It has been suggested that testosterone may enhance the attractiveness of the dominant male to females by behavioural (e.g. posture) and/or non-behavioural (e.g. coat quality and odour) means, although experimental evidence for either suggestion is lacking.

A similar consequence of the dominance hierarchy is seen in the females of the group. The dominant female receives more sexual attention from the dominant male than does the subordinate female(s), and hardly any aggressive behaviour from other members of the group (Fig. 5.30). An intriguing consequence of this state of affairs is seen if both the dominant and subordinate females are challenged with an oestradiol surge. Only dominant females display an LH surge. If plasma prolactin levels are measured, it is seen that they are much higher in the subordinate females (hyperprolactinaemia). Thus, it seems likely that dominant females have the capacity for positive feedback, show normal menstrual cycles and are therefore potentially fertile. Not so subordinate females, who appear to lose the capacity to respond to an oestrogen surge with an LH surge, therefore having anovulatory and irregular cycles and probably amenorrhoea. This is clearly reminiscent of the situation seen in some women with hyperprolactinaemia. That prolactin is the key to this change in positive feedback capacity was demonstrated by lowering plasma prolactin concentrations in subordinate females using a dopamine receptor agonist, and raising it in dominant females using a dopamine receptor antagonist. These treatments effectively reversed the LH response to an oestradiol surge in the two types of female. These two examples emphasize the important and far-reaching consequences of social interaction in determining levels of sexual activity as well as endocrine and reproductive status.

SUMMARY

The critical event underlying ovulation in the menstrual or oestrous cycle is an oestradiol-induced gonadotrophin surge. The oestradiol surge is not preceded by an obvious endocrine or neural trigger, and appears to

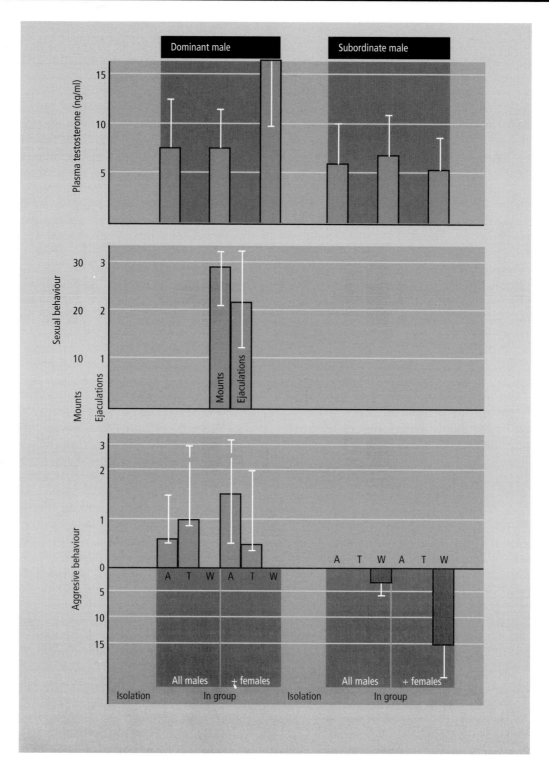

Fig. 5.29 Changes in sexual and aggressive behaviour and plasma testosterone levels in dominant and subordinate male talapoin monkeys during a 12-week period in isolation, a 6-week period in an all-male group, and a 7-week period in a social group with oestradiol-treated females. Note that only the dominant male's plasma testosterone increases in the social group, and that only he is sexually active and being aggressive towards, but not receiving aggression from, other males. The subordinate male receives, but does not give, aggression and withdraws more frequently, especially when females are in the group. A: attacks, T: threats (both measures of aggressive behaviour). W: withdrawals (a measure of submissive behaviour). (Mounts and ejaculations are both measures of sexual behaviour).

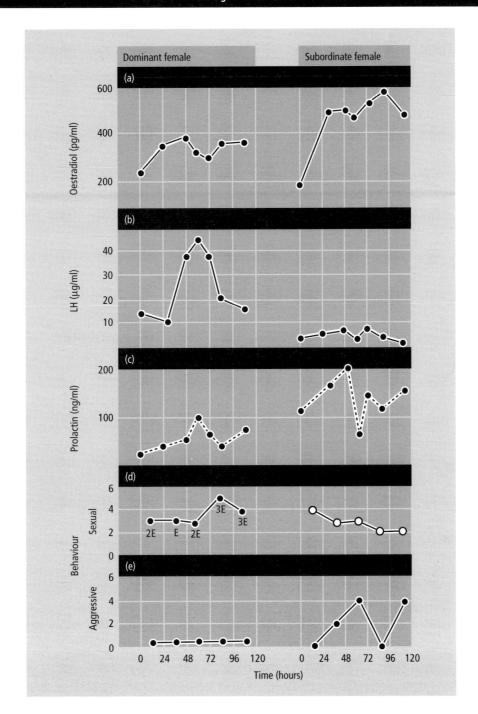

Fig. 5.30 Sexual and aggressive behaviour received by the dominant and subordinate female talapoin monkeys in a social group, and changes in plasma LH and prolactin levels when challenged with an oestradiol surge. (a) Increase in plasma oestradiol concentrations following the oestradiol surge. (b) Changes in plasma LH concentrations, which follow the oestradiol surge. (c) Plasma prolactin concentrations in the females. (d) Sexual interaction with males: ○--○ indicates mounts without ejaculation; ●--● indicates mounts with ejaculations (E). (e) Level of aggressive behaviour received by the females. Note that the dominant female receives high levels of sexual behaviour and low levels of aggression, has low plasma levels of prolactin and shows a surge of LH in response to an oestradiol surge. The converse is true of the subordinate female.

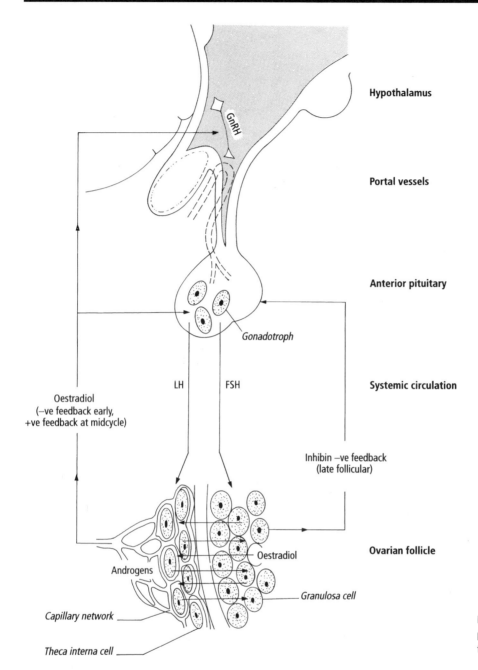

Fig. 5.31 Summary of hypothalamo-pituitary–ovarian interactions during the follicular phase of the cycle.

arise spontaneously within the ovary, being determined by the dynamics of follicular growth. This leads to the conclusion that the ovary, and not the hypothalamus or pituitary, determines cycle length. However, the oestrogen surge from the ovary acts on both the pituitary and the hypothalamus to induce the release of an ovulatory surge of LH (Fig. 5.31). This capacity to respond to an oestradiol surge with an LH surge is lacking in male

non-primate species. However, in both males and females, similar negative feedback effects of gonadal hormones on the pituitary and hypothalamus are observed (Figs 5.32 & 5.33). The feedback effects of gonadal hormones are adequate, in themselves, to explain all the basic features of the reproductive patterns in both males and females. However, the hypothalamo-pituitary–gonadal axis is not a closed system.

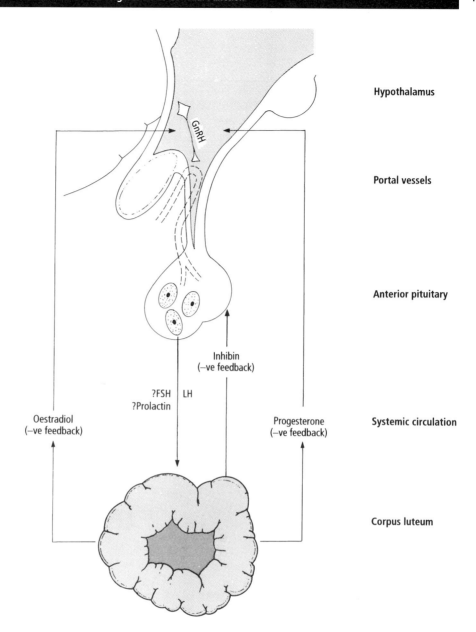

Hypothalamus

GnRH

Portal vessels

Anterior pituitary

Inhibin
(−ve feedback)

?FSH | LH
?Prolactin

Oestradiol
(−ve feedback)

Progesterone
(−ve feedback)

Systemic circulation

Corpus luteum

Fig. 5.32 Summary of hypothalamo-pituitary–ovarian interactions during the luteal phase of the cycle.

External factors clearly modulate its activity to render the basic reproductive pattern susceptible to environmental influences, such as the time of day or year, and proximity of a potential reproductive partner or rival. By such mechanisms, the efficiency of the reproductive process is increased and the survival of the species promoted.

FURTHER READING

Baird DT & Smith KB (1993) Inhibin and related peptides in the regulation of reproduction. *Oxford Reviews in Reproductive Biology* **15**, 191–232.

Ciba Foundation Symposium 62 (1979). *Sex, Hormones and Behaviour.* Excerpta Medica (New Series).

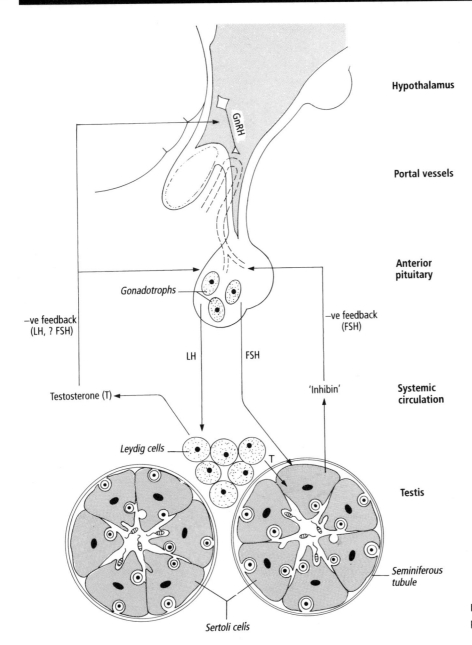

Fig. 5.33 Summary of hypothalamo-pituitary–testis interactions in the male.

Everitt BJ & Keverne EB (1986) Reproduction. In: *Neuroendocrinology* (Eds Lightman SL & Everitt BJ). Blackwell Scientific Publications, Oxford.

Everitt BJ, Meister B & Hökfelt T (1992) The organization of monoaminergic neurons in the hypothalamus in relation to neuroendocrine integration. In: *Neuroendocrinology* (Ed. Nemeroff CB), pp. 87–129. CRC Press, Boca Raton.

Kreiger DT & Hughes JC (Eds) (1980) *Neuroendocrinology*. Sinauer Associates Inc. Sunderland.

Yen SSC & Jaffe RB (Eds) (1978) *Reproductive Endocrinology*. Saunders Philadelphia.

Maturation of the Hypothalamo-Pituitary–Gonadal Axis

C O N T E N T S

In earlier chapters, we have seen how the gonads and genitalia develop during fetal life, mature at puberty and function in the adult. The regulation of adult function is complex, as we saw in the last chapter. Now we return to the subject of puberty and examine how the adult pattern of endocrine activity is established. Our account will concentrate mainly on human puberty.

HORMONAL CHANGES AT PUBERTY

In Fig. 6.1, we summarize the physical changes occurring through puberty described in Chapter 1. They are largely controlled by gonadal and adrenal steroids, together with some direct involvement of pituitary hormones. It is important to re-emphasize that although the absolute tempo of these changes may vary individually, their sequence is remarkably consistent: so much so that, should it deviate from the pattern shown, a clinical investigation is often warranted. We will now examine the endocrinology of the process of gonadal activation, which results in the adult pattern of regulation described in Chapter 5.

Surprisingly, plasma levels of gonadotrophins may be in the adult range soon after birth, rising and falling intermittently for the first year or two of life. They then decline and remain very low during childhood until the initiation of events leading to puberty. From this time,

mean levels of follicle stimulating hormone (FSH) and luteinizing hormone (LH) rise gradually to reach adult levels (Fig. 6.2). Prolactin concentrations also increase in late puberty in girls, but not in boys, in whom the plasma levels of this hormone are already similar to those seen in men. This sex difference may be attributed to the rising oestradiol that enhances prolactin secretion in females (see Chapter 5).

Probably the most intriguing and dramatic change in gonadotrophin secretion occurs in early puberty at night during sleep. In men and women, there is no evidence of a circadian rhythm of FSH and LH secretion (Fig. 6.3d). The same is true in prepubertal children (Fig. 6.3a), who have uniformly low plasma levels. From early to mid-puberty, however, a striking increase in the magnitude, and possibly also the frequency, of LH pulses occurs, which reflects a *sleep-augmented LH secretion* (Fig. 6.3b). In late puberty, daytime LH pulses also increase (Fig. 6.3c), but are less than those still occurring at night, until the adult pattern of higher basal levels with no daily variation is achieved.

Testosterone levels in plasma follow the gonadotrophins. Thus, they are less than 0.1 ng/ml in boys and girls (except during the first 3–5 months after birth in boys when levels similar to those at puberty are found). In early puberty, testosterone levels in boys rise at night when LH secretion becomes elevated (Fig. 6.4). Later in puberty, blood samples taken during the day also show

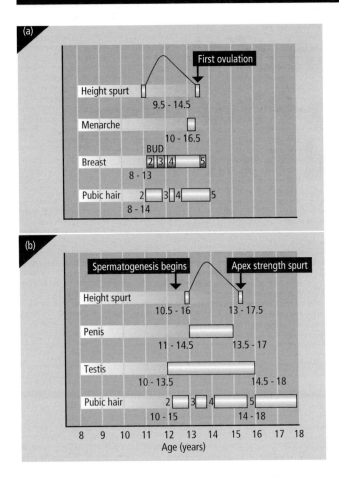

Fig. 6.1 Summary of the sequence of events during puberty in (a) girls and (b) boys. The figures below each symbol represent the range of ages within which each event may begin and end. The figures within each symbol refer to the stages illustrated in Figs 1.15, 1.16 & 1.17.

increases, the greatest changes appearing during pubertal Stage 2, when testosterone concentrations may change from 0.2 to 2.4 ng/ml. There are smaller, but none the less consistent, increases in plasma testosterone concentrations in girls between pubertal Stages P1 and P4.

Oestrogen plasma concentrations are extremely high in both male and female fetuses at birth (5000 pg/ml) because of the conversion of fetal and maternal C19 steroids by the placenta (see Chapter 10). Indeed, newborn infants may display breast budding and even milk secretion ('witches milk') as a consequence of these high oestrogen levels. Oestradiol and oestrone levels soon drop to 7 and 20 pg/ml, respectively, and remain low until puberty. In girls, oestradiol levels rise consistently through the stages of puberty to reach the concentrations seen in mature females (Fig. 6.2). In boys,

plasma concentrations of oestrone are higher than oestradiol, but both are considerably lower than in girls at comparable stages of puberty. In males, about half the oestradiol is derived from extraglandular aromatization of testosterone, and a quarter, or less, from testicular secretion.

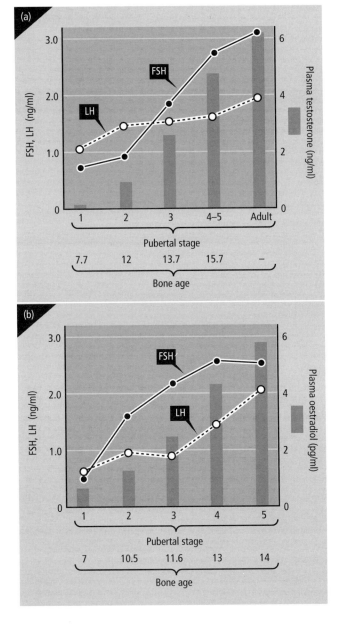

Fig. 6.2 Plasma concentrations of gonadotrophin and steroid hormones during various pubertal stages in: (a) boys; and (b) girls. Bone age is assessed by examining radiographs of hand, knee and elbow, and comparing them with standards of maturation in a normal population. It is an index of physical maturation, and better correlated with the development of secondary sexual characters than chronological age.

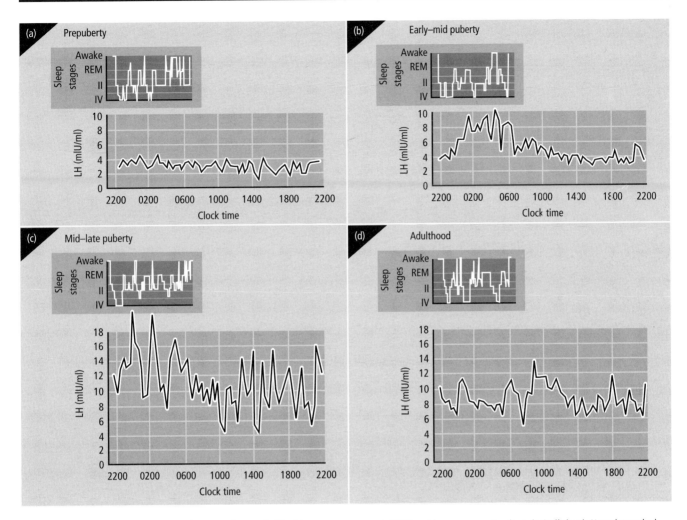

Fig. 6.3 Plasma LH concentrations throughout a 24-hour period in: (a) a prepubertal girl (9 years); (b) an early pubertal boy (15 years); (c) a late pubertal boy (16 years); and (d) a young adult male. The sleep pattern for each nocturna sleep period is depicted in the top left hand corner of each graph (REM: rapid eye movement or 'paradoxical' sleep). Note the marked daily rhythm in (b), with sleep-augmented LH secretion, and the overall higher LH concentrations in (d) compared with (a), but no clear daily rhythm in either.

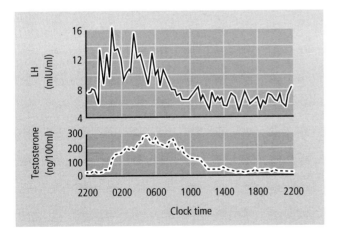

Fig. 6.4 A similar figure to Fig. 6.3(b) showing plasma LH and testosterone concentrations in a 12-year-old boy. Note the night-time rise in testosterone concentrations, coincident with the rise in LH.

Surprisingly, the *earliest* detectable endocrine change, preceding those of gonadotrophins and gonadal steroids, is a progressive increase in the plasma concentration of *adrenal androgens*, notably dehydroepiandrosterone and its sulphate. This adrenal maturation is selective, in that glucocorticoid and mineralocorticoid secretion does not increase at the same time. It is called *adrenarche*, and starts around 8 years of age (6–8 years skeletal age), continuing until 13–15 years. The circulating levels of these weak androgens are orders of magnitude higher than those of the gonadal sex steroids, yet their significance in relation to hypothalamo-pituitary–gonadal activation is obscure. The only clear *somatic* function of these adrenal androgens appears to be the *promotion of pubic and axillary hair growth*. The adolescent growth spurt does not depend on them. Neither hypersecretion nor hyposecretion of adrenal androgens seems to be associated consistently with either early or delayed puberty, and there is little evidence to suggest that these hormones are concerned with timing the onset of puberty in normal children.

In summary, the peripubertal period is associated with the activation of the gonads (and adrenals), which results in elevated steroid secretion. These events are, in turn, dependent on increased trophic stimulation by FSH and LH. In the next three sections, we will examine the control of these trophic stimuli, the nature and timing of the trigger that induces them, and the regions of the central nervous system (CNS) involved.

MECHANISMS UNDERLYING PUBERTY

There are two distinct, but not mutually exclusive, hypotheses concerning the process of pituitary–gonadal activation. The oldest of these, often called the *gonadostat hypothesis*, has ascribed special importance to a progressive maturation of the feedback action of oestradiol or testosterone on gonadotrophin secretion and to changes in the pituitary responsiveness to gonadotrophin releasing hormone (GnRH). The second ascribes a central maturational role to the CNS and, in particular, the hypothalamus. The balance of evidence has shifted in favour of the second of these models.

The gonadostat hypothesis

This model is illustrated in Fig. 6.5. The proposition is that the negative feedback regulation of FSH and LH secretion, prepubertally, operates at a very low threshold or *set point* and is therefore very sensitive to low levels of steroids, and that this set point then increases to become less sensitive at puberty. This process would lead to rising concentrations of gonadotrophins and sex steroids in the circulation during puberty. While it is clear that changes in pituitary responsiveness to steroids do occur, and that steroid-mediated changes in pituitary responsiveness to GnRH also occur, it is probable that these represent *secondary responses* to pubertal change rather than their *driving force*.

As far as positive feedback is concerned, it does seem that the *capacity to evoke a gonadotrophin surge* develops only late in the pubertal process, and then inefficiently, because well over half the cycles occurring in early postpuberty are anovulatory, a proportion that decreases to one-fifth after 5 years or so. The reason for this delayed appearance of positive feedback capacity is not fully understood. However, as we saw in Chapter 5, the anterior pituitary depends critically on continued oestrogen exposure in order to synthesize and store sufficient amounts of gonadotrophins to respond to an oestrogen surge. It may simply be, therefore, that plasma concentrations of follicular oestradiol adequate for this purpose are not achieved until pubertal Stage 5. Therefore, the capacity to respond to an oestrogen surge with an LH surge will likely be a late event in the pubertal process. These observations are also explicable, therefore, on the hypothalamic maturational model.

Hypothalamic maturation

The hypothesis that pubertal activation requires only an increased output of hypothalamic GnRH to drive the system places emphasis on the CNS and its final common pathway of communication with the pituitary, the hypothalamic GnRH neurons, rather than on progressive shifts in the sensitivity of negative or positive feedback interactions between sex steroids, the hypothalamus and the anterior pituitary.

Experimental evidence supporting the view that puberty is driven by a primary change in the hypothalamic output of GnRH comes from experiments in young monkeys, which show that the decrease in pulsatile LH release seen at the end of the neonatal period, and the subsequent rise in pulsatile LH release at the normal time of puberty, occurs even in the absence of the gonads. Thus, male rhesus monkeys gonadectomized at birth show pulsatile LH and FSH secretion and levels that vary within the adult range for the first 10 weeks of life (Fig. 6.6a). Thereafter, concentrations fall and remain at low or undetectable levels for the next 2.5 years or so (Fig. 6.6a). Following this period of gonadotrophic quiescence, often called the *juvenile hiatus* in gonadotrophin secretion, the pulsatile secretion of LH and FSH recommences and levels rise to the adult range, thereby heralding the onset of puberty. As we have established in Chapter 5, pulsatile gonadotrophin secretion is driven by pulsatile GnRH secretion. Therefore, the search for the mechanisms underlying puberty must inevitably focus on why the GnRH pulse generator becomes inactive during the juvenile hiatus and what overcomes it at the time of onset of pulsatile GnRH secretion.

Complementary experiments in intact immature female rhesus monkeys (about 2 years old) make a similar point. They were provided with an externally situated pump, which delivered pulses of GnRH intravenously at 1–1.5-hour intervals (as described in Chapter 5). As can be seen in Fig. 6.6(b), ovulatory menstrual cycles, complete with oestradiol surges, LH surges and luteal progesterone peaks, were initiated

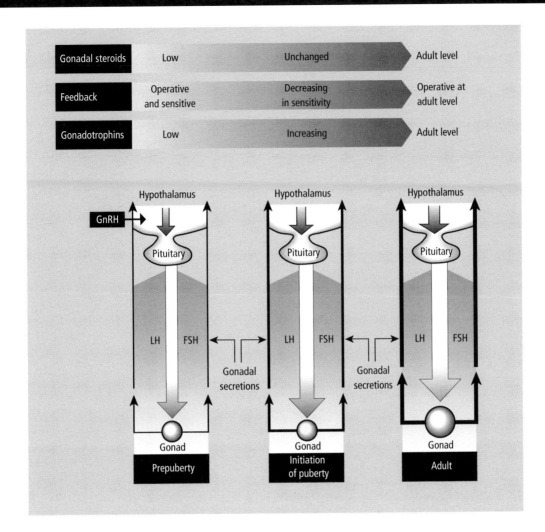

Fig. 6.5 The so-called 'gonadostat' theory of puberty. This summary diagram illustrates the proposed decreasing sensitivity of the negative feedback mechanism to sex steroids during puberty claimed to underlie increasing FSH and LH output. The latter then drives gonadal steroid production to reach adult levels. The evidence supporting this hypothesis is not particularly conclusive and it seems more likely that these changes in feedback sensitivity, which do indeed occur, are secondary to the increased drive to the hypothalamo-pituitary–gonadal axis provided by enhanced GnRH secretion (see Fig. 6.6).

and maintained in these females (Fig. 6.6b). The data have been interpreted to suggest that a most important event in the initiation of puberty is activation of the hypothalamic mechanism, which delivers GnRH pulses to the anterior pituitary. Having done this, the pituitary and ovary are able to respond instantly, and maintain their steroid-mediated negative and positive feedback interactions with the GnRH and gonadotrophin secretory mechanisms. In other words, all that the pituitary–gonadal unit requires for an adult pattern of functioning is the pulsatile secretion of GnRH from the hypothalamus. Figure 6.6(b) also shows rather dramatically that switching off the GnRH pump is followed by re-entry into the immature, prepubertal state. This result strongly indicates that exposure of the hypothalamus to adult levels of circulating steroids does not contribute to the attainment of a 'mature' pattern of functioning. This experiment represents a convincing demonstration that puberty could arise solely as a consequence of a maturational event within the CNS, translated to the pituitary–gonadal system as a stream of GnRH pulses.

The outcome of clinical studies confirms these experimental results. Comparable to the monkey, in the human, the first 6 months or so of life are associated with fluctuating, often high plasma levels of FSH, LH, testosterone (in boys) and oestradiol (in girls, see above). By about 1–2 years, the system has quietened down and gonado-

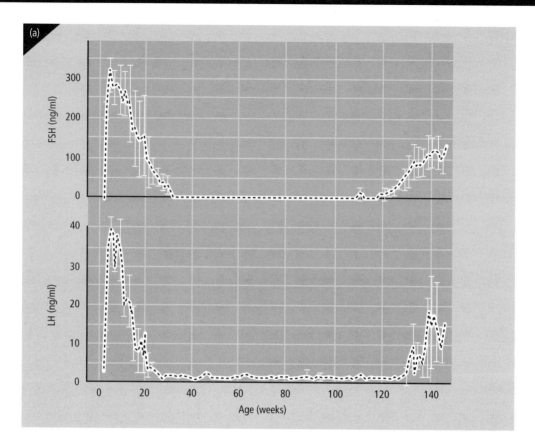

Fig. 6.6 (a) Circulating FSH and LH levels in the plasma of three male rhesus monkeys that were bilaterally orchidectomized at approximately 1 week of age. Blood samples were collected immediately before castration and thereafter at weekly intervals. Data for the first 142 weeks of postnatal development are shown (0 day of birth). It can be seen that LH and FSH pulses occur during the first 20 weeks or so of life when plasma levels are in the adult range. Thereafter LH and FSH secretion ceases and levels remain low or undetectable until about week 120. The first visible change is an increase in FSH secretion followed soon after by increasing, pulsatile secretion of LH. It is important to emphasize that these changes occur in the absence of testosterone in the circulation. Thus, altered gonadotrophin secretion is a function of altered hypothalamic GnRH output independent of any alteration in steroid feedback regulation of this hypothalamo-pituitary system.

trophin and steroid concentrations are very low prepubertally. Gonadotrophin levels in agonadal children, for example those with Turner's syndrome (see Chapter 1), increase around the expected time of puberty in the absence of gonadal steroid influences, similar to the observations on immature castrate monkeys (Fig. 6.6a). Furthermore, precocious puberty may occur in children as young as 2 years of age. Here it is also apparent that the pituitary and gonads can function in an adult manner when activated pathologically, often as a result of a CNS tumour (see below). In such cases, however, there is little opportunity of investigating other maturational changes that might occur in the pituitary or gonads, as the clinician is usually presented with the fact of precocious puberty, rather than early indications of it.

Thus, on balance, the genesis of puberty must be sought not in interactions between gonad and the hypothalamic–pituitary axis, but within the CNS. What might the trigger be, and does it lie within the CNS or impinge upon the CNS from outside?

THE TIMING OF PUBERTY

Secular trend towards earlier puberty

The factors responsible for triggering and timing the onset of puberty have proved elusive. One phenomenon illustrated in Fig. 6.7 may, however, be important in directing our attention towards particular factors. Thus, although the age at which girls first menstruate shows a considerable range within the population, there has been a clear secular trend towards an earlier menarche in girls

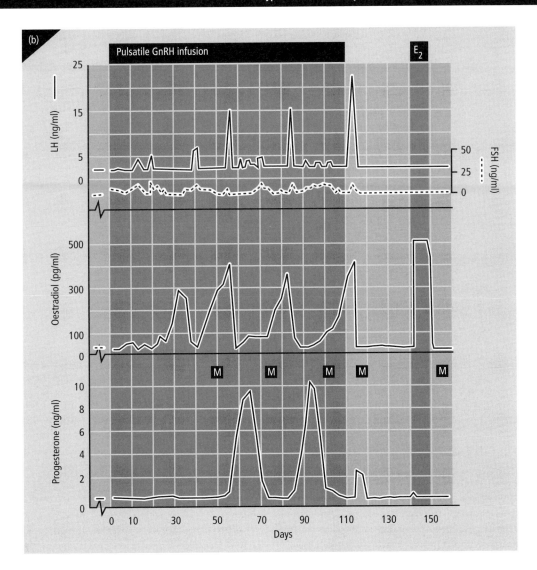

Fig. 6.6 *(continued)* (b) Induction of ovulatory menstrual cycles in an intact immature rhesus monkey by the infusion of GnRH (1 μg/min for 6 minutes once every hour). The period of GnRH infusion is shown by the horizontal bar (Day 0–110); levels of LH, FSH, oestradiol and progesterone were undetectable in blood samples prior to GnRH infusions. Note that the first oestradiol surge did not elicit a full LH surge (*c.* Day 25). However, subsequent oestradiol surges induced both LH surges and evidence of corpus luteum formation (progesterone peaks). These LH surges, as well as ensuring menstrual periods, occurred at 28-day intervals. Cessation of GnRH infusions (immediately after the LH surge, *c.* Day 112) was followed by prompt re-entry into a non-cyclic, prepubertal state. Implantation of an oestradiol-containing silastic capsule subcutaneously (between 140 and 150 days, as indicated by bar) to produce surge levels of oestradiol in blood did not induce an LH surge in the absence of exogenous GnRH. M: menstruation.

and puberty in boys in Western Europe and the USA over the past century. What factors have changed that might have contributed to the earlier attainment of sexual maturity and do they give any insight into the mechanisms controlling the initiation of puberty?

Environmental factors

Clearly, there may be more than one answer to this question. Health care and personal health have improved during this time, along with living conditions and socio-economic standards. Undoubtedly, these have contributed to the attainment of earlier puberty and our longer life expectancy in general: indeed, the latter ensures that the majority of women now survive to experience the menopause (see Chapter 14). But from clinical and experimental studies, two factors have been implicated more consistently in this secular trend to earlier puberty, and in the mechanisms that normally underlie the onset of puberty. These are photoperiod and nutrition.

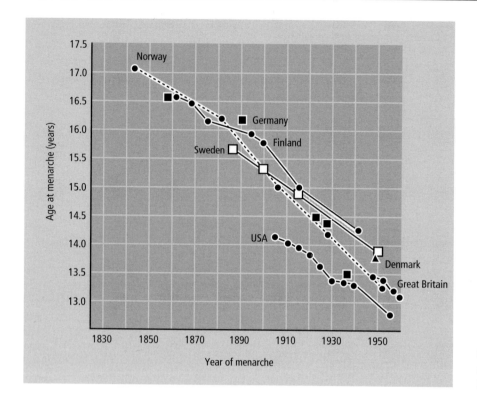

Fig. 6.7 Secular trend towards an earlier age at menarche in girls from Western Europe and the USA.

Photoperiod

In Western society, we are able to control our physical environment, for example by using electricity to extend artificially the length of the day. Indeed, daily dark periods may be consistently as short as 7 hours all the year round, as if we were in constant 'long days' or a summer photoperiod. As we discussed in Chapter 5, in non-primate species, such changes in day length may have a major impact on reproductive status. However, the intensity of domestic lighting probably is not high enough to influence photosensitive neural mechanisms that might regulate GnRH secretion. Moreover, there is little direct evidence that photoperiod affects either reproductive maturation or adult reproductive activity in the human.

Nutrition

It seems reasonable to suggest that nutritional factors might have important effects on sexual maturation. Indeed, examination of cultures, such as the nomadic Lapps, that have not experienced such major improvements in living standards and nutrition, show that between 1870 and 1930 there was little or no trend towards an earlier menarche. Experimental studies that emphasize the importance of nutrition on reproduction in the adult are abundant. For example, maintaining female rats on a low-protein diet, such that their body weight is held consistently at 80% of normal, results in

the cessation of oestrous cycles. Subsequent sudden exposure of these animals to high-protein food restores the ability to discharge LH after oestradiol challenge. The practice of 'flushing' sheep, exposing them to rich pasture, increases the ovulation rate and also induces an earlier onset of oestrus and lambing, a phenomenon exploited by farmers. Adolescent girls, with the complicated syndrome of anorexia nervosa, exhibit a very low food (particularly carbohydrate) intake and body weight, have irregular menstrual cycles or amenorrhoea. These examples suggest that food intake, or a reflection of it — body weight — is highly correlated with reproductive efficacy in the adult. Is it similarly associated with the onset of puberty?

Body weight

In Fig. 6.8, it can be seen that although age at menarche has changed considerably during the past 100 years, the body weight at menarche has remained surprisingly constant at about 47 kg. Similar constancy is seen in the weight at onset of the adolescent growth spurt (Fig. 6.8). These data have led to the suggestion that, at least in girls, a critical weight must be attained before the activation of the hypothalamo-pituitary–gonadal axis and the growth spurt can occur. According to this view, body weight, or more correctly a critical metabolic mass

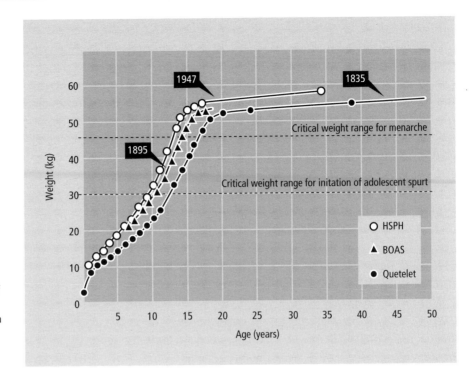

Fig. 6.8 Age plotted against body weight in three populations of girls in 1835, 1895 and 1947. Note the constant weights at initiation of the growth spurt (30 kg) and at menarche (47 kg). (Populations: Belgian girls in 1835 from data of Quetelet, 1869; American girls in 1895 from data of BOAS, and in 1947 from data of Reed and Stuart, 1959.)

related to body weight, may trigger and therefore time the onset of puberty. A similar suggestion for a critical weight of 55 kg underlying sexual maturation in boys has also been made. The earlier occurrence of puberty today, compared to a century ago, may, therefore, be explained by earlier attainment of a critical weight due to improvements in nutrition, health care and social living conditions. Evidence in support of this view is, at first sight, abundant. Moderately obese girls experience an earlier menarche than lean girls. Malnutrition is associated with delayed menarche. Primary amenorrhoea is extremely common in ballet dancers at professional schools who are in the very low range of weight for height and relative fatness for their age. The amenorrhoea frequently seen in girls with anorexia nervosa has its onset when body weight drops somewhat below 47 kg, while in some anorexic girls who begin to re-feed, the reoccurrence of menstruation is associated with attainment of a 47-kg body weight.

This is an impressive array of data supportive of what, teleologically, may be regarded as a very logical signal for the onset of puberty, namely, attainment of a body size sufficient to cope with the demands placed on it by adult reproductive activities, such as pregnancy. However, much controversy surrounds this hypothesis, particularly that part of it which assumes a causal relationship between a critical metabolic mass, derived from assessment of lean body weight, body fat and total body water, and the time of onset of puberty. First, the studies quoted are retrospective rather than prospective, and depend on derived, rather than direct, measurements. It has subsequently proved quite difficult to predict age of menarche from a knowledge of an individual's body weight and weight gain. Second, many anorexic girls who re-attain their critical body weight do not begin having menstrual cycles, although this may simply reflect the fact that anorexia nervosa is a far more complicated syndrome than just a disorder of food intake and consequent weight loss. Third, menarche is a *rather late* event in puberty, and may, therefore, be much removed from the critical factors that determine the *onset* of those endocrine changes described above, even if menarche itself is often related to body weight.

Summary

The concept of a metabolic signal, conveying information about the state of development of the body to the CNS, thereby acting as a trigger to the onset of puberty, is an attractive one. However, there are no conclusive data as to the form that the information might take nor as to what that trigger might be. Some humoral reflection of the body fat : lean ratio or a substance synthesized by developing bone have been suggested, but the evidence is not conclusive. One thing does seem clear, however: the trigger, whatever it is, induces a change in the activity of GnRH neurons in the hypothalamus. It is not certain whether this reflects the removal of an inhibitory influence that resulted in the juvenile hiatus in gonadotrophin secretion, or the induction of an excitatory input to GnRH neurons. These are not mutually exclusive mechanisms, as the inhibitory mechanism that holds the GnRH pulse generator activity in check during prepuberty may not operate directly on the GnRH neurons themselves, but on the excitatory input to GnRH neurons, which many consider to be glutamergic. What do we know of the areas of the CNS concerned with puberty?

CNS AND PUBERTY

As the evidence on the nature of the trigger for puberty is vague, it is perhaps not surprising that the neural sites through which they act are also poorly defined, save that the hypothalamic GnRH neurons are the final common pathway. The latter point is emphasized by an intriguing clinical condition known as *Kallman's syndrome*. This syndrome is characterized by delayed puberty and anosmia. The delayed puberty is the result of an isolated gonadotrophin deficiency, which is itself the result of a GnRH deficiency that reflects a primary deficit in the formation of GnRH neurons in the olfactory placode and their migration into the developing brain, including the medial preoptic area. Puberty can be initiated in these subjects by giving exogenous GnRH, which causes the release of FSH and LH from the pituitary.

Other clinical cases of advanced or delayed puberty associated with an underlying neuropathology, especially tumours (Table 6.1), provide just about the only additional information on areas of the CNS concerned with puberty. However, understanding the mechanisms by which such lesions in the CNS affect the onset of puberty is far from simple. Generally the *site* rather than the *type* of tumour determines its effects on puberty, unless the tumour is gonadotrophin-producing. In the hypothalamus, lesions of the anterior area are associated with delayed puberty. Lesions of more posterior areas, from the median eminence to the mammillary bodies, are consistently associated with precocious puberty. Tumours involving the pineal gland were once thought to affect the onset of puberty indirectly, by mechanical compression of the underlying hypothalamus (Fig. 5.1). However, this now seems unlikely as such tumours may be associated with advanced or delayed puberty depending upon the involvement of pineal non-glandular or glandular tissue, respectively. To what extent these effects are related to the production of gonadotrophins (particularly by teratomas) or melatonin (by glandular tumours of the pineal) is unclear. Chromaphobe adenomas (hyperplastic lactotrophs) producing high levels of prolactin are associated with a delay of puberty, which may be induced subsequently with bromocriptine: the dopamine D_2 receptor

Site of lesion	Puberty Precocious	Delayed	Type of lesion
1 Hypothalamus			
(a) Anterior		✓	Hamartoma
(b) Middle and posterior			Germinoma
including mammillary bodies	✓		Teratoma
			Third ventricle cyst
2 Pineal gland			
(a) Parenchymatous		✓	(a) Glandular tissue tumour (rare)
(b) Non-parenchymatous	✓		(b) Non-glandular tissue tumour
3 Pituitary gland			
		✓	{ Craniopharyngioma { Chromaphobe adenoma
	✓		Gonadotrophin-producing

Explanation of terms:
  hamartoma: hyperplastic growth formed of nerve cells, fibres and glia;
  germinoma: tumour of germ cell origin;
  teratoma: tumour of partially developed embryonic tissues;
  craniopharyngioma: tumour of Rathke's pouch originating from the pituitary stalk;
  chromaphobe adenoma: prolactin secreting tumour of the anterior pituitary.

Table 6.1 Neurological lesions associated with advanced or delayed puberty in humans

agonist (see Chapter 5); however, it seems unlikely that this bears on the normal pubertal triggering mechanism.

The problem is that the lesions are often large and of complicated and widespread origin, so accurate interpretation of their effects is difficult unless the tumours involved are gonadotrophin- or prolactin-producing. The identification of those areas of the CNS critically involved in integrating the endocrine changes associated with the initiation of puberty is an important and active area of research in reproduction.

FURTHER READING

Adams LA & Steiner RA (1988) Puberty. *Oxford Reviews of Reproductive Biology* **10**, 1–52.

Bullough VL (1981) Age at menarche: a misunderstanding. *Science* **213**, 365–366.

Delamarre-van der Waal HA, Plant TM, van Rees GP & Schoemaker J (1989) *Control of the Onset of Puberty.* Excerpta Medica International Congress Series, No 861 (Amsterdam).

Everitt BJ & Keverne EB (1986) Reproduction: In: *Neuroendocrinology* (Eds Lightman SL & Everitt BJ). Blackwell Scientific Publications, Oxford.

Foster DL (1994) Puberty in sheep. In: *The Physiology of Reproduction* (Eds Knobil E & Neill JD), Volume 2, 2nd edition, pp. 411–452. Raven Press, New York.

Grumbach MM, Grave GD & Mayer FE (1974) *The Control of the Onset of Puberty.* John Wiley, Chichester.

Plant TM (1994) Puberty in primates. In: *The Physiology of Reproduction* (Eds Knobil E & Neill JD), Volume 2, 2nd edition, pp. 453–486. Raven Press, New York.

Tanner JM (1978) *Foetus into Man; Physical Growth from Conception to Maturity.* Open Books, Wells.

Tanner JM (1986) *Growth at Adolescence.* Blackwell Scientific Publications, Oxford.

Yen SC & Jaffe RB (Eds) (1978) *Reproductive Endocrinology.* (Particularly Chapter 10.) WB Saunders & Co, Philadelphia.

CHAPTER 7

Actions of
Steroid Hormones
in the Adult

CONTENTS

In the foregoing chapters we have established the pivotal role of the gonads and their steroid secretions in reproductive events. We have been concerned mainly with steroid action in two areas: first, in the generation and maintenance of sexual differentiation during fetal, neonatal and pubertal life; and second, in the regulation of gonadotrophin secretion by the hypothalamic–pituitary axis. In this chapter, we will examine in more detail those remaining actions of the steroids in the adult male and non-pregnant female, which ensure the attainment of full reproductive capacity both physically and behaviourally.

The effects of sex steroids may conveniently be thought of as falling into two broad categories: some steroid actions are *determinative*, others are *regulatory*. Determinative actions involve essentially qualitative changes, which are irreversible or only partially reversible. Examples of this type of action are provided by the effect of androgens on the development of the Wolffian ducts and the generation of male external genitalia, the mild enhancement of these sexually distinct features by the low prepubertal androgen levels in males, and the changes in hair pattern, baldness, voice tone, penile and scrotal size and bone growth that occur at puberty, as well as those in brain structure and physiology in non-primates. These actions constitute part of a progressive androgenization, which establishes a clear and distinctive male phenotype. It represents the completion of a process initiated with the expression of the SRY (sex-determining region on the Y chromosome) gene. In females, an active determinative role for steroids first occurs during the prepubertal and pubertal period, when body growth, size and shape and growth of secondary sex hair are stimulated.

In contrast, the regulatory actions of steroids are reversible, and can involve both quantitative and qualitative changes to established accessory sex organs and tissues. These actions are not concerned with establishing the individual as a male or female. They are concerned with ensuring that their reproductive tracts and genitalia function effectively in the reproductive process. In males, the regulatory actions of testicular steroids influence the activity of the accessory sex glands, metabolism, erectile capacity and, in some species, the more exotic secondary sexual characters, such as antlers in deer. These actions may be continuous or show seasonal variations. In females, it is the regulatory action of oestrogens and progestagens that results in the external manifestations of the menstrual and oestrous cycles. These external changes are accompanied by cyclic changes in the vagina, cervix, uterus, oviducts and, in some species, for example the ferret, unusual swelling of the vulva or, in various female primates, the sexual skin.

ANDROGENS AND THE MALE REPRODUCTIVE SYSTEM

In Chapter 3, we saw that testosterone was essential for the maintenance of spermatogenesis. Neutralization of testosterone by an antibody, or by synthetic anti-androgens, such as cyproterone, blocks or reduces the effectiveness of sperm production. Testosterone deprivation also has profound and immediate effects on the accessory sex glands of the male's genital tract (Fig. 7.1). After castration, the prostate, seminal vesicles and epididymides, or their equivalents in various species (Table 7.1), involute, their epithelia shrink and secretory activity ceases (Fig. 7.2). Direct measurement of their metabolic and synthetic activity shows a dramatic fall, and seminal plasma is no longer produced. If the castrate animal is provided with exogenous testosterone, the involuted organs are fully restored, both in size and secretory activity. This reversible regression of acces-

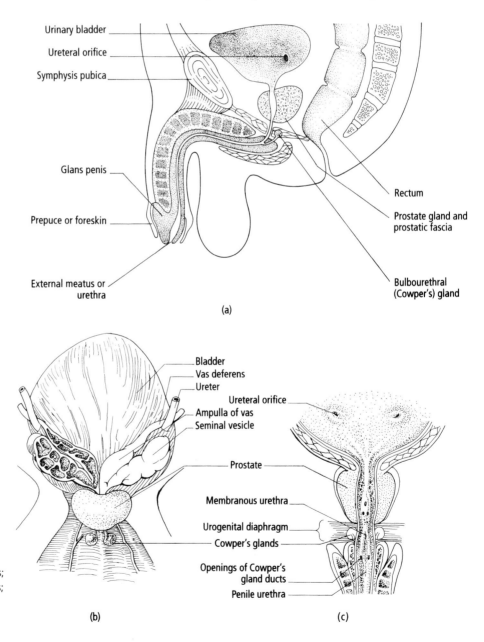

Fig. 7.1 View of human male accessory sex glands: (a) mid-sagittal section through pelvis; (b) posterior view of dissected pelvic contents; (c) coronal section through (b) viewed anteriorly.

Table 7.1 Relative size of principal male accessory sex glands

Species	Prostate	Seminal vesicle	Ampulla	Cowper's or Bulbo-urethral glands
Human	+++	++	+	±
Bull	+	+++	++	±
Dog	+++	-	-	-
Boar	++	+++	-	++
Stallion	++	+	++	±
Ram	++	+	++	++

sory sex gland activity occurs naturally in seasonally active males, such as sheep and deer. The seasonal appearance of secondary sexual characteristics, such as antlers, and the behavioural interactions during which they are used, are similarly dependent on the actions of androgens.

Not surprisingly, these target organs for androgen activity are found to possess androgen receptors. Although the testicular androgen that normally mediates this activity is testosterone, the enzyme 5α-reductase is present and active in these target tissues. Therefore, most androgenic stimulation is provided by the more active 5α-dihydrotestosterone. In Chapter 5, we mentioned that the local activity of androgens is enhanced by prolactin. This relationship is reciprocated, as androgen-dependent prolactin receptors are present in both prostate and seminal vesicles, and prolactin itself is detectable in seminal plasma at higher levels than in blood.

In addition to their effects on secondary sex organs, androgens also have anabolic or myotrophic effects, a reflection of which is the characteristically more muscular appearance of males that progressively establishes itself after puberty. Androgens also increase kidney and liver weight, depress thymus weight and stimulate erythropoiesis.

OESTROGENS, PROGESTAGENS AND THE FEMALE REPRODUCTIVE SYSTEM

In the female, the shifting balance of hormones during the ovarian cycle affects oviducal, uterine, cervical and vaginal activity, as well as exerting more generalized physiological actions. The effects of steroids on these target organs may be studied by correlating changes either in a normal cycle with changing steroid levels, or after the injection of exogenous steroids into intact or castrate females.

The oviduct (Fallopian tube)

The oviduct is a thin muscular tube covered externally with serosal tissue and peritoneum. A ciliated, secretory, high columnar epithelium overlies the stromal tissue internally. The oviduct is the site of fertilization and therefore the oocyte passes along it from the fimbriated ostium towards the spermatozoa, which pass in the opposite direction from the isthmic junction

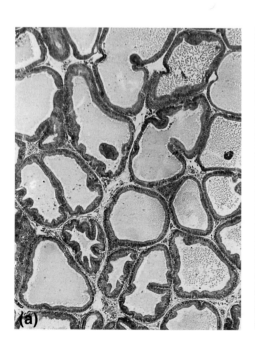 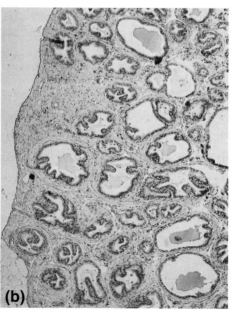

Fig. 7.2 Sections through prostate from: (a) intact rat; and (b) rat 18 days postcastration. Note reduction in luminal content, reduced height of luminal secretory epithelium and relative expansion of proportion of connective tissue.

with the uterus (Fig. 4.1). A few days later, the fertilized zygote reverses the path taken by the spermatozoa to enter the uterus (see Chapters 8 & 9). Clearly then, the oviduct has a role in gamete transport. It must also provide a suitable environment for fertilization and the early growth of the conceptus. After ovariectomy, oviducal cilia are lost, secretion ceases and muscular activity declines, indicating an important steroidal influence on the oviduct. Subsequent injections of oestrogen restore both the ciliated, high columnar epithelium and secretory activity, and increase spontaneous muscle contractions. When progesterone is imposed on this oestrogen background, the numbers of cilia decline, and the quantity of the oviducal secretion also declines, the small volumes of fluid that are produced having a lower sugar and protein content. Raising the progesterone to oestrogen ratio may also exert a mildly depressant effect on oviducal musculature, particularly relaxing the sphincter-like muscle at the utero-tubal junction (although data on this are somewhat conflicting).

There is considerable evidence to suggest that sudden changes in the oestrogen : progesterone ratio, as achieved, for example, by administration of exogenous pharmacological doses of steroids during a cycle, can cause disturbances of gamete and conceptus transport. Thus, if high doses of the synthetic oestrogen, stilboestrol, are taken within 72 hours of fertilization, the pregnancy rate is reduced, probably by causing *premature expulsion* of the conceptus from the oviduct via the uterus into the vagina. However, the same dose given earlier (just before or immediately after ovulation) can result in *tube-locking* of the conceptus, in which prolonged spasm of the oviducal musculature *prevents* passage of the conceptus to the uterus. Pregnancy may none the less ensue, but frequently it is not uterine, but occurs in the oviduct (*ectopic*).

The uterus

The uterus shows even more prominent steroid-dependent cyclic changes in structure and function than the oviduct (Fig. 7.3). During each cycle, the uterus first prepares to receive and transport the spermatozoa from

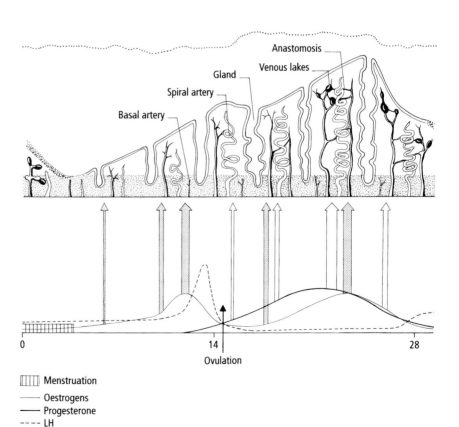

Fig. 7.3 Changes in human endometrium during the menstrual cycle. Underlying steroid changes are indicated below and basal body temperature is indicated above. Thickness of arrows (oestrogens: stippled; progestagens: white) indicate strength of action.

Basal artery
Spiral artery
Gland
Venous lakes
Anastomosis

0   14   28
Ovulation

Menstruation
Oestrogens
Progesterone
LH

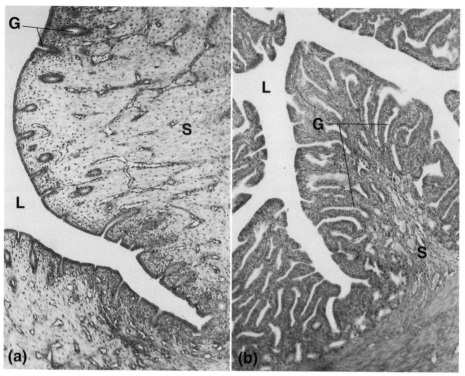

Fig. 7.4 Sections through rabbit endometrium during: (a) oestrus; and (b) progestational phase. Note the dramatic increase in invaginations of the glandular epithelium (G) from luminal surface (L) into stromal tissue (S).

the cervix to the oviduct, and subsequently it prepares to receive the conceptus from the oviduct and nourish it. The uterus consists of an outer peritoneal and serosal investiture, over a thick myometrium of smooth muscle arranged in distinctly oriented layers. Internally, the endometrium consists of a stromal matrix over which lies a simple low columnar epithelium with glandular extensions penetrating into the stroma (Fig. 7.4). After ovariectomy, the uterus hypotrophies, and its blood supply is reduced. Administration of oestrogen reverses this effect, with massive increases in RNA and protein synthesis, cellular division and growth. In the normal cycle, the period during which oestrogen rises rapidly coincides with the latter half of the follicular phase of the menstrual cycle or, in non-primates, with the whole of the abbreviated follicular phase. During this phase a similar *uterotrophic* effect of oestrogen is observed. The myometrium increases both its contractility and its excitability, showing increased spontaneous activity. Meanwhile in the endometrium, stromal thickening occurs, partly due to stromal cell proliferation (so that the *uterine cycle* equivalent of the ovarian *follicular* cycle is often called the *proliferative phase*) and partly due to oedema. The surface epithelium increases in surface area and metabolic activity. In primates and in large farm animals, this also involves an increase in the numbers, and size, of the glandular invaginations of the stroma;

this is less marked, however, in rabbits and rats. The oestrogen-primed epithelial cells secrete a fluid of a characteristically watery constitution, which contain a range of proteins, including proteolytic enzymes. These changes reach a maximum at the time of the oestrogen surge.

The oestrogens act by binding to oestrogen receptors present in abundance in uterine tissue. One of the most crucial actions of oestrogens over this period is to induce the synthesis of intracellular receptors for progesterone. At the beginning of the oestrogenic phase of the cycle, progesterone-binding receptors are at a low level, and progesterone therefore has little effect on the uterus. However, after the oestrogen surge and ovulation have occurred, the uterus is primed to bind progesterone, and so the progestagenic or *secretory phase* of the uterine cycle begins (corresponding to the luteal phase of the ovarian cycle). In the rabbit and mouse, there is a rapid extension of the epithelial proliferation into glandular regions at this time (Fig. 7.4).

Progesterone stimulates the synthesis of secretory material by the glands of most species so that they become distended with a thick secretion rich in glycoprotein, sugars and amino acids. In some, but probably not all, species, the release of this glandular secretion into the lumen requires, or is facilitated by, a secondary peak of luteal-phase oestrogen imposed upon the progesterone background (see also Chapter 9). Stromal proliferation also increases under the influence of progesterone, the stromal cells becoming larger and plumper. This is particularly marked in rodents and primates. Within the stromal tissues of primates, characteristic spiral arteries become fully developed (Fig. 7.3). Progesterone also acts on the myometrium causing further enlargement of cells, but, in contrast to oestrogens, progesterone *depresses* the excitability of the uterine musculature. It is important to re-emphasize that these actions of progesterone will only occur in an oestrogen-primed uterus, another example of receptor regulation (see Chapter 2).

With the withdrawal of steroid support at the end of the luteal phase of the cycle, the elaborate secretory epithelium collapses, with evidence of apoptotic cell death. In most mammals, the endometrium is resorbed, and a thin stromal layer overlain with epithelium replaces it, ready for entry into a new uterine cycle as oestrogens rise. In humans, apes and Old World monkeys, the endometrial tissue is shed as the menses via the cervix and vagina, together with blood from the ruptured arteries. The spiral arteries contract to reduce bleeding.

The cervix

The cervix is traversed by the spermatozoa at coitus and the neonate at parturition (see Chapters 8 & 12). In many species, including humans, the spermatozoa must actively swim through the cervix (see Chapter 8). The properties of the cervix show marked steroid-dependent changes during the cycle and these can be crucial to normal fertility. During exposure to oestrogen in the follicular phase, the muscles of the cervix relax and the epithelium becomes secretory. However, during the luteal phase, when progesterone levels are elevated, secretion is reduced and the cervix is firmer.

Cervical mucus may be collected for examination during the human cycle and tested in a variety of ways for its steroid-dependent properties. The test of greatest functional significance is that of sperm penetration, in which the capacity of spermatozoa to swim into and through a smear of mucus on a slide is assessed. Characteristically, sperm penetration is low in the luteal phase of the cycle, and reaches a maximum around the time of ovulation (Fig. 7.5). These effects can be mimicked by administration of exogenous steroids: oestrogens enhance sperm penetration while progesterone, even in the presence of oestrogens, depresses penetration. Thus, continuous administration of progestagens throughout the cycle, or the local release of progestagens from cap-

sules placed in the uterus, suppresses sperm penetration even at the time of ovulation and the oestrogen surge. Use is made of this property in the low-dose progestagenic contraceptives. The steroids act via an effect on the amount and nature of the glycoproteins secreted by the cervical epithelium. Under progestagen dominance, small volumes of thick mucus are secreted and strands of mucus can only be stretched a short length before the threads snap: a low *spinnbarkeit* (Fig. 7.5). If mucus from oestrogenic cervices is allowed to dry on a slide, its distinctive molecular composition results in a characteristic pattern known as *ferning* (Fig. 7.5). These tests of cervical mucus are important, as a hostile, impenetrable mucus will reduce sperm progress towards the oviduct and thus fertility.

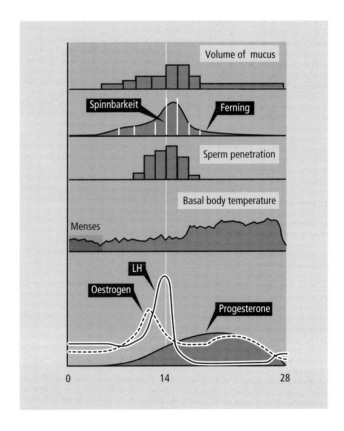

Fig. 7.5 Schematic view of changes in properties of cervical mucus at various days of the human cycle (blood hormone levels and basal body temperature shown). Parameters changing under influence of high oestrogen and low progesterone are: volume of mucus; spinnbarkeit of mucus (bar height illustrates relative length to which mucus thread can be stretched before snapping); ferning (curve illustrates proportion of crystallized mucus that shows a ferning pattern when dried on a slide); and *in vitro* tests of the ability of spermatozoa to penetrate mucus. Note luteal oestrogen does not induce changes due to elevated progesterone at the same time. Progestagenic contraceptives prevent normal periovulatory changes.

The vagina

The vagina shows marked structural changes during the cycle in some species. In the guinea-pig, for example, a membrane completely closes the vaginal *os* throughout most of the cycle, and only breaks down under the influence of rising plasma oestrogen concentrations at oestrus. In many mammals, including humans, oestrogens induce an increased mitotic activity in the columnar epithelium of the vagina, with a tendency to keratinize. This change is particularly marked in rodents, in which the stage of the oestrous cycle can be assessed quite accurately by examination of the different cell types

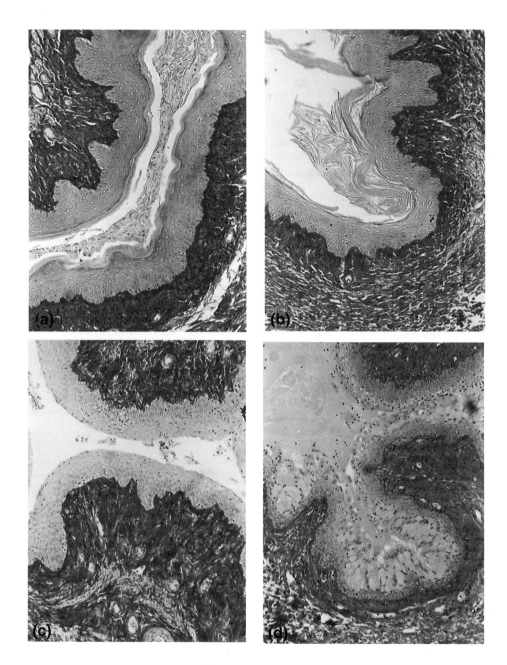

Fig. 7.6 Sections through rat vagina. (a) Leading up to oestrus (pro-oestrus), rising oestrogen causes nucleated cells to be shed into the lumen and a keratin layer to develop on the surface epithelium. (b) This process is completed at oestrus with a heavily keratinized layer and some desquamation of keratin. (c) During the short pre-gestational phase (dioestrus), with a low oestrogen : progesterone ratio, the epithelial cells are nucleated and non-keratinized, and leucocytes are visible in the lumen. (d) In pregnancy, the surface epithelium is glandular and leucocytes abound.

present in smears from the vaginal epithelium (Fig. 7.6). The fluids within the vagina also change during the cycle, and one effect of this is to vary the metabolic substrates available to the bacterial flora there. Cyclic changes in the vagina, induced by oestrogen and progesterone, result in the generation by bacteria of differing proportions of volatile aliphatic acids. These give distinctive odours to vaginal secretions and may have marked behavioural consequences, as will be seen later in this chapter.

Other tissues

Several other features of a female's anatomy and physiology may change with the ovarian cycle. Oestrogens have general effects on the cardiovascular system and metabolism that may be revealed cyclically or in women taking oral contraceptives. Thus, oestrogens depress appetite, are mildly anabolic and maintain bone structure. They also appear to be involved in the reduced capillary fragility, higher levels of low and high density lipoproteins, and thus in the ability to bind cholesterol, and reduced incidence of thrombosis seen in premenopausal women when compared to men. Oestrogens may also increase the ability of the cardiovascular system to withstand high blood pressures. The mechanisms by which oestrogens act in this way are unclear. However, maintained elevated levels of oestrogens, such as occur in pregnancy or during treatment with some contraceptive pills, may actually cause hypertension and, via effects on lipid metabolism, an increased blood clotting rate (see Chapter 14).

Progesterone, in contrast to oestrogens, is mildly catabolic in humans, and has been suggested to increase appetite. It also has two general actions that may become manifest during the menstrual cycle. Progesterone elevates basal body temperature (Fig. 7.5); this is not a consequence of increased catabolism or stimulation of the thyroid, but appears to involve a direct action on hypothalamic areas concerned with thermoregulation. The rise in temperature occurs only *after* ovulation and thus is only useful as a basis for establishing the regularity of a woman's cycle if the rhythm method of contraception is contemplated (see Chapter 14). Progestogenic steroids also have anaesthetic properties, probably mediated by actions within the brainstem reticular formation, and this may be related to the increased tiredness and sleepiness of some women even early in pregnancy.

Progesterone shows some affinity for aldosterone receptors in the kidney, presumably because of similarities of stereochemical structure. However, after binding by progesterone, the receptor is not activated and, in consequence, progesterone acts as an inhibitor. Natriuresis ensues. A compensatory rise in aldosterone output occurs to restore sodium retention. Sodium retention in women may also be enhanced in the luteal phase by a direct stimulatory effect of luteal oestrogen on angiotensinogen production. The consequence of these events may be a net retention of sodium and water towards the end of the luteal phase, which contributes to some symptoms characteristic of the premenstrual period, for example, heavy, tender breasts.

These examples of the widespread cyclic changes in structure and function of many of the somatic tissues of the female, and not just her reproductive organs, show how the ovary and its secretions play such a dominant part in normal day-to-day physiology. Ovarian steroids also have profound and cyclic effects on the behaviour and mood of the females of some species, and this, together with comparable effects of testicular hormones in males, is the subject of the next section.

HORMONES AND THE REGULATION OF SEXUAL BEHAVIOUR

We have already discussed in Chapter 1 how hormones may affect the developing brain and so influence or even determine the types of sexually dimorphic behaviour observed later in life. These same hormones can also influence how and whether these types of behaviour are actually expressed in the adult.

Stimuli capable of eliciting sexual behaviour surround most of us most of the time, but sexual interaction occurs only sporadically. What determines when these sexual stimuli induce sexual activity? The answers to this question, particularly in primates and humans, are complex, but if we examine non-primate species we see rather clearly that hormones are very important. A castrated male or female rat, cat or dog does not display sexual activity when placed with a member of the opposite sex. Treatment with the appropriate hormones results in the activation of sexual behaviour. The sexual stimuli become effective again in inducing sexual responses. Thus, in these species, hormones increase the probability that an appropriate set of sexual stimuli will elicit sexual activity. In primates, including humans, this profound controlling influence of sex hormones has been modified considerably, such that social and volitional factors become increasingly important. Undoubtedly, this is one reflection of the increasing size and complexity of the brain, particularly the neocortex, as the phylogenetic scale is ascended. We will first discuss the ways in which hormones affect sexual behaviour in various species, and then go on to consider how and where they exert their effects. Although much remains to be discovered, an important message will become apparent. Hormones alter the expression of sexual behaviour by acting *both* within the *brain* and on the *genitalia*.

Masculine sexual behaviour

Non-primates

In adult non-primate males, sexual behaviour is clearly hormone-dependent. It declines after castration and this decline is reversed by treatment with testosterone (Fig. 7.7). However, one feature of this relationship between testicular androgens and sexual behaviour remains difficult to explain. Although plasma testosterone is virtually undetectable within hours of castration, the decline in sexual behaviour takes several weeks to reach its nadir (Fig. 7.7). Similarly, treatment with testosterone after long-term castration will restore sexual behaviour, but does so with a long latency (Fig. 7.7). Moreover, if the interval between castration and the initiation of replacement therapy is short, much less hormone is required to restore elements of sexual behaviour than if several weeks elapse (Fig. 7.8). Furthermore, not all behaviours are equally affected by removal of testosterone. Thus, mounting persists long after cessation of intromission and ejaculation.

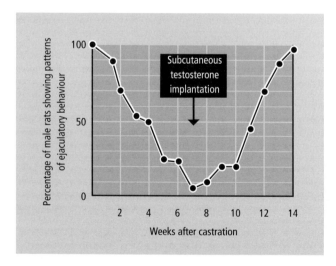

Fig. 7.7 Effects of castration and testosterone replacement on patterns of ejaculatory behaviour in the male rat. (Ejaculatory behaviour consists of a prolonged intromission and a characteristic manner of terminating the intromitted mount.) Note that appreciable levels of the behaviour persist for some weeks after castration and that a comparable time was required for restoration of the behaviour following the subcutaneous implantation of testosterone.

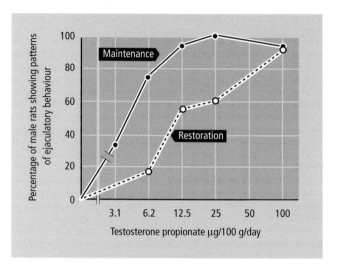

Fig. 7.8 Differential testosterone requirements for 'maintaining' ejaculatory behaviour in newly castrated rats, and 'restoring' the same behaviour in long-term castrates. Note that much more testosterone proprionate was required in the latter group; treatment, as in Fig. 7.7, was given for several weeks to cause this change.

These results suggest that a 'memory trace' of the hormone's presence lasts for some time in the target tissue(s) after hormone removal, that sexual behaviour is only slowly restored, and that perhaps there are degrees of 'dependence' of different sexual responses on the hormone. What are these target tissues and do different tissues influence different responses and have also different sensitivities to testosterone? Clearly, the brain might be a target, affecting the male's motivational state *directly*. But so also might the genitalia be a target, the peripheral effects of testosterone on their sensitivity to tactile stimuli or on the capacity for penile erection contributing *indirectly* to changes in sexual behaviour.

Some insight has been gained into this complicated problem as a result of the fortuitous finding in rats that testosterone exerts its *behavioural effects on the brain* only after aromatization to oestradiol (see Chapters 1 & 2). After castration, therefore, small amounts of oestrogen administered to male rats will reverse the central or motivational deficits, such as reduced mounting behaviour, but regression of seminal vesicles, prostate and cornified spines on the penis continues as do the attendant deficits in related sexual responses, such as erection and intromission. Conversely, when dihydrotestosterone, which cannot be aromatized to oestrogens in the brain, is given to castrate male rats, it has *potent stimulatory effects on sex accessory glands, penis and erectile responses*, but little or no effect on mounting behaviour.

Combining extremely small amounts of oestradiol with dihydrotestosterone results in identical effects on behaviour and on genitalia to those seen after testosterone treatment. These findings suggest that testosterone serves as a *prehormone*. Its androgenic effects on somatic structures depend on reduction to dihydrotestosterone and its effects on the brain depend on aromatization to oestradiol. Unfortunately, this precisely worked out chain of events is not universally applicable, for example in primates aromatization does not seem to represent an obligatory event for the central actions of androgens, so the same dissection of behaviour cannot be applied.

Primates

Castration of male monkeys or men may eventually be followed by a reduction in sexual activity after months (in monkeys) or years (in men), but there is considerable interindividual variability, and few show total loss. Among those monkeys losing sexual activity, testosterone treatment effectively restores it to previous levels. Even in monkeys, there are enormous individual differences in the types of response to castration. For example, a male monkey that has a history of separate sexual interactions with each of two females may, after castration, continue copulating more or less unchanged with one female, but lose his sexual interest completely in the other. Clearly, while testosterone has an important influence on sexual activity, its presence is not obligatory and other factors are also important.

Testosterone therapy has been used successfully to treat low levels of sexual arousal and activity in *hypo- or agonadal* men, and withdrawal of the hormone is followed reliably by a reduction in arousal and activity (Fig. 7.9). In contrast, treating intact (i.e. *eugonadal*) men with testosterone, generally does *not* result in an increase in sexual activity. Testosterone is not an aphrodisiac! It seems that treatment is only beneficial at restoring or stimulating sexual activity and interest *in men with low testosterone levels*. There is a critical range of low plasma testosterone concentrations over which a clear positive sexual response to treatment can be expected. But once the adult range of testosterone levels is reached, that relationship is lost. The situation may be different in female primates, including women, as will be discussed further below. Conversely, antiandrogens, such as cyproterone, may reduce indices of sexual arousal in eugonadal men, which has led to their use clinically to treat individuals with antisocial patterns of sexual behaviour, such as exhibitionism or paedophilia.

If the nature of the processes affected by androgens in these studies of hypogonadal men is to be discovered, it is important to measure more than just the frequency of sexual acts. In the study summarized in Fig. 7.9, the frequency of sexual thoughts has been measured and this has proven to be a sensitive index of the effects of testosterone in hypogonadal men. Penile erections, both during sleep (*nocturnal penile tumescence* or NPT) and in response to sexual stimuli, have also been measured. Only nocturnal erections are clearly testosterone-dependent, an observation that has led to the differential diagnosis of organic and psychogenic impotence. Thus, preservation of NPT in men who lose erectile ability during sexual interactions is strongly indicative both of psychogenic impotence and the likely ineffectiveness of testosterone therapy. Moreover, while it is the case that the low levels of plasma testosterone seen in hypogonadal men are highly correlated with low frequencies of sexual thoughts and erections induced by sexual fantasy (internal stimuli), erections in response to external stimuli, such as erotic films, do not appear to be so testosterone-dependent (Fig. 7.10). As with studies on animals, the nature of the stimulus and the response

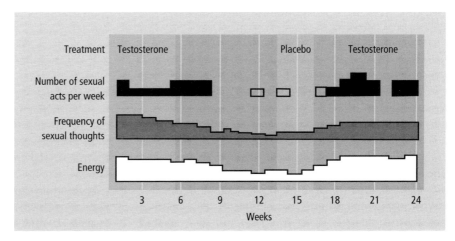

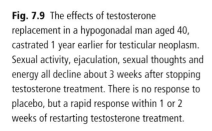

Fig. 7.9 The effects of testosterone replacement in a hypogonadal man aged 40, castrated 1 year earlier for testicular neoplasm. Sexual activity, ejaculation, sexual thoughts and energy all decline about 3 weeks after stopping testosterone treatment. There is no response to placebo, but a rapid response within 1 or 2 weeks of restarting testosterone treatment.

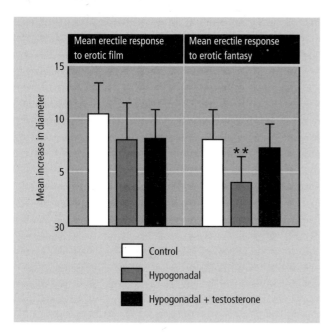

Fig. 7.10 Erectile response (measured as increase in penile diameter) to erotic film and fantasy in hypogonadal men with and without testosterone replacement. The hypogonadal men did not differ from controls in their response to the film, but their response to fantasy was significantly lower than controls when they were androgen-deficient but improved with testosterone replacement. The latency of their erectile response was significantly reduced after hormone replacement. ** $P < 0.01$.

must be considered carefully when interpreting these data. An interesting issue in studies of sexual behaviour in men is whether erections occur as a result of sexual arousal, or whether they contribute to its development. Is penile erection a response or a stimulus? Erections may be such an important indicator to men of sexual excitement that a decrease in erectile ability may contribute to a much higher threshold for sexual arousal and hence decreased sexual activity. A clear separation of genital and motivational responsiveness to androgens in hypogonadal men has simply not been achieved, but actions at each site are probable.

Feminine sexual behaviour

Non-primates

In Chapters 4 and 5, we described how sexual receptivity or 'heat' occurs cyclically, and is closely coordinated with the period of ovulation, which ensures that copulation occurs at a time when fertilization is most likely. (Reflex ovulators, for example the cat, rabbit and ferret have a somewhat different mechanism for achieving this end: the female comes into heat and stays in that state until she copulates, and this event itself triggers ovulation, see Chapters 4 & 5.) Ovariectomy in these species is followed by a prompt and usually complete abolition of *receptivity* (they will no longer accept mounts by the male) and *proceptivity* (they no longer 'solicit' males). Restoration of these elements of feminine sexual behaviour is achieved equally rapidly by treatment with oestradiol and, in some species, progesterone, which must be given by injection in appropriate sequence with oestradiol so as to mimic the hormonal events seen during the oestrous cycle (Fig. 5.26 for an example of a progesterone rise at oestrus in the rat). These behavioural events may be affected additionally by a circadian rhythm (e.g. in the rat, oestrus occurs in the dark phase of the day–night cycle) and/or a seasonal rhythm (e.g. in the ewe, oestrous cycles begin in the autumn), events discussed in more detail in Chapter 5.

Similarly, dramatic changes in behaviour are seen in large domestic animals, like the sheep and pig. Thus, the active nudging, blocking and nibbling responses displayed by oestrous ewes, or the immovably solid posture displayed by sows in heat, disappear after ovariectomy. Reinstatement of oestrous levels of these behaviours follows appropriate treatment with ovarian hormones, for example, progesterone followed by oestradiol in the ewe, mimicking the end of the luteal phase and the short follicular phase leading up to oestrus in the natural cycle.

So like non-primate males, sexual behaviour in females seems to be totally dependent on their hormonal state. But, as for primate males, the situation for female primates is more complex.

Primates

The strict relationship between ovarian hormones and sexual behaviour seen in non-primate female mammals is largely lost, or at least takes a rather different form, in female primates. Superficially, this does not appear to be the case as, if sexual interaction between a male and female monkey is measured, it is often seen to follow a cyclic pattern (Fig. 7.11a). Ovariectomy is followed by a major reduction in sexual interaction, which is restored by treating females with oestradiol. However, careful measurement of the sexual responses of males and females shows that it is mainly the *male's behaviour* that declines markedly after ovariectomy and that increases after oestradiol therapy. Oestradiol is somehow changing the sexual 'attractiveness' of the female to the male. How?

Numerous experiments have pointed to the vagina as the site where oestradiol exerts these effects (Fig. 7. 11b, panels D), and have shown that it affects the female's

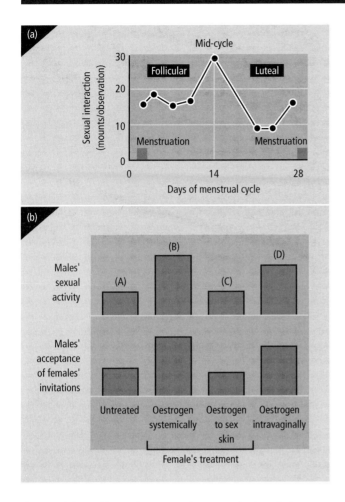

Fig. 7.11 (a) Sexual interaction, measured as mounts by the male, between pairs of male and female rhesus monkeys during the menstrual cycle. Note the high follicular-phase levels of interaction, which peak at mid-cycle and fall in the luteal phase. The premenstrual rise is characteristic. This pattern of interaction seems to be more due to fluctuating interest in the female by the male than *vice versa*, as is revealed more clearly in (b). (b) A block diagram summarizing the results of experiments which demonstrate that: (A) ovariectomy of the female depresses the male's sexual activity and reduces his acceptance of her sexual invitations; (B) treating the female with oestradiol, by subcutaneous injections, increases both these parameters of the male's behaviour; (C) the effect is lost if the oestradiol is smeared, as a cream, on the female's perineum; (D) if it is placed in the vagina, an effect identical to that seen in (B) occurs. These data indicate that oestradiol increases sexual attractiveness by an action on the vagina.

odour. Male rhesus monkeys rendered reversibly anosmic fail to discriminate between oestrogen-treated and untreated ovariectomized females. They copulate with the latter until their anosmia is reversed, at which point they usually cease promptly until oestrogen treatment (systemic or intravaginal) is resumed.

Analysis of vaginal secretions using gas chromatography and mass spectrometry has revealed the presence of a mixture of simple aliphatic acids (acetic, propionic, isobutyric, butyric and isovaleric), the concentrations of which vary during the menstrual cycle, which disappear after ovariectomy and are restored after oestrogen treatment. Vaginal lavages taken from oestrogen-treated female monkeys and placed on the perineum of untreated females stimulate the sexual interest of males, as indeed do vaginal secretions from women and other species of monkey. Not surprisingly, therefore, the same aliphatic acids have been found in human vaginal secretions and they also vary in concentration during the menstrual cycle. Proof of the involvement of the aliphatic acids comes from the observation that a synthetic mixture in the correct proportions can achieve the same male interest (Fig. 7.12). The effectiveness of this mixture can be enhanced by the addition of phenolic compounds (phenylpropanoic and para-hydroxyphenylpropanoic acids), also present in vaginal secretions, but which are ineffective when applied alone. Thus, the female monkey's vagina clearly produces odour cues that are dependent on oestradiol. The acids are not a glandular product, but result from microbial action on vaginal secretions, blockable by use of penicillin.

Progesterone can decrease the sexual attractiveness of female monkeys and does so by reducing the sex-attractant properties of vaginal secretions. This observation explains the decrease in sexual interaction during the luteal phase of the menstrual cycle. Clearly vaginal secretions can both 'turn on' and 'turn off' the male's sexual interest.

Finally, it must be emphasized that these effects of odours on the sexual activity of males are not all or none. Some males are oblivious to changes in the female's odour, others seem to have their sexual activity completely regulated by them. This situation differs enormously from the stereotyped behavioural responses of other mammals, and particularly insects, to 'pheromones'. These highly species-specific substances 'release' patterns of aggressive or sexual (and other) behaviours in an invariant way, very different from the effects of aliphatic acids in primates.

Do ovarian hormones affect sexual behaviour in women? There is now some consensus that coital activity does vary predictably during the menstrual cycle. Most studies have reported mid-cycle (periovulatory) peaks, luteal troughs and often secondary increases in sexual activity prior to menstruation (Fig. 7.13). It is generally accepted, however, that ovariectomy in women does not normally result in loss of libido, although oestrogen-replacement therapy may be necessary to ensure vaginal lubrication and to prevent atrophic changes in the reproductive tract. As with males, then, direct effects of sex steroids on the genitalia in maintaining levels of sexual

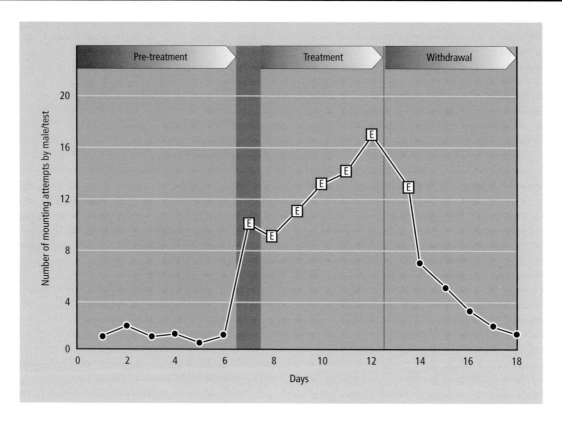

Fig. 7.12 Summary of effects on three male rhesus monkeys of a synthetic mixture of aliphatic acids applied to the perineal region (sexual skin) of an ovariectomized female rhesus monkey. Mounting behaviour is markedly stimulated during treatment, and ejaculations (E) occur consistently. Withdrawal of the mixture is followed by a reversal of these changes in the male's sexual behaviour.

activity must always be taken into account before making assumptions about changes to motivational states.

It has been suggested, however, that ovarian hormones may exert effects on sexual interaction between men and women in a manner comparable to that described above in rhesus monkeys. In Fig. 7.13, data collected using a self-report, diary technique showed that coital activity varied typically during the menstrual cycle in this population of subjects. But when initiation of the sexual interaction was analysed, the cyclic variation was accounted for by changes not in the women's behaviour, but in the behaviour of their male partners. This was interpreted as evidence of cyclic fluctuations in sexual attractiveness, and suggested to be the result of olfactory communication between the couples similar to that seen in monkeys (see above). Even more interesting was the finding that oral contraceptives were associated with the absence of a luteal trough in sexual initiation by the male partners in this study (Fig. 7.13). This phenomenon was explained by the absence of endogenous progesterone production in women taking oral contraceptives, so that its anti-oestrogenic (attractiveness-reducing) actions on the vagina did not occur.

In fact, there is no direct evidence for this interpretation nor, indeed, that sexual activity in men is affected by cyclic variation in vaginal odours. But the data summarized in Fig. 7.13 indicate how relatively little we understand about the influence of ovarian hormones on human sexual interaction. The fact that synthetic forms of these hormones are taken by large numbers of women to control their fertility emphasizes the importance of investigating their behavioural effects. Indeed, loss of libido has been reported to be a major symptom among women taking oral contraceptives, being four times more common than in controls. The mechanism underlying this change in sexual behaviour, apparently at variance with the data shown in Fig. 7.13, is also obscure, although it will be further discussed below.

A logical conclusion from the majority of data is that ovarian hormones have little effect on the sexual activity of female monkeys and women, even though they may have pronounced effects on sexual interaction. In monkeys at least, this is because ovarian hormones influence the sexual interest in females by males. Clearly, this situation is very different from the strict dependence of oestrous behaviour on ovarian hormones in female

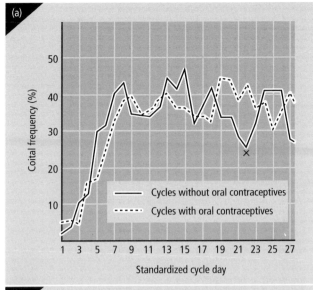

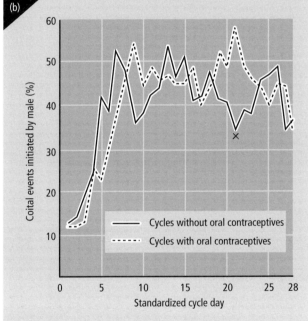

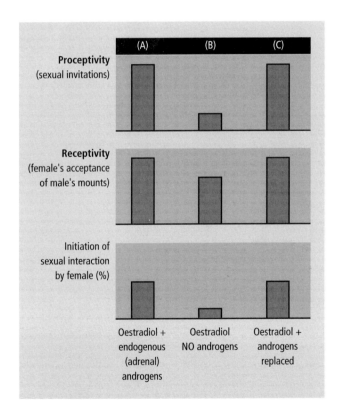

sexual behaviour. Thus, as we have seen, oestrogen-treated, ovariectomized female monkeys are both attractive to males and willing to solicit and accept their attempts to copulate. However, removing the remaining source of circulating androgens (by suppression or removal of the adrenal cortex) or passive immunization against testosterone, results in a marked reduction of proceptive and receptive behaviours (Fig. 7.14), despite the continuing presence of oestradiol. These changes in the female's behaviour are reversed by treatment with testosterone (Fig. 7.14) or an androgen not bound by the testosterone antiserum, such as androstenedione. Although the decreases in sexual behaviour in female monkeys deprived of androgens are considerable, it should be emphasized that they may still accept mounting attempts by males (Fig. 7.14), just as androgen-

Fig. 7.13 (a) Changes in coital frequency in a group of women during menstrual cycles without and with oral contraceptives. The luteal trough at X occurring in normal cycles does not occur in cycles during oral contraceptive treatment. (b) When the percentage of coital events initiated by the male is plotted, a similar picture is obtained, men initiating sexual activity less in the luteal phase of the menstrual cycle. Oral contraceptives prevent this reduction.

Fig. 7.14 A block diagram summarizing the results of experiments on adrenal androgens and proceptive and receptive behaviour in female rhesus monkeys. (A) All females were ovariectomized and received oestradiol benzoate throughout the experiment, ensuring their attractiveness to the males. (B) Removal of endogenously secreted androgens from the adrenal by suppression with dexamethasone or adrenalectomy with glucocorticoid replacement, caused large decreases in proceptive behaviour and initiation of sexual interaction by the female, and smaller decreases in receptive behaviour. (C) These decreases were reversed by treatment with testosterone proprionate or androstenedione.

non-primates. But this does not necessarily mean that proceptive and receptive behaviour in female primates are independent of hormonal influences. Several studies in monkeys and women have indicated that *androgens* of adrenal and ovarian origin do influence female primate

deprived males may continue to display sexual interest in females. It is the *incidence* of such acceptances and interest which falls. Thus, the strict dependence of sexual behaviour on steroid hormones seen in female non-primates has no obvious parallel in primates.

Anecdotal evidence from clinical studies suggests that androgens affect sexual activity in women. For example, women taking oral contraceptives have low urinary

levels of androgen metabolites and may also report loss of libido (see above). A recent controlled study of the effects of androgens and oestrogens on sexual behaviour in ovariectomized women has provided clear evidence. Three groups of women were studied after surgery, one receiving an oestrogen–androgen preparation intramuscularly once a month, a second receiving oestrogen alone, and the third group remaining untreated. Women who received both androgen and oestrogen reported higher rates of sexual desire, sexual arousal and numbers of fantasies (Fig. 7.15) than those who received oestrogen alone or were untreated. Equally impressive was a significant correlation between these measures and plasma testosterone, but not plasma oestradiol. These data strongly suggest that androgen may be critical for the maintenance of optimum levels of sexual functioning in women, perhaps especially after the menopause or after ovariectomy.

The finding that androgens so readily increase measures of sexual arousal and interest in women, but only do so in hypogonadal men has led to the suggestion that males are 'over-determined' and females 'under-determined' so far as androgenic influences on sexual behaviour are concerned. Figure 7.16 illustrates this

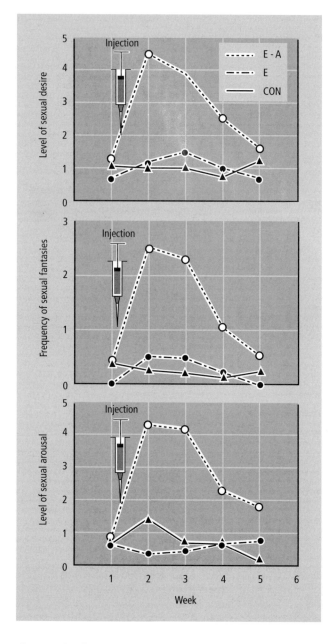

Fig. 7.15 The effects on women who had undergone ovariectomy of replacing oestradiol (E), oestradiol plus androgen (E-A), or no hormone (CON). Levels of sexual desire, the frequency of sexual thoughts and level of sexual arousal are all significantly greater in the combined treatment group, indicating the impact of androgenic steroids on sexuality in women.

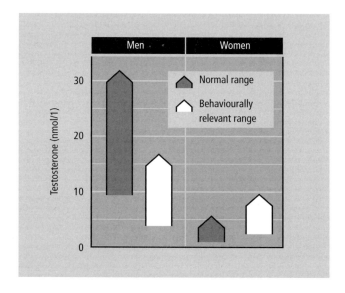

Fig. 7.16 The hypothetical relationship between the normal circulating levels of testosterone and the levels between which changing testosterone can affect sexual behaviour in men and women (behaviourally relevant range). It is postulated that the behaviourally relevant range is smaller than the normal range in men, which explains why clear sexual effects of testosterone are only seen in hypogonadal subjects and not in intact eugonadal adults. By contrast, the behaviourally relevant range is larger than the normal range in women and this may explain why administration of testosterone to women more readily results in changes in sexual behaviour, for example like those illustrated in Fig. 7.15.

notion whereby the normal range of plasma testosterone levels in men exceeds the range within which there is a clear relationship between the hormone and indices of sexual activity. Only when starting from a low baseline, as in hypogonadal men, is this relationship seen. However, in women, it is hypothesized that the range over which testosterone can influence sexual responsiveness overlaps and exceeds the normal range of plasma levels.

Summary

This account of steroid hormones and sexual behaviour should reveal two important points. First, there is convincing evidence that ovarian (or adrenal) and testicular hormones affect sexual behaviour in primates, but that they exert a less dramatic, controlling effect than in other mammals. Such control as there is may be profoundly modified by the dynamics of the social group in which monkeys live; for example, as we saw in Chapter 5, subordinate male monkeys tend not to engage in sexual interactions, yet their plasma levels of testosterone are clearly adequate. These variations are even more obvious in the case of human sexual behaviour where, although steroids undoubtedly influence it, other factors, such as mood, time, place, partner preference and novelty, are also important determinants. Second, this is a complicated and controversial field of study, which, so far as human sexual behaviour is concerned, has only become socially, scientifically and ethically acceptable during the recent past. Unqualified assertions as to the importance, or otherwise, of hormones on human sexual behaviour should be viewed, therefore, with caution. We now turn our attention to an investigation of the possible sites of hormone action in the brain where the expression of sexual behaviour is controlled or influenced.

CENTRAL ACTIONS OF HORMONES IN THE CONTROL OF SEXUAL BEHAVIOUR

The techniques available to study the neuroendocrine mechanisms underlying sexual behaviour include the stereotaxic placement of lesions, hormones and also drugs at discrete neural loci; electrical stimulation in these areas and autoradiographic methods designed to localize the target neurons bearing receptors for behaviourally relevant hormones. Although these techniques are sophisticated, each brings with it particular sorts of interpretational problems. For example, electrolytic lesions destroy not only the neuronal cell bodies in a part of the brain, but also axons traversing that area. Changes in sexual behaviour that follow such lesions

may, therefore, be independent of damage to the neuronal population under study and reflect instead incidental damage to the passing fibre system. Implanted or injected hormones may diffuse away from the original site (although this can be controlled). While advances in techniques and their application have increased our understanding of some neuroendocrine mechanisms controlling sexual behaviour, it should be borne in mind that sex hormones are taken up in many areas of the brain, but it is not clear if they are all involved in reproductive functions there, and, if not, what functions they may serve at such sites.

Males

It has been a consistent finding that lesions, including those specific to neuronal cell bodies, placed in the medial preoptic area (including the sexually dimorphic area, see Chapter 1) and adjacent anterior hypothalamus (Figs 5.2 & 5.3) of males of many non-primate and some primate species, severely impair the ability to copulate. The testes do not atrophy as a result of these lesions, indicating that the pituitary–gonadal axis is unaffected, and the behavioural changes are not simply secondary to a decrease in testosterone secretion. Indeed, treatment with testosterone does not restore sexual behaviour in males with lesions in the preoptic area.

Conversely, implantation of testosterone into the preoptic–anterior hypothalamic areas, which are rich in androgen receptors, restores sexual behaviour in castrate male rats (Fig. 7.17). These dramatic effects of testosterone in the anterior hypothalamus (which in rodents depend upon aromatization to oestradiol) argue strongly that this area is of primary importance in the hormonal regulation of sexual behaviour. This does not in any way minimize the important peripheral effects of androgens, but instead points to the fact that neuroendocrine integration in this behavioural system requires the actions of the hormone at different levels. Thus, copulation in males not only requires the action of testosterone in the hypothalamus, but also in the periphery (and in the spinal cord), presumably to facilitate the transduction and transmission of sensory stimuli to behaviourally relevant areas of the central nervous system (CNS) or to enable the motor responses to these stimuli.

It has become clear that, while both preoptic area lesions and removal of testosterone by castration decrease the display of sexual behaviour, they do so in different ways. Castrate male rats do not copulate and show no interest in females in heat. By contrast, preoptic area-lesioned male rats show high levels of interest in oestrous females and make repeated attempts to mount.

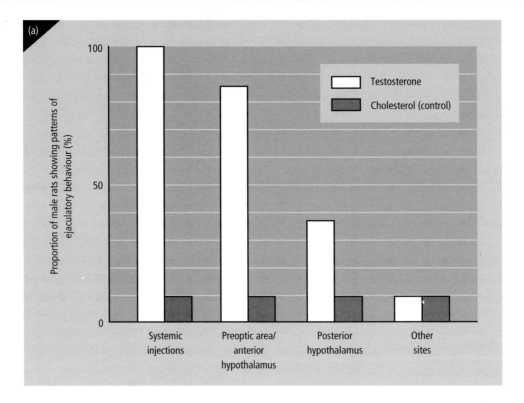

Fig. 7.17 (a) The effects of testosterone implanted in the CNS on the sexual behaviour of castrated male rats. (a) Testosterone placed in the preoptic–anterior hypothalamic continuum induced high levels of ejaculatory patterns, comparable to those observed following systemic treatment with the hormone. Testosterone placed in the posterior hypothalamus or elsewhere in the brain was without significant effect. Important controls were provided by the fact that cholesterol had no effect on sexual activity in any site, so arguing for some degree of hormonal specificity. More recently it has been shown that, at least in rats, oestradiol has similar effects.

However, these mounting attempts are often ill-directed: they are not associated with pelvic thrusting and these males are unable to intromit. This observation suggests a dissociation in the neural substrates underlying the appetitive behaviour that precedes copulation (behaviour that serves the purpose of bringing the male and female into close proximity) from that influencing the performance of copulatory reflexes. The preoptic area would appear to be especially important for the latter and also provides the key site of action for steroids in facilitating the appearance of these reflexes in sexual contexts. A similar conclusion follows the fascinating observation that male rhesus monkeys with lesions in the preoptic area make little attempt to copulate with females, but masturbate to ejaculation at other times. Clearly, the capacity for sexual arousal and the performance of copulatory reflexes are separable neurally in primates as well.

This pattern of results leaves open some important questions. For example, where does testosterone exert its motivational effects? As castrate males show low levels of sexual interest in females, but preoptic area lesioned males retain their interest, it follows that the medial preoptic area is unlikely to be the only site at which testosterone exerts effects on sexual behaviour. An extra-hypothalamic site central to the regulation of appetitive behaviour, including pre-copulatory behaviour, is the dopamine-dependent area of the nucleus accumbens. Dopamine receptor antagonists, for example, when infused directly into this site can selectively impair appetitive aspects of sexual behaviour, but leave the ability to copulate unaffected.

Females

In female non-primate mammals, notably the rat, it is the ventromedial nucleus of the hypothalamus that is the principal site of action of oestradiol (Figs 5.2 & 5.3). Implanting hormone in this area, sufficient to saturate a proportion of the oestrogen receptors, increases markedly the receptive behaviour of ovariectomized females. This treatment is quite adequate as a back-

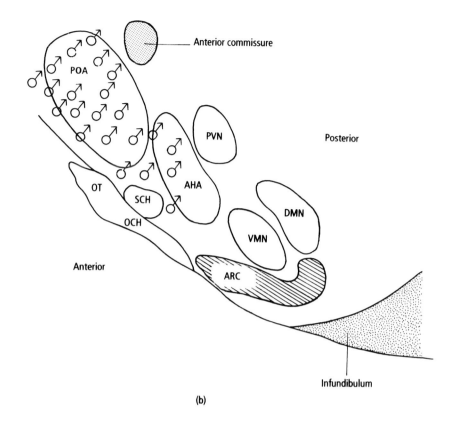

Fig. 7.17 (*continued*) (b) Sites (♂) in the hypothalamus at which testosterone induced increases in sexual behaviour in male rats. AHA: anterior hypothalamic area; ARC: arcuate nucleus; DMN: dorsomedial nucleus; OT and OCH: optic tract and chiasm; POA: preoptic area; PVN: paraventricular nucleus; SCH: suprachiasmatic nucleus; VMN: ventromedial nucleus.

(b)

ground 'priming' for the subsequent actions of progesterone in those species, such as the rat, which require it together with oestradiol to induce oestrous levels of proceptive and receptive behaviour. The site of action of progesterone has proved more elusive to define, but a number of studies now point to the ventromedial hypothalamic area as most likely. It is rich in progesterone receptors, which are actually induced by prior treatment with oestradiol. This probably explains why the sequence of oestradiol followed by progesterone is so critical in this species for oestrous behaviour to occur.

As we have seen, androgens rather than oestrogens seem to underlie proceptivity in female monkeys and they also seem to act via the anterior hypothalamus. Thus, implanting testosterone in an area extending from the ventromedial nucleus to the preoptic area reverses the decrease in sexual activity that follows androgen deprivation in female monkeys (Fig. 7.18). No comparable effects of intrahypothalamic oestradiol have been reported in monkeys.

How do hormones affect behaviour?

Relatively little is known about the ways in which hormones alter neural activity to bring about changes in sexual behaviour. Experiments on rats have revealed that oestradiol, for example, exerts its effects on receptivity after a delay of 26–48 hours. The finding that inhibitors of protein synthesis completely prevent these actions of oestradiol has led to the widely held view that the hormone, which is clearly accumulated in neuronal nuclei, modulates neural activity via an action involving gene expression. However, the nature of the products of this action has not been determined, although they may include progesterone receptors in females. Nor is it clear whether the passage of time between exposure to oestradiol and its behavioural effects is necessary for materials synthesized in the neuronal cell body (e.g. enzymes, receptor molecules or peptidergic transmitters themselves) to be transported to other regions of the neurone (e.g. terminals, dendrites).

There is considerable interest in the nature of the neurochemical mechanisms in the hypothalamus with which sex steroids interact when exerting their behavioural effects. Hypothalamic neurons containing GnRH have been suggested to be one indirect target of steroid action (see Chapter 5). Thus, GnRH infused into the dorsal midbrain, a site to which preoptic GnRH neurones project, enhances the display of receptive lordosis postures in the female rat. Conversely, GnRH antibodies

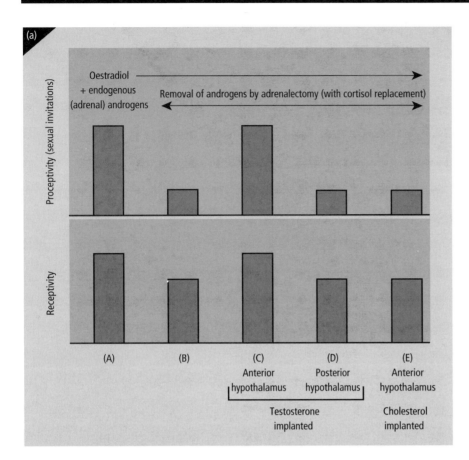

Fig. 7.18 (a) A block diagram summarizing the results of experiments in which testosterone propionate was implanted in the CNS of androgen-deprived female rhesus monkeys. All females were ovariectomized and received injections of oestradiol benzoate throughout the experiment, so that they remained attractive to the males. Adrenalectomy was followed, as shown in Fig. 7.14, by decreased levels of proceptive and receptive behaviour (compare A with B). These changes in sexual activity were reversed by placing testosterone in the anterior hypothalamus (C) but not in the posterior hypothalamus (D). Cholesterol in the anterior hypothalamus was without effect (E).

infused into the same site have the opposite effect. However, despite some early positive demonstrations, GnRH generally has little or no effect on the sexual behaviour of male rats or monkeys of either sex, so its precise importance remains uncertain.

Neurons in the arcuate nucleus containing proopiomelanocortin-derived peptides, particularly β-endorphin, richly innervate the medial preoptic area (see Fig. 5.2), and infusion of β-endorphin into the latter site profoundly inhibits copulation in male rats. Levels of β-endorphin in the preoptic area vary markedly with the steroidal environment, suggesting that this neural system may be important in mediating some of the behavioural effects of testosterone withdrawal. The fact that the same peptide reduces GnRH secretion in both males and females has been taken to indicate that it may have an important, pivotal role in mediating the inhibition of reproduction in a number of different situations, for example during stress, which is known to activate this population of neurons and increase intracerebral levels of β-endorphin.

Summary

A great deal is known about the strict, controlling role of hormones in the sexual behaviour of non-primate species and about the hypothalamic sites where their actions are exerted. But even in these species, many aspects of the neural regulation of sexual behaviour are not understood: for example, what are the functions of structures in the limbic forebrain, such as the amygdala and septum, which bind large amounts of steroid hormones in males and females and are well known to be concerned with the expression of motivated and emotional behaviour? In primates, although the effects of hormones and their sites of action in males and females have been described to some extent, it is clear that much remains to be understood about the ways in which social interactions modify, and are modified by, these basic neuroendocrine mechanisms. Clearly, gonadal hormones are not the only determinants of sexual activity. Furthermore, social interaction, as we

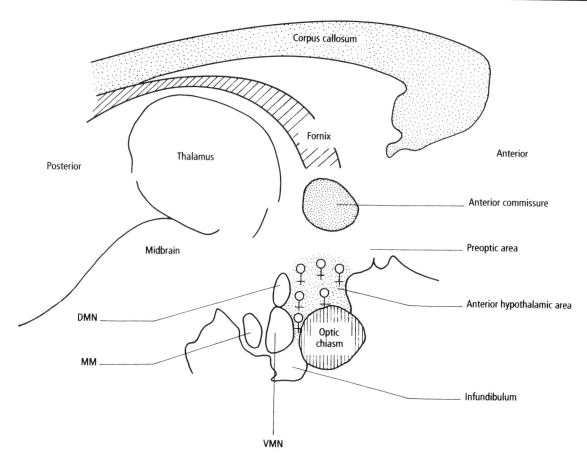

Fig. 7.18 (b) Diagram of a sagittal section of the rhesus monkey's hypothalamus to show sites in the anterior hypothalamic area where testos- terone implants reversed the behavioural effects of adrenalectomy in females (♀). Abbreviations as in Fig. 7.17; MM: mammillary nucleus.

have seen in Chapter 5, may profoundly influence the activity of the hypothalamo-pituitary–gonadal axis and hence fertility as well. If such factors are complicated and poorly understood in non-human primates, it is hardly surprising that we only have limited information concerning the interactions between hormones (which do influence human sexual activity), social environment and individual variables (such as personality, mood and early history), which determine patterns of human sexual behaviour.

FURTHER READING

Bancroft J (1989) *Human Sexuality and its Problems,* 2nd edition. Churchill Livingstone, Edinburgh.

Bermant G & Davidson J (1974) *Biological Bases of Sexual Behaviour.* Harper & Row.

Ciba Foundation Symposium 62 (1979) *Sex, Hormones and Behaviour.* Excerpta Medica (New Series).

Everitt BJ (1990) Sexual motivation: a neural and behavioral analysis of the mechanisms underlying appetitive copulatory responses of male rats. *Neuroscience and Biobehavioral Reviews* **14,** 217–232.

Everitt BJ & Bancroft J (1991) Of rats and men: the comparative approach to male sexuality. In: *Annual Review of Sex Research* (Eds Bancroft J, Davis CM & Ruppel HJ Jr), Volume 2, pp. 77–118. Society for the Scientific Study of Sex.

Everitt BJ & Keverne EB (1986) Reproduction. In: *Neuroendocrinology* (Eds Lightman SL & Everitt BJ), Chapter 18. Blackwell Scientific Publications, Oxford.

Hafez ES & Evans TN (Eds) (1973) *Human Reproduction.* Harper & Row.

Knobil E & Neill JD (Eds) (1994) *The Physiology of Reproduction,* Volume 2, 2nd edition. Raven Press, New York. (Particularly Chapters 10, 18, 23 & 24.)

Kreiger DT & Hughes JC (Eds) (1980) *Neuroendocrinology.* Sinauer Associates Inc., Sunderland.

Meisel RL & Sachs BD (1994) The physiology of male sexual behavior. In: *The Physiology of Reproduction* (Eds Knobil E &

Neill JD), Volume 2, 2nd edition, pp. 3–107. Raven Press, New York.

Pfaff DW, Schwartz-Giblin S, McCarthy MM & Kow L-M (1994) Cellular and molecular mechanisms of female reproductive behavior. In: *The Physiology of Reproduction* (Eds Knobil E &

Neill JD), Volume 2, 2nd edition, pp. 107–221. Raven Press, New York.

Sherwin BB & Gelf MM (1987) The role of androgen in the maintenance of sexual functioning in oophorectomized women. *Psychosomatic Medicine* **49,** 397–409.

Coitus and Fertilization

In the foregoing chapters, we have considered reproductive function in the male and the non-pregnant female. A central concept underlying all of our discussion is the coordinating role played by the gonadal hormones in the production of mature gametes and the conditioning of reproductive function and sexual behaviour. In this way the chance of fertilization is maximized. If successful fertilization occurs, then cyclic sexual patterns must be replaced by pregnancy and nidatory patterns. In this chapter, we will consider how the oocytes and spermatozoa travel from their sites of production to the site of fertilization in the oviduct. Then we will consider the events of fertilization itself.

TRANSPORT OF SPERMATOZOA TO THE FEMALE

In Chapter 3, we left the spermatozoa in the lumina of the seminiferous tubules. Human spermatozoa, a few microns in length, must travel through some 30–40 cm of male and female reproductive tract, or more than 100 000 times their own length, to reach the oviduct. During this long and hazardous journey, several major obstacles must be overcome, including *transport between*

individuals at coitus. Fewer than one in a million of the spermatozoa produced ever complete the journey. It is not just that the journey itself is difficult, but also the spermatozoa must successfully undergo a series of changes in both the male and female genital tracts before they gain full fertilizing capacity. These changes are termed *maturation* in the male tract, and *capacitation* and *the acrosome reaction* in the female tract.

Maturation

Spermatozoa are released from their close association with the Sertoli cells into a fluid secreted by these cells such that a continuous flow rich in spermatozoa washes towards the *rete testis* (Fig. 1.13). As the fluid passes through the rete testis, the composition of its ions and small molecules changes, probably mainly by diffusional equilibration through the tubule walls, as the absence of inter-Sertoli cell junctions renders the blood–testis barrier much less complete. The spermatozoa are then carried through the short and delicate *vasa efferentia* into the *epididymis* (Fig. 8.1). If the vasa efferentia are ligated, then fluid outlet is blocked and the seminiferous tubules literally 'blow-up' with accumulating fluid (Fig. 8.2).

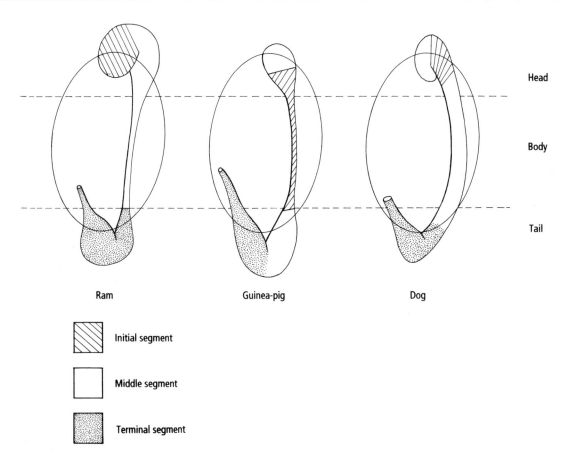

Fig. 8.1 Excurrent ducts of ram, guinea-pig and dog (see Fig. 1.13 for human) to illustrate variation. Vasa efferentia (varying in length and from 10 to 20 in number between the species) connect the rete testis to a long, highly convoluted tube: the epididymis. Anatomically, the coils of the tube form a head (caput), body (corpus) and tail (cauda). Three histological segments, which usually do *not* correspond with the anatomical divisions, are present: an initial segment of high, ciliated epithelium and smallish lumen; a middle segment with wider lumen and shorter cilia; and a wide terminal segment of low cuboidal, relatively poorly ciliated epithelium with more smooth muscle in the underlying stroma. Micropinocytosis and fluid absorption occur in the initial and middle segments, and possibly also some secretion in the more terminal parts of the middle segment. The terminal segment is not clearly absorptive but stores spermatozoa at a high density in the lumen.

Within the epididymis, most of this fluid is reabsorbed, concentrating the spermatozoa 100-fold. In addition, the epididymis adds secretory products, including carnitine, glycerophosphorylcholine, fructose and glycoproteins, the latter coating the surface of the spermatozoa. Passage through the vasa efferentia and epididymis takes around 6–12 days in most species and affects spermatozoal behaviour profoundly. Thus, spermatozoa entering the vasa efferentia are quite incapable of movement (beyond an infrequent twitch) and, when inseminated into females, cannot attach to and fertilize an oocyte. However, by the time they arrive in the cauda epididymidis, spermatozoa have acquired the capacity to fertilize oocytes and to swim progressively (although they do not swim actively *in vivo*, but only after their release from the male tract when they commence swimming and are said to be *activated*). These maturational changes in functional capability are accompanied by changes in the biochemistry and morphology of the spermatozoa (Table 8.1). This whole process of maturation is crucially dependent upon adequate stimulation of the epididymis by androgens.

If the androgens are removed by castration, the epididymis hypotrophies. Injection of testosterone restores activity. Most of the androgen stimulating epididymidal function is derived not from the circulation but from the lymph and rete testis fluid. Thus, ligation of the vasa efferentia to impair these flows results in considerable functional and structural regression of the epididymis. Within the testis fluid, testosterone is bound to androgen binding protein and reaches concentrations approaching those of testicular venous blood (between 30 and 60 ng/ml; dihydrotestosterone is also present at about half this level). Within the epididymis, intracellular receptors take up the androgens, and 5α-reductase converts testosterone to dihydro-

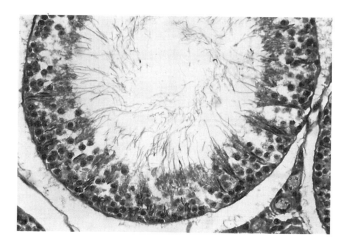

Fig. 8.2 Section through a rat seminiferous tubule 2 days after ligation of vasa efferentia. Note the distended lumen full of fluid and trapped spermatozoa (inapropriate to this stage of the cycle of the seminiferous epithelium: compare with those shown in Fig. 3.9). The fluid accumulation has forced apart adjacent Sertoli cells adluminally, revealing clearly their close association with spermatogenic cells and the Sertoli cells. If this fluid pressure persists, it damages the epithelium, disrupts the blood–testis barrier and causes aspermatogenesis and tubular hypotrophy.

testosterone to yield very high tissue levels of this more active androgen. There is some evidence to suggest that the epididymis may even engage in a little androgen synthesis itself.

The spermatozoa pass from the tail of the epididymis into the *vas deferens* as a very densely packed mass, and their transport is no longer a result of fluid movement but due to the muscular activity of the epididymis and vas deferens. The absence of a massive fluid flux is important, as patients or animals with a ligated vas deferens (*vasectomy*) do not accumulate masses of fluid behind the ligature as occurred with ligation of the vasa efferentia (see Fig. 8.2). Spermatozoa, however, do build up behind the ligature and these must be removed either by phagocytosis within the epididymis or by leakage through the epididymidal wall. The normal non-ligated vas deferens serves as a storage reservoir of spermatozoa. In the absence of ejaculation, spermatozoa dribble through the *terminal ampulla* of the vas deferens into the urethra and are washed away in the urine.

Semen (spermatozoa and seminal fluid)

The seminal fluid within which spermatozoa are carried to the female tract is derived largely from the major accessory sex glands (Fig. 7.1), with only a small contribution from the epididymis. Different species have bewilderingly different patterns of accessory sex gland structure (see Table 7.1) and function (Table 8.2). Knowledge of the origin of some of the main constituents of seminal fluid, as shown in Table 8.2, can help

Property	Details of changes
Concentration	100-fold; 50×10^6/ml entering suspended in fluid; dense packed at 50×10^8/ml on leaving
Completion of sperm modelling	Nuclear condensation and acrosomal shaping completed; cytoplasmic drop 'squeezed' down tail and shed
Metabolism	Cholesterol and phospholipids selectively metabolized, shifting lipid balance towards diacylglycerol, unsaturated fatty acids and desmosterol Increased dependence on external fructose for glycolytic energy production; little oxidative metabolism pH rises
Mobility	Increase in disulphide linkages between proteins in outer dense fibres of tail, yielding a more rigid flagellum with a stronger potential beat cAMP content of tail rises Acquires capacity for forward motion
Membrane	Coated with glycoproteins Rise in surface charge (due to sialic acid increase) and change in profile of surface proteins Membrane fluidity changes, as does its lipid composition

Table 8.1 Changes to spermatozoa in the epididymis

to diagnose deficiencies of function, in particular accessory sex glands. Seminal fluid cannot be essential for effective sperm function, as spermatozoa taken directly from the vas deferens can fertilize oocytes. However, spermatozoa require a 'fluid vehicle' for their normal transport and the seminal plasma supplies this, either exuberantly in half-litre volumes in the boar, or with a more conservative 3 ml or so in the human. In addition to providing a transport medium, the seminal plasma also provides nutritional factors (fructose, sorbitol), and protective factors (buffering capacity to alkalinize the acid pH of vaginal fluids, plus reducing agents, such as ascorbic acid, hypotaurine and ergothioneine, to protect against potential oxidation following exposure of spermatozoa to atmospheric oxygen). It has been proposed, but not proved, that prostaglandins in semen might stimulate muscular activity in the female tract.

Semen does not only carry spermatozoa and substances to assist the maintenance of sperm fertility. Large numbers of leucocytes may be present in seminal plasma, as well as potentially infective agents. Sexual interaction between individuals provides one occasion at which these genitourinary infectious agents can be transmitted. Of particular concern is the presence of large amounts of free hepatitis B virus and human immunodeficiency virus (HIV) in the semen of infected men, as these agents cause seriously debilitating and/or potentially fatal diseases for which treatment is limited. The chance of transmitting these viral infections during vaginal intercourse depends on: the virulence of the viral strain; the viral load of the infected individual; concurrent infectious or inflammatory conditions in the male's genital tract and the recipient tissues of his partner; and the general health of the partner. In general, rates of transmission of hepatitis B virus are about 10-fold those of HIV, and there is an increasing risk of transmission with oral, vaginal and anal intercourse. Potential transmission is, of course, not one way, although infected

Table 8.2 Composition of ejaculate of man and domestic animals

Constituent	Species	Concentration range (mM)	Principal source	Function
Spermatozoa (concentration expressed as no./nl; ejaculate volume in ml in brackets)				
	Boar (150–500)	20–300		
	Bull (2–10)	300–2000		
	Dog (2–15)	60–300		
	Human (2–5)	50–150	Testis	
	Ram (1–2)	2000–5000		
	Stallion (30–300)	30–800		
Fructose	Human, ram, bull	8–37	Seminal vesicle and ampulla	Anaerobic fructolysis
	Boar, dog, stallion	<0.5		(Sorbitol also used in ram)
Inositol	Human, bull, stallion, ram	1–3	Testis and epididymidal fluid	Preserves seminal osmolarity?
	Boar	28	(seminal vesicle - bulls and boar)	
Citric acid	Bull, ram	15–45	Seminal vesicle prostate	Ca^{2+} chelator (limits rate of
	Human	5–73	(stallion, ram, boar, bull)	$Ca^{2\pm}$ dependent coagulation
	Stallion	0.5–2.5	prostate (man and dog)	to prevent seminal 'stones'?)
	Boar	2.5–10		
Glycerylphosphorylcholine	Human, stallion	2–3	Epididymis	See below
	Ram	58–73		
	Bull	4–18		
	Dog, boar	5		
Acid phosphatase (expressed in activity units/ml)	Human	2470	Prostate	Cleaves choline from glycerylphosphorylcholine for use in phospholipid metabolism
	Bull	6		
	Boar	2		

pH = 7.2–7.8; also contains significant amounts of various prostaglandins, especially 19–hydroxylated PGE_1 and $_2$ (humans, monkeys); PGE_1 and E_2; and 19–OH PGFs and PGF_{1a} and $_{2a}$ in humans. Role unclear.

females seem to transmit to males during vaginal intercourse at about half the reciprocal rate. Effective prophylactic vaccination is available only against hepatitis B. Proper use of the right type of condom, lubricant and viricide can provide effective protection against HIV transmission, if, of course, a pregnancy is not desired (see Chapter 14).

Coitus, genital reflexes and sexual responses

In mammals, fertilization is internal and the male gametes must be deposited in the female tract at coitus. Coitus itself is of variable duration (minutes in man, hours in the camel) and is accompanied by extensive physiological changes not just in the genitalia but also in the body as a whole.

It is only since the mid-1960s that research into human sexual physiology has become an accepted part of the study of reproduction. Masters and Johnson, from their studies on human heterosexual interaction and masturbation, proposed a widely accepted model for sexual responses in men and women. Their so-called 'EPOR' model describes: (a) an initial *excitement phase* (E) during which psychogenic or somatogenic stimuli raise sexual arousal; (b) the *plateau phase* (P) during which arousal becomes intensified; (c) the *orgasmic phase* (O), which is reached if the level of stimulation is adequate and entails the few seconds of involuntary *climax* in which sexual tension is relieved, usually in an explosive wave of intense pleasure; and (d) the *resolution phase* (R) during which sexual arousal is dissipated and pelvic haemodynamics resolve to the unstimulated state. The specific physiological changes occurring during these phases will be discussed below. In the male, an *absolute refractory period* occurs after orgasm during which time sexual re-arousal and orgasm is impossible. Its duration depends somewhat on age, being abbreviated in boys, and also on a variety of situational factors, such as novelty of partner or context. The female is generally said not to experience an absolute refractory period and may, therefore, be repeatedly aroused and orgasmic.

The male

Penile *erection* can be elicited by *psychogenic stimuli*, such as visual cues and erotic imagery. These are integrated in the brain, presumably involving mechanisms within the *limbic system*, which is then able, via descending projections to the spinal cord, to influence somatic and autonomic efferents to the genitalia. These same efferents can be activated reflexly by tactile stimulation of the penis and adjacent perineum, this being a most effective means of inducing erection. Data from animals and from men with spinal cord transection reveal that the afferent

limb of the reflex is carried by the *internal pudendal nerves*. Three efferent outflows influence erection: (1) the *pelvic nerve* (parasympathetic outflow) promotes erection; (2) the *hypogastric nerve* (sympathetic outflow) carries fibres that depress erection and possibly some also promote it; and (3) the *pudendal nerve* (somatic) promotes erection. In man, erection is achieved entirely by *haemodynamic changes*, which in other species (e.g. the bull and macaque) may be complemented by the relaxation of a retractor muscle (which pulls the penis back into the prepuce in the flaccid state) and/or by an *os penis* attached to the capsule of the corpora cavernosa (e.g. in the dog, macaque, mink and sea lion). It is the *corpora cavernosa* (Fig. 8.3) (trabeculated sinus spaces surrounded by a tough fibrous capsule) that provide the main erectile tissue.

The likely sequence of events underlying the change from *flaccidity* through *tumescence* to *erection* and then *detumescence* of the penis is as follows. In the flaccid state, myogenic tone within the smooth muscle fibres of the cavernous trabeculae and of the arteries supplying the penis is maintained by the sympathetic outflow of the hypogastric nerve. Use of *papaverine* or prostaglandin E1, or its synthetic analogue alprostadil (smooth muscle relaxants) or of α-adrenergic blockers can reduce this tone and lead to tumescence. During a natural erection, it is stimulation of the parasympathetic outflow that counters this sympathetic effect to reduce myogenic tone in the arterial smooth muscle (causing arterial dilatation and an increased blood flow into the corpora cavernosa) and in the cavernous trabecular muscle (decreasing intracavernous resistance and expanding cavernous volume). Additionally, arteriovenous shunts, which bypass the sinuses of the corpora cavernosa when the penis is flaccid, now direct blood into them. Finally, the venous outflow from the corpora cavernosa is reduced, probably simply by compression of the veins by the rapidly developing turgor. This sequence of events changes the intracavernous space from a low-volume, low-pressure system into a large-volume, high-pressure one. In man, all this is achieved by increasing the *inflow* of arterial blood but not its *through flow*. The corpus spongiosum (Fig. 8.3) also increases in turgor, but not as much as the corpora cavernosa, thereby avoiding compression of the urethra.

Three neurotransmitters have been implicated in the erectile activity of the parasympathetic outflow and they may work cooperatively: (a) *acetylcholine* is present at some parasympathetic termini; (b) *vasoactive intestinal polypeptide* (VIP) is found both terminally and in the ganglionic plexus; and (c) *nitric oxide* production is elevated during vasodilatation. The parasympathetic outflow also relaxes the retractor penis muscle in those species that have one. *Impotence*, the failure to achieve a complete or

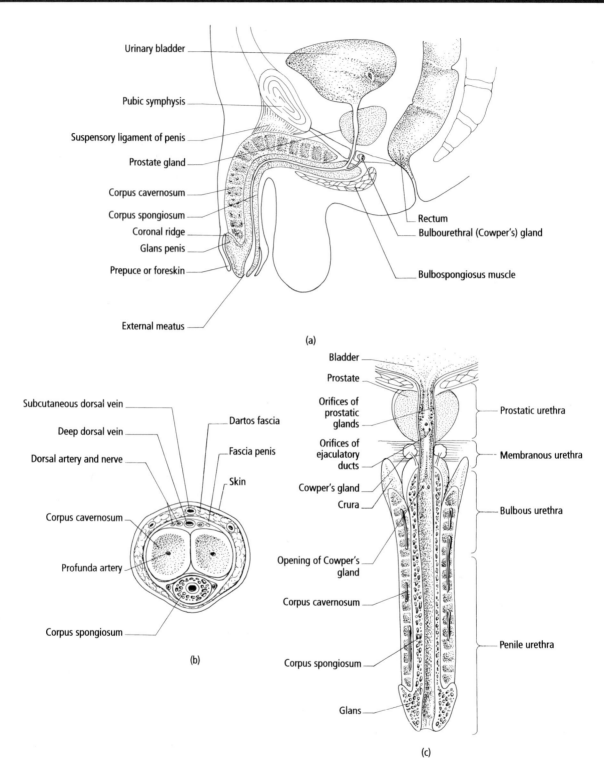

Fig. 8.3 Structure of the human penis: (a) mid-sagittal section through pelvis; (b) transverse section through shaft of penis: note fibrous sheaths enclosing corpora cavernosa, which allow generation of hydrostatic pressure required for erection; tears in the fibrous capsule lead to failure of erection; (c) coronal section showing entry of prostatic, ejaculatory (vasa deferentia and seminal vesicles) and Cowper's (bulbo-urethral) gland ducts. Penile secretory glands of Littré open into the roof of the penile urethra. The internal pudendal arteries supply both the dorsal arteries to the glans and the pudendal arteries to the corpora cavernosa. There is considerable species variation in glans structure: mushroom-shaped in man, corkscrew-shaped in boar and bull, spiny in cat and rat, and having thin urethral veriform processes in goat and ram.

enduring erection, may be caused by: tears in the fibrous capsule of the corpora cavernosa; obstructions to the vessels supplying the penis; use of drugs which antagonize neurotransmitters mediating tumescence; psychogenic factors; or a mixture of these. Pharmacological stimulation of erection can be achieved by the intracavernosal injection of papaverine. The importance of testosterone in erectile function is not entirely clear. Nocturnal erections, which occur during each episode of *rapid eye movement (dreaming or paradoxical) sleep* are known to be testosterone-dependent, whereas erections in response to visual erotic stimuli, for example, are much less dependent upon testosterone and occur readily in hypogonadal men. It has been suggested, therefore, that measuring nocturnal penile tumescence (NPT) provides a method of assessing the capacity for sexual arousal independently of the complicating cognitive factors that may compromise sexual function, resulting in impotence.

As ejaculation approaches, the turgor in the penis increases further and the penile circumference at the *coronal ridge* increases, largely achieved by the action of the pudendal somatic outflow to the perineal striated muscle fibres surrounding the *corporeal crura* (Fig. 8.3). Coincidentally, the testes are drawn reflexly towards the perineum and may increase their volume by as much as 50% due to vasocongestion. The scrotal skin thickens and contracts due to activity in the dartos muscle. With further stimulation, a sequence of contractions of the muscles of the prostate, vas deferens and seminal vesicle is induced, and the components of the seminal plasma, together with the spermatozoa, are expelled into the urethra. This process of *emission* is mediated largely by sympathetic fibres via the hypogastric plexus, and administration of drugs that interfere with the α-adrenergic system (as in treatment for hypertension) lead to '*dry orgasms*'; erection is not impaired (and may be assisted: see earlier), but emission is.

Ejaculation, whereby semen is expelled from the posterior urethra, is achieved by contraction of the smooth muscles of the urethra and striated muscles of bulbocavernosus and ischiocavernosus. It is usually associated with contractions of the pelvic floor musculature innervated by the pudendal nerves. The passage of semen back into the bladder is normally prevented by contraction of the vesicular urethral sphincter; failure of this can lead to *retrograde ejaculation* into the bladder. The composition of the early and late fractions of the human ejaculate reflects the sequential nature of the contractions and the relative lack of mixing of the various seminal components within the urethra. The early fraction is rich in acid phosphatase (prostatic), the mid-fraction in spermatozoa (vas deferens) while the late fraction is rich in fructose (seminal vesicle).

Concomitant with penile erection in men, erection of the nipples and increases in heart rate and blood pressure occur as sexual excitement rises. Immediately prior to ejaculation, skin rashes may develop over the epigastrium, chest, face and neck together with involuntary muscle spasms. At ejaculation, the cardiovascular changes, skin rash and muscle spasms intensify and are often accompanied by hyperventilation, contractions of the rectal sphincter and vocalizations. Associated with ejaculation is orgasm whereby sexual tension and arousal are released and an intense sensation of pleasure occurs. Penile detumescence is achieved via the activity of the pelvic nerve sympathetic outflow restoring smooth muscle tone.

The female

At coitus, frictional stimulation of the glans penis is provided by movements against the external genitalia and vaginal walls of the female (Fig. 8.4). The human female undergoes a remarkably similar sequence of reflex responses to that observed in the male, and its elicitation is also dependent on tactile and psychogenic stimulation. Tactile stimulation in the perineal region, and in some women particularly on the *glans clitoris*, provides the primary afferent input reinforced by vaginal stimulation after penile penetration. A vascular response may cause engorgement of the corpora of the clitoris, with consequent clitoral erection, although the extent of clitoral involvement varies considerably. Vaginal lubrication occurs by *transudation* of fluid through the vaginal wall, which vasocongests and becomes purplish red. The vagina expands and the labia majora become engorged with blood. With increased stimulation, the width and length of the vagina increase further and the uterus elevates upwards into the false pelvis lifting the cervical os to produce the so-called *tenting effect* in the mid-vaginal plane. At orgasm, frequent vaginal contractions occur and uterine contractions beginning in the fundus spread towards the lower uterine segment.

Systemically, the female may experience increased heart rate and blood pressure and manifest skin flushes, vocalizations and muscle spasms (notably rhythmic contractions of the pelvic striated musculature), and intense sensations of pleasure. Post-orgasmically, clitoral erection is lost, the labia and the vaginal os detumesce, and the uterine and vaginal walls relax to their original positions. It has not proved possible to differentiate between the orgasms that follow clitoral or vaginal stimulation in terms of their physiological manifestations.

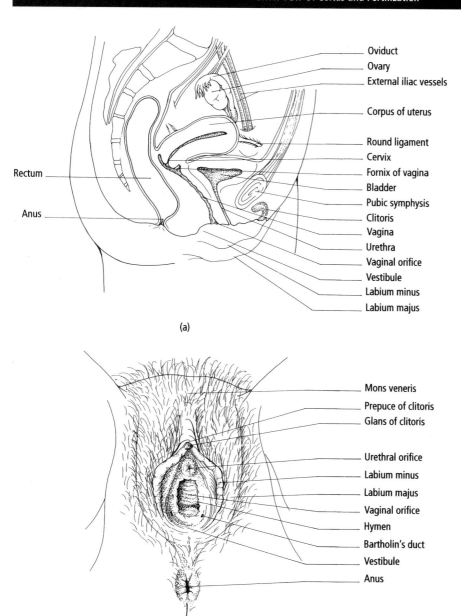

(a)

Oviduct
Ovary
External iliac vessels
Corpus of uterus
Round ligament
Cervix
Fornix of vagina
Bladder
Pubic symphysis
Clitoris
Vagina
Urethra
Vaginal orifice
Vestibule
Labium minus
Labium majus
Rectum
Anus

(b)

Mons veneris
Prepuce of clitoris
Glans of clitoris
Urethral orifice
Labium minus
Labium majus
Vaginal orifice
Hymen
Bartholin's duct
Vestibule
Anus

Fig. 8.4 Structure of human female genitalia; (a) mid-saggittal section through pelvis: note ventroflexure of uterus, which tends to lift upwards during sexual stimulation, moving the cervical os anteriorly in the tenting effect. The smooth muscular wall of the vagina is overlain by extensive highly vascularized stroma covered in a stratified squamous epithelium. It is not a secretory epithelium: vaginal fluids are derived either by transudation, or from secretions of cervix and Bartholin's glands, as well as minor vestibular glands at the vaginal vestibule; (b) view of genitalia from exterior. In some species, the clitoris contains a small bony structure, the os clitoris, homologous to the os penis in the male.

The general descriptions of orgasm recorded by men and women are so similar as to be indistinguishable. The time course of sexual responses in the female is generally longer than in the male. Indeed, ejaculation and orgasm are the end result of coitus within minutes in men (an average of 4 minutes in the Western World according to one study). Proportionately, fewer women appear to achieve orgasm from coitus, various surveys giving figures of 30–50%. Larger numbers achieve orgasm by clitoral stimulation, but even then a significant number, 10–20%, do not achieve orgasm despite being highly sexually aroused. It has been suggested that orgasm in women is not a reflex action, as it appears to be in men, but is learned. Cross-cultural studies support such a view; for example, women are more likely to enjoy sex and have orgasms in societies where this is expected than in societies where it is not.

The ability to communicate sexually and to give and receive sexual pleasure within a relationship is very often a cornerstone of its success in modern Western Society. Social and physical factors affect the capacity to respond sexually to a partner by experiencing orgasm and, when communication and the relationship itself are becoming poor, efforts to define at least some of these factors can have considerable value.

Semen deposition

At coition, the semen is ejaculated within the vagina and onto the cervical os and perhaps into the cervical canal (e.g. in the human, sheep or cow). In some species (e.g. the pig, dog, horse, mouse or rat), there is either a direct deposition in the cervix and uterus or, more likely, a very rapid and effective transport of most vaginally deposited semen into the uterus. In many species studied, the semen *coagulates* rapidly during or immediately after deposition. The coagulation may become gelatinous (e.g. in the human, pig and horse), fibrous (e.g. in the guinea-pig) or calcareous (e.g. in the mouse) and results from an enzyme–substrate interaction. In humans, coagulating enzymes derived from the prostate interact with a fibrinogen-like substrate derived from the seminal vesicle. The coagulum may act to retain spermatozoa in the vagina preventing their physical loss or perhaps buffering them against the hostile acidity of the vaginal fluids (pH 5.7). In some cases of ejaculatory disorder, coagulation takes place within the urethra or (in *retrograde ejaculation*) within the bladder, and obstruction to urinary flow can result. In normal circumstances in the human female tract, the coagulum is dissolved within 20–60 minutes by progressive activation of a proenzyme derived from the prostatic secretion of the ejaculate.

GAMETES IN THE FEMALE GENITAL TRACT

Spermatozoal transport

Within a minute or so of mating in all species studied, spermatozoa can be detected in the cervix or the uterus. It is not at present clear how human spermatozoa

Table 8.3 Estimates of the survival of viable fertile gametes in the female genital tract

Species	Spermatozoa (hours)	Oocyte (hours)
Human	28–48	6–24
Cow	30–48	8–12
Sheep	30–48	15–25
Horse	75–120	6–8
Pig	25–50	8–10
Mouse	6–12	6–15
Rabbit	30–36	6–8

deposited in the vagina actually enter the cervix. Perhaps the ciliated surface of the cervical os wafts them towards the cervical canal. However, in the human over 99% of spermatozoa do not enter the cervix and are lost by leakage from the vagina. The few successful spermatozoa may survive for many hours deep in the cervical crypts of the mucous membrane (Table 8.3). Here they are nourished by mucoid secretions, their further progress to the uterus depending on its consistency. Only in the absence of progesterone domination does the mucus permit sperm penetration (see Chapter 7), and even then morphologically abnormal spermatozoa are prevented from passing further up the female tract.

Studies on the transport of spermatozoa through the uterus to the oviduct are difficult to undertake. It seems clear that the vaginal, cervical and uterine movements often present in the pre-orgasmic and orgasmic phases are not *required* for effective sperm transport but may assist it. Neither is it likely that the prostaglandins present in the semen are required as stimulants to the female tract. Probably, the spermatozoa move through the uterus and into the ampulla of the oviduct by their own propulsion and in currents of fluid set up by the action of uterine cilia. This conclusion fits with the time scale of sperm migration, as, although some sperm are found in the oviducts within minutes of coitus, they are few and often dead. The earliest that *living* spermatozoa are recovered from the oviduct is 2–7 hours post-coition. The number of living oviducal spermatozoa detected during the early stages of fertilization can be measured in tens or hundreds, and even subsequently the total number present at any one time rarely exceeds several hundred. It is not at all clear how the flow of sperm to the oviduct is regulated. The cervical crypts may act as a reservoir slowly releasing sperm into the uterus. Additionally, the utero-tubal junction seems to regulate entry to the oviduct by its action as an intermittent sphincter.

Having reached the isthmus of the oviduct, spermatozoa tend to linger and become immotile, perhaps binding temporarily to oviductal epithelial cells. Only at ovulation do spermatozoa re-acquire motility and pass to the ampullary–isthmic junction (Fig. 4.1) and the site of potential fertilization. How this passage is regulated is unknown, although recently experiments *in vitro* have suggested that the oocyte-cumulus mass might produce a chemotactic factor. The nature of such a chemo-attractant is unclear, although the examination of spermatozoa for potential 'chemo-attractant' sensing receptors has revealed the presence of chemoreceptors similar to those found in the olfactory epithelium, as well as those involved in mediating chemo-attraction of neutrophils. However, it remains to be proved that such a system of chemo-attraction actually functions *in vivo*.

Oocyte transport

While the spermatozoa are arriving in the oviduct, the ovulated oocyte(s), with enclosing cumulus cells, is picked up from the surface of the ovary in the peritoneal cavity by the fimbriated ostium of the oviduct (Fig. 4.1), and swept by oviducal cilia along the ampulla towards the junction with the isthmus (Fig. 4.1). Oocyte transport is affected adversely if cumulus cells are lacking and/or if oviducal cilia are malfunctional. As for the uterus, there is no evidence that muscular activity in the oviduct is essential for transport along the ampulla, although smooth muscle in the mesosalpinx and tubo-ovarian ligaments is involved in the ostial pick-up of oocytes. Furthermore, as we saw in Chapter 7, pharmacological doses of steroids may adversely affect oocyte transport in the oviduct. The oocytes and the spermatozoa come together in the ampulla and it is here that the events leading up to fertilization occur.

FERTILIZATION

Sperm capacitation

If mature spermatozoa recovered at ejaculation are placed with oocytes *in vitro*, fertilization either does not occur or does so only after a delay of several hours. In contrast, spermatozoa recovered from the uterus or oviduct a few hours after coitus are capable of immediate fertilization. This attainment of a fertilizing capacity within the female tract is called *capacitation*. The process includes a stripping from the spermatozoal surface of the coating of glycoprotein molecules acquired during passage through the epididymis and after contact with seminal plasma, resulting in changes to the surface charge and macromolecular organization of the spermatozoal membrane. The process is experimentally reversible; thus, if capacitated spermatozoa are reincubated in epididymal fluid or seminal plasma, they become *decapacitated*. An oestrogen-primed uterus or oviducal isthmus is optimal for capacitation, critical features being the proteolytic enzymes and high ionic strength provided by their secretions. However, it is possible to mimic these conditions *in vitro* by preparation of suitable culture media.

The process of capacitation has two elements: (a) a change in the movement characteristics of the spermatozoa to a *hyperactivated motility pattern* in which the regular undulating wave-like flagellar beats are replaced by stronger, wide amplitude or *'whiplashing' beats* of the tail that push the spermatozoa forwards in vigorous

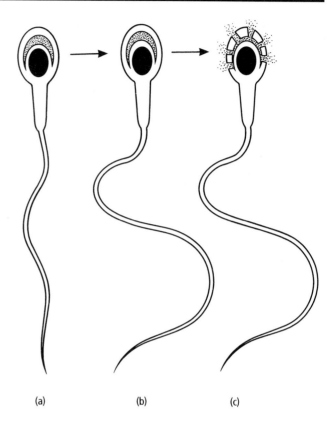

(a)   (b)   (c)

Fig. 8.5 (a) Schematized spermatozoon prior to capacitation; a consequence of capacitation is (b) hyperactivated tail movements, and the capacity subsequently to undergo (c) the acrosome reaction, in which multiple sites of fusion between the plasma membrane and the outer acrosomal membrane occur, first at the tip of the acrosome and then at the equatorial region. As a result of the acrosome reaction, the plasma membrane remaining in the equatorial and post-acrosomal regions acquires the potential to fuse with the plasma membrane of the oocyte.

lurches (Fig. 8.5b); and (b) a change in the surface membrane properties that renders the spermatozoon responsive to signals encountered in the immediate vicinity of the oocyte, which then induce a further change in the spermatozoa called the *acrosome reaction* (see next section). These two elements of capacitation are associated with the acquisition by the spermatozoa of four properties. First, the stability of the surface membrane is reduced; this is important for the subsequent acrosome reaction (see below). It may also be responsible for acquisition of the second property, namely an increased calcium permeability of the sperm membrane, leading to a modest rise in internal calcium levels. Additionally, among the proteins stripped from the sperm surface are calmodulin binding proteins, and their removal may make the spermatozoa more responsive to the effects of calcium. The development of hyperactivated motility is sensitive to calcium levels. Third, and as a result of elevated calcium, adenyl cyclase activity within the spermatozoa increases, leading to elevation of cAMP levels and increased cAMP-dependent phosphorylation of spermatozoal proteins. Sperm motility is enhanced by addition of either exogenous cAMP or inhibitors of phosphodiesterase (such as pentoxyfylline, thereby preventing destruction of cAMP). Fourth, the activity of a spermatozoal tyrosine kinase is increased, leading to its autophosphorylation as well as to the presumptive phosphorylation of tyrosines in intracellular proteins. If this autophosphorylation is prevented, the ability of spermatozoa to undergo a subsequent acrosome reaction is impaired, and thus the event seems to be a critical component of the capacitatory process.

The acrosome reaction

Perhaps the most dramatic morphological transformation of the capacitated spermatozoa is the acrosome reaction (Fig. 8.5c). During this process, the acrosome swells, its membrane fuses with the overlying plasma membrane, a vesiculated appearance is created, and both the contents of the acrosomal vesicle and the inner acrosomal membrane become exteriorized in a process of *exocytosis*. The acrosome reaction is associated with further large increases in intracellular calcium and cAMP. The calcium elevation seems central to the acrosome reaction as is shown by the failure of activation in the absence of calcium and by the induction of activation prematurely by use of calcium ionophore to enhance its entry. It also seems clear that a rise in intracellular pH from about 7.1 to 7.5 occurs, and may work synergistically with the calcium. What stimulates the calcium entry and how? A number of agents have been shown to increase calcium influx. The problem has been to discriminate which of them is biologically relevant. Two strong candidates have emerged.

It is now generally accepted that the agent responsible for the acrosome reaction is a constituent of the zona pellucida called *ZP3*. Capacitated spermatozoa bind specifically to a short, galactose-containing oligosaccharide component of this large acidic glycoprotein in a process that is species specific. The oligosaccharide can be isolated and, when added to suspensions of spermatozoa, binds to them and inhibits their attachment to the zona pellucida. The sperm binding site is located on the plasma membrane overlying the acrosome, and may have *galactosyl transferase* activity. Having bound, other oligosaccharide and protein regions of the ZP3 glycoprotein then interact with the sperm plasma membrane and, via an action involving a G protein, stimulate calcium influx and a rise in pH. The acrosome reaction follows. As might be expected, blocking the calcium rise prevents

the acrosome-inducing effect of ZP3. As much of the sperm receptor for ZP3 is shed on the vesiculating membrane during the acrosome reaction, sperm binding to ZP3 is short lived. In addition, the acrosome releases β-hexosaminidase B, which digests away any local ZP3 receptor, so preventing further binding.

Recently, it has been suggested that, in humans, progesterone can also stimulate an influx of calcium into spermatozoa, leading to the stimulation of adenyl cyclase and a rise in cAMP levels. The follicular production of progesterone rises just prior to ovulation (see Chapters 4 & 5), and it is possible that progesterone is carried in the cumulus mass of granulosa cells surrounding the oocyte. Maybe *in vivo* progesterone acts on spermatozoa arriving at the cumulus mass to make their acrosomes leaky, releasing the intra-acrosomal enzyme *hyaluronidase* onto the surface of spermatozoa and into the fluids around them. This enzyme can digest the intercellular matrix of hyaluronic acid, which holds together the cumulus cells around the oocyte, so facilitating the passage of spermatozoa through the cumulus towards the zona pellucida of the oocyte, the interaction with ZP3 and the definitive acrosome reaction (Fig. 8.6a). Progesterone may also be involved in the capacitation process, stimulating calcium entry.

The exposure of the inner acrosomal membrane reveals a new set of binding sites, this time specific for a second zona glycoprotein, ZP2. These binding sites hold the spermatozoa and zona in contact, and their action is essential for the passage of the spermatozoon through the zona pellucida towards the oocyte (see next section).

It is important that the acrosome reaction occurs close to the oocyte, because acrosome reacted spermatozoa have a very short life span. Herein probably lies the explanation for the rather complex series of changes that spermatozoa undergo in the female tract. Hours (or even days, in the human) may intervene between deposition of the ejaculate into the female tract and the ovulation of an oocyte. High fertility may be ensured by establishing a reservoir of spermatozoa, releasing a gradual trickle through the capacitating uterus and oviduct where their final fertilizing capacity can only be realized in the presence of an oocyte. If all spermatozoa were fully competent to fertilize immediately after their release into the female tract, much stricter time limits on fertility would operate.

Gamete fusion

Only spermatozoa that make it to the zona with the acrosome intact can bind to ZP3 via the receptors on the sperm surface. The ZP3 induces the acrosomal reaction to expose the ZP2 binding sites on the inner acrosomal

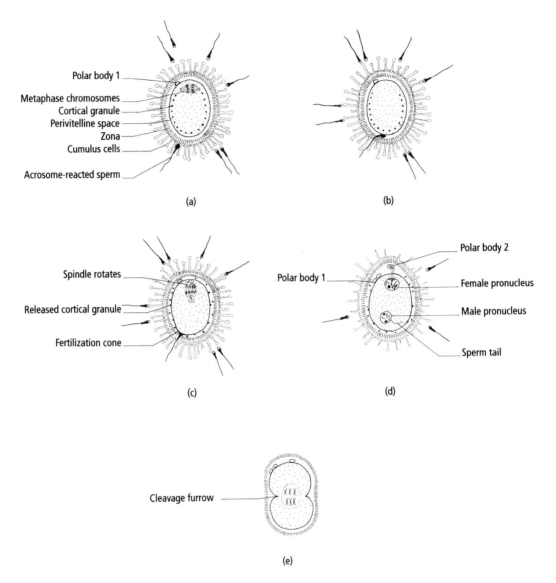

Fig. 8.6 (a) Spermatozoa approach an ovulated mouse oocyte, which has its first polar body, chromosomes on the second metaphase spindle, cortical granules, and perivitelline space between the oolemma of the oocyte and the zona pellucida. One capacitated spermatozoon binds to ZP3 on the zona pellucida, acrosome reacts, and then binds to ZP2, before (b) penetrating the zona to lie in the perivitelline space, where it (c) fuses with the plasma membrane via its equatorial and post-acrosomal membrane, thus activating a calcium wave (see Fig. 8.7). At the point of fusion, the swelling of the fertilization cone occurs. The calcium induces release of cortical granules and resumption of meiosis. (d) Meiotic division is completed, yielding the second polar body and female pronucleus. Meanwhile the sperm nucleus decondenses and the male pronucleus forms, the sperm tail being adjacent. (e) Chromosomes duplicate their DNA, pronuclei migrate together, their membranes break down as the first mitotic spindle forms, chromatids separate and move apart, and a cleavage furrow develops. By this stage the cumulus cells have dispersed.

membrane. Also exposed on this membrane are proteolytic enzymes which digest a path through the zona, along which the spermatozoon passes aided by the whiplash forward propulsion of the hyperactivated tail. Zona penetration takes between 5 and 20 minutes. The spermatozoon eventually comes to lie tangential to the oocyte surface between the zona pellucida and the oocyte membrane (*oolemma*) in the *perivitelline space*, where microvilli on the surface of the oocyte envelop the sperm head (Fig. 8.6b).

The surface membrane overlying the middle and the posterior half of the spermatozoal head (the equatorial and post-acrosomal region) is the site of fusion with the surface membrane (oolemma) of the oocyte. However, *only capacitated spermatozoa that have undergone an acrosome reaction are capable of fusion.* An increased fusibility of the membrane in this region is a critical outcome of the acrosome reaction. The fusion process may involve an interaction between an integrin-like molecule in the oolemma and a disintegrin-ligand on the spermatozoon. However, a number of other candidate molecules have also been implicated in the fusion process, and it is entirely possible that several are involved. This is an area of intensive research. Calcium is required for the fusion process, and its action may be mediated by calmodulin, as antagonists to this molecule can block sperm–oocyte fusion. Once fusion has occurred, the spermatozoon dramatically ceases to move and its nucleus (together with variable parts of the midpiece and tail contents, depending on the species) passes into the ooplasm. This first phase of fertilization, from entry into the cumulus mass to fusion, is completed.

The ensuing events of fertilization last some 20 hours or so and are concerned with two distinctive types of activity. First, the diploid genetic constitution of the embryo must be ensured. Second, the developmental programme of the embryo must be initiated.

Establishment of diploidy

The newly fertilized oocyte confronts two immediate problems. First, it must prevent further spermatozoa from fertilizing it (the *block to polyspermy*) as this would cause *triploidy* (for one extra sperm) or *polyploidy* (for several). Polyploidy of this sort is described as *androgenetic*, as the extra sets of haploid chromosomes are derived from the male. Second, the oocyte was ovulated and arrested in its second meiotic division; it is thus not yet haploid. If it is to transmit only one set of chromosomes to the next generation, it must complete its second meiotic division, dispatching one haploid set of chromosomes to the *second polar body*, and enter interphase of the first cell cycle. The

failure to jettison one of the two haploid sets of female chromosomes would lead to *gynogenetic triploidy*.

Immediately after fusion of the oocyte with the spermatozoon, there is a dramatic increase in the level of free intracellular calcium, due largely to the release of calcium from internal stores (Fig. 8.7). The rise lasts about 2–3 minutes and does not occur synchronously over the whole oocyte, but rather sweeps in a wave across the oocyte starting from the point of sperm entry. This first rise is followed by a series of calcium spikes, each spike being of 1–2 minutes duration, becoming more synchronous over the whole oocyte and occurring every 3–15 minutes depending on the individual oocyte (Fig. 8.7). This spiking activity can last for several hours. The rise in calcium is associated with a change in the pattern of protein phosphorylation in the oocyte. How the spermatozoon causes the initial calcium spike or the changes in phosphorylation is uncertain. One view suggests that the binding of a sperm to its receptor is followed by the activation of an enzyme. Two enzyme systems have been implicated but neither definitively proved to be involved (Fig. 8.8). Perhaps a tyrosine kinase is activated to phosphorylate intracellular targets, including perhaps calcium release channels on internal calcium stores or phospholipase Cγ. Alternatively, the enzyme *phospholipase Cβ* may be activated via a G protein, thereby stimulating release of the second messengers *inositol triphosphate* (IP_3) and *diacylglycerol* (DAG). The former can activate calcium release from internal stores, while the latter activates *protein kinase C* (PKC) to stimulate its phosphorylating activity. An alternative view suggests that the fertilizing spermatozoon releases a protein into the oocyte at fusion and that this protein initiates the release of internal calcium stores (Fig. 8.8). What is clear is that the long-term consequences of the rise in free calcium and the increased phosphorylating activity are crucial both for euploidy and the development of the zygote.

Fig. 8.7 Plot of the calcium spiking pattern of a fertilized zona-free mouse oocyte inseminated with spermatozoa at the time indicated by the arrow. After 12 minutes, a spermatozoon fused with the oocyte, generating a transient five- to 10-fold rise in internal calcium. Thereafter, calcium pulses follow at 3–15-minute intervals (characteristic for each oocyte) and can last for several hours.

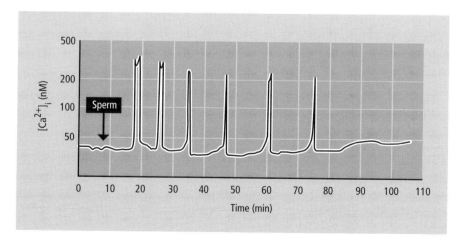

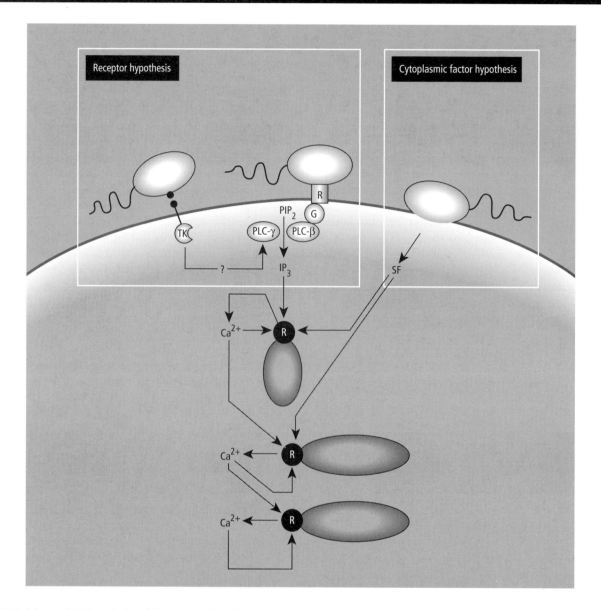

Fig. 8.8 Models to explain the activation of the oocyte. Perhaps the spermatozoon fuses with a receptor, either an integrin-like receptor, which activates a tyrosine kinase (TK), and so phospholipase Cγ activity (PLCγ), or a G-protein-linked receptor (R+G) and so phospholipase Cβ activity (PLCβ). The consequence is elevated synthesis of IP$_3$ from PIP$_2$. The IP$_3$ then binds to intracellular receptors (IP$_3$R) to stimulate calcium release. The calcium, in turn, activates deeper stores by binding to receptors (RyR or IP$_3$R), thereby spreading the calcium wave across the oocyte. An alternative model suggests release of an as yet unidentified sperm cytoplasmic factor (SF) from the fusing spermatozoon, which might increase the sensitivity of receptors located on internal calcium stores.

The calcium has two important actions, which can be demonstrated clearly by blocking the calcium rise experimentally. First, the elevated calcium results in fusion of the *cortical granules* in the cortex of the zygote with the overlying oolemma, thereby releasing their contents into the perivitelline space (*the cortical reaction*) (Fig. 8.6c). Among the contents of these granules are enzymes that act on the zona pellucida to prevent, or impair, further binding and penetration by spermatozoa (*the zona reac-*

tion). These enzymes have at least three actions. A protease cleaves glycoprotein ZP2, and β-hexosaminidase B digests the oligosaccharide receptor on glycoprotein ZP3. In this way, the ZP sperm binding properties are removed, and sperm binding to the zona ceases. In addition, tyrosine residues on adjacent ZPs are cross-linked, rendering the zona indissoluble to acrosin and so impenetrable by spermatozoa. A final consequence of sperm–oocyte fusion, the mechanism for which is not yet understood, is a reduction in the sperm binding properties of the oolemma itself. All these events occur rapidly to reduce the chances of androgenetic polyploidy.

The avoidance of gynogenetic triploidy also depends on the calcium rise. At fertilization, the oocytes are arrested in second meiotic metaphase (M phase). It is now clear that this arrest depends upon the continuing presence of two types of cytoplasmic activity called *maturation promoting factor* (MPF) and *cytostatic factor* (CSF). MPF has been shown to consist of a complex of at least two proteins: a kinase (phosphorylating) enzyme called *pp45* and a smaller protein called *cyclin B*. Only when this complex is present will M phase persist. CSF seems to act by stabilizing the complex. However, this complex is very sensitive to calcium, such that a rise in calcium tilts the equilibrium away from complex synthesis towards

destruction. Almost certainly the calcium acts by stimulating another enzyme to destroy cyclin B, so that, insufficient of it is available to complex with pp45. As a result, the oocyte exits from M phase, a rather complicated process.

In the mouse, the metaphase spindle lies just under and parallel to the surface, from which it is separated by actin-containing microfilaments (Fig. 8.9b–d). These rotate the spindle ensuring that, on completion of meiosis, a large zygote and a small, second polar body are generated. However, in the human and the sheep, rotation is not required, the oocyte being ovulated with the spindle perpendicular to the surface. Actin-containing microfilaments also seem to be important in the incorporation of the sperm nucleus itself at the site of the *fertilization cone* (Figs 8.6c & 8.9d–f). It is important to

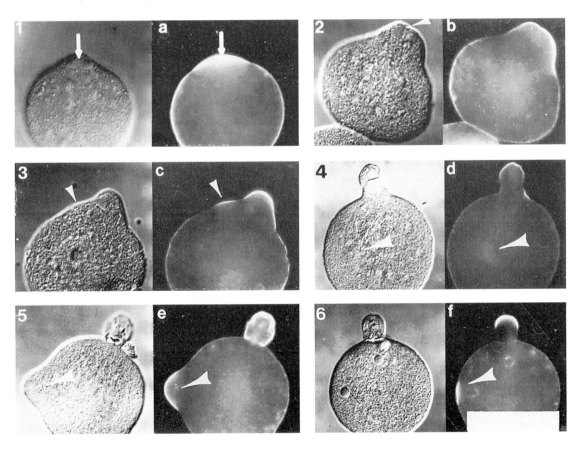

Fig. 8.9 Mouse oocytes at different stages after fertilization examined by differential interference microscopy (1–6) or, after visualizing the distribution of actin, by use of a fluorescent antibody (a–f). (1 & a) The unfertilized oocyte: note the second metaphase spindle and the heavy staining for submembranous actin that overlies it (arrow). (2 & b) Within minutes of fertilization the membrane over the spindle indents at the equator of the spindle (arrow) yielding a 'double-bump' structure over the spindle. (3 & c) One of the actin-lined 'bumps' appears to contract down (arrow) and as a result the spindle rotates. (4 & d) The second 'bump' now starts to restrict at

its base, leading to the formation of the second polar body. Notice also the fertilizing spermatozoon (arrow) now being pulled into the oocyte at the fertilization cone at which actin is evident (arrow). (5 & e) A slightly later stage: the second polar body is now pinched off, but the fertilization cone is large and actin-lined (arrow). (6 & f) By about 4 hours, male and female pronuclei can be seen clearly. The female pronucleus is smaller and adjacent to the actin-positive second polar body. The male pronucleus is still close to its actin-positive entry point (arrow), but the marked fertilization cone is now gone. (Scale bar in panel is 60 μm.)

note that spermatozoa do not bind to the oocyte membrane immediately overlying the second metaphase spindle, probably due to the absence of microvilli in this region (Fig. 8.10). This makes sense biologically, as the coincidence of a spermatozoon entering the oocyte and a second polar body being ejected could lead to mutual interference and *aneuploidy* (deviation from the normal euploid chromosomal number). There is evidence that the extrusion of the polar body requires the activity of PKC in addition to the triggering calcium rise. The whole process takes about 30–45 minutes.

With the successful attainment of one maternal and one paternal set of haploid chromosomes, a euploid *zygote* is formed. Aneuploidy causes embryonic abnormality. Fortunately, most aneuploid conceptuses die relatively early, although a few survive to term and beyond. Others may develop into tumours, such as hydatidiform mole, a benign trophoblastic tumour, or choriocarcinoma, its malignant derivative. Thus, aneuploidy, apart from being of potential danger to the

mother, is reproductively wasteful and can be distressing. Failure of the normal mechanisms for establishing euploidy is increased under certain conditions; for example, the fertilization of prematurely ovulated oocytes (see Chapter 4) is associated with defective cortical granule release and poor sperm head incorporation. The fertilization of 'aged' oocytes that have been ovulated and passed into the oviduct hours before sperm arrival can also be problematic (Table 8.3). In most mammalian species in which ovulation and behavioural oestrus are closely synchronized, the problem of ageing oocytes will not be acute. In humans, and some higher primates and cetaceans, however, where coitus and ovulation are not necessarily, or even usually, associated, the fertilization of old oocytes may be much commoner. Could the high incidence of genetic abnormality (50% or more of all conceptions) reported for human embryofetuses reflect a genetic price to be paid for the increased social and sexual freedoms derived by dissociating mating from fertility? (See Chapter 14.)

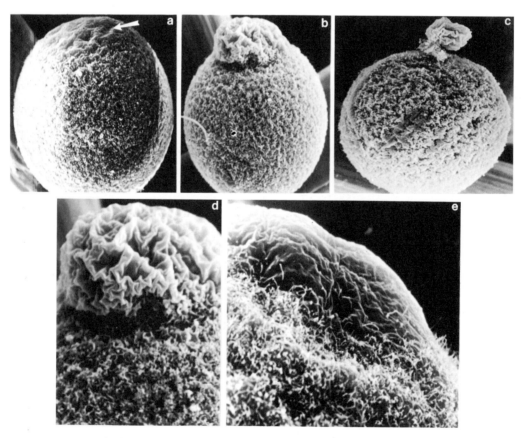

Fig. 8.10 (a) Scanning electron micrograph of a newly ovulated mouse oocyte from which zona pellucida has been removed. Note the absence of microvilli at one end (arrow) marking the site of metaphase spindle, as shown in (b) where polar body extrusion is underway after fertilization (sperm tail evident). (c) Note the small size and absence of microvilli on the polar body. (d, e) Higher power pictures of b & a in the region demarcating the microvillous and non-microvillous zones.

Initiation of development: oocyte activation

During oocyte–sperm fusion and second polar body expulsion, the cytoplasmic contents of the sperm cell membrane (now fused with the oocyte membrane) pass into the oocyte cytoplasm. The sperm nuclear membrane breaks down and the highly condensed chromatin starts to swell, releasing filamentous strands of chromatin into the cytoplasm. The protamines that created such compressed chromatin are released and replaced by normal histones. This decondensation is actively induced by factors in the oocyte cytoplasm that develop in the terminal phases of intrafollicular maturation, such as high levels of the reducing agent *glutathione* (see Chapter 4).

Between 4 to 7 hours after fusion, the two sets of haploid chromosomes each become surrounded by distinct membranes and are now known as *pronuclei* (Figs 8.6d,e & 8.9f). The male is usually the larger of the two. Both pronuclei contain several nucleoli. During the next few hours, each pronucleus gradually moves from its subcortical position to a more central and adjacent cytoplasmic position. During this period, the haploid chromosomes synthesize DNA in preparation for the first mitotic division, which occurs about 18–24 hours after gamete fusion. The pronuclear membranes around the reduplicated sets of parental chromosomes break down (Fig. 8.6e), the mitotic metaphase spindle forms and the chromosomes assume their positions at its equator. The final phase of fertilization has been achieved: *syngamy* (or coming together of the gametic chromosomes) has occurred. Immediately, the first mitotic anaphase and telophase are completed, the *cleavage furrow* forms, and the one-cell zygote becomes a two-cell conceptus.

Parthenogenetic activation

Although the spermatozoon induces many remarkable changes in the oocyte, it does not appear to be essential for many of them! An oocyte may be activated *parthenogenetically* by a variety of bizarre stimuli, such as electric shock, exposure to various enzymes or to alcohol. Activation by these stimuli is especially easy in oocytes that have been ovulated for several hours. The stimuli induce a calcium rise, thereby mimicking the act of spermatozoal fusion. As a result, cortical granules exocytose, meiotic metaphase is resumed and the oocyte's developmental programme is activated. The *parthenote* so formed may undergo cleavage to form a blastocyst and (in the mouse) implantation and development to a stage where a beating heart, somites and forelimbs are present. However, most parthenotes die fairly early on in this sequence and none survive to term (see also discussion on asexual reproduction in Chapter 1). The reason for this lack of survival resides in some extraordinary events that normally take place during oogenesis and spermatogenesis.

It seems that during the packaging of the chromosomes for transmission to the zygote, the environment in which the chromosomes find themselves influences the organization of small numbers of their genes in a way that affects their ability to become transcriptionally active subsequently in the conceptus. It is not the actual genetic code (base sequence) itself that is changed but the way in which the genes are wrapped up in associated proteins to form chromatin, although the precise details are unknown. This process is called *imprinting*, because it leaves an imprint on the genes that is 'remembered' and affects later expression. The important point is that the genes affected in the spermatogenic lineage differ from those in the oogenetic lineage. In other words, the imprinting pattern is *parentally specific*. These maternal and paternal imprinting processes mean that, although the oocyte and the spermatozoon each contribute one complete set of chromosomes and genes to the conceptus, each set is not on its own fully competent to direct a complete programme of development. Only when a set of genes from an oocyte is combined with a set of genes from a spermatozoon is a fully functional genetic blue-print achieved. A parthenote lacks access to some crucial genetic information, which, although present in its chromosomes, cannot be accessed because of the maternal imprinting to which it was subjected during oogenesis. In the normal zygote, this information would be provided by genes on the paternally derived set of chromosomes.

Parental imprinting has profound implications for mammals, because it compels us to reproduce sexually. There are costs to obligatory sexual reproduction, as was discussed in Chapter 1. The time and energy taken up with seeking a sexual partner, courting and mating is costly. Males cannot be dispensed with but consume resources. Asexual reproduction has many advantages for rapid and efficient propagation of the species. Sexual reproduction may be most advantageous at times of stress when maximum genetic flexibility is required for the species to survive. While most organisms retain an option on asexual reproduction, even where sexual reproduction occurs, mammals have lost that option during evolution. One of the great puzzles of mammalian biology is to determine why.

SUMMARY

The interval between the departure of the gametes from the gonads and the successful formation of a two-cell conceptus encompasses an extraordinarily complex sequence of events involving biochemical, behavioural, endocrine, physiological and genetic components. Not surprisingly, aspects of this process often go wrong and infertility results (see Chapter 14). Fortunately, recent biomedical advances mean that most of these events can now be carried out *in vitro*. Mature oocytes can be aspirated directly from the follicles, ejaculated spermatozoa can be capacitated and activated in defined media, fertilization and the formation of a conceptus can be achieved outside the body. These techniques have provided help for many otherwise infertile couples. Considerable controversy has surrounded the ethics of use of these procedures; in particular, the decisive role that fusion of an oocyte and a spermatozoon might play in the establishment of a human life has been stressed. It is important to be aware that scientific evidence does not support the view that one single event is decisive for the creation of an individual life. The process is a continuum which starts with the growth of the oocyte and the synthesis of the maternal cytoplasmic inheritance, and continues with the formation and development of the zygote, the signalling by the conceptus to the mother of its presence, and the subsequent acquisition of a distinctive human form capable ultimately of independent existence. The natural losses in mammals are massive over the early events in this sequence, particularly from oocyte atresia and pre-implantation death. Human pre-implantation losses are increased further by our changed social and sexual habits and the use of various forms of contraceptive. Biology cannot provide the answer to the question 'When does a life begin?' because the question is based on false premises. Biologically, life is a continuum. Ethical decisions cannot therefore be taken on the basis of biological observation alone.

FURTHER READING

Austin CR (1968) *Ultrastructure of Fertilization.* Holt, Rinehart & Winston.

Bancroft J (1989) *Human Sexuality and Its Problems.* Churchill Livingstone, Edinburgh.

Bock G. & O'Connor. M. (eds) (1986) *Human Embryo Research: Yes or No?* Tavistock Publications.

Creed KE, Carati CJ & Keogh EJ (1991) The physiology of penile erection. *Oxford Reviews in Reproductive Biology* **13,** 72–95.

Drobius EZ & Overstreet JW (1992) Natural history of mammalian spermatozoa in the female reproductive tract. *Oxford Reviews in Reproductive Biology* **14,** 1–46.

Edwards RG (1982) *Conception in the Human Female.* Academic Press.

Fraser LR (1994) Sperm functional changes from ejaculation to fertilization. In: *Male Factors in Human Infertility* (Ed. Tesarik J). Aeres-Serono Symposia Publications, Rome.

Johnson MH (1989) Did I begin? *New Scientist* 9 December, pp. 39–42.

Masters WH & Johnson VE (1966) *Human Sexual Response.* Churchill, London.

Masters WH & Johnson VE (1970) *Human Sexual Inadequacy.* Churchill, London.

Money J & Musaph H (Eds) (1977) *Handbook of Sexology.* Elsevier, Armsterdam.

Surani MAH (1994) Genomic imprinting: control of genetic expression by epigenetic inheritance. *Current Topics in Cell Biology* **6,** 390–395.

Wagner G & Green R (1981) *Impotence: Physiological, Psychological, Social Diagnosis and Treatment.* Plenum Press, New York.

Whitaker M & Swann K (1993) Lighting the fuse at fertilisation. *Development* **117,** 1–12.

Yanagimachi R (1994) Mammalian fertilization. In: *The Physiology of Reproduction* (Eds Knobil E & Neil JD), 2nd edition. Raven Press, New York.

Implantation and the Establishment of the Placenta

The conceptus remains at the oviducal site of fertilization for a further few days (Table 9.1, column 2). It is then transferred through the isthmus of the oviduct to the uterus. Transfer is facilitated by the changing endocrine milieu of the early luteal phase with its rising ratio of progesterone to oestrogen, which affects the oviducal and uterine musculature and relaxes the isthmic sphincter (see Chapter 7). It is probable, however, that the cilia, rather than the musculature of the genital tract, are the primary active transporters of the

Species	Cleavage to four cells	Conceptus enters uterus	Formation of blastocyst	Time of attachment	Expected time of luteal regression if mating infertile	Duration of pregnancy
Invasive						
Mouse	1.5–2	3	3	4.5	10–12	19–20
Rat	2–3	3	4.5	4.5–5.5	10–12	21–22
Rabbit	1–1.5	3.5	3.5	7–8	12	28–31
Human	2	3.5	4–5	7–9	12–14	270–290
Non-invasive						
Sheep	4	2–3	6–7	16	16–18	144–152
Pig	1–3	2	5–6	18	16–18	112–115
Cow	2–3	3–4	7–8	30–45	18–20	277–290
Horse	1.5–2	5–6	6	30–40	20–21	330–345

Table 9.1 Times (in days) after ovulation at which various developmental and maternal events occur

conceptus. Thus, pharmacological inhibition of oviducal muscle does not prevent transfer of the conceptus; furthermore, if a segment of oviduct is excised, turned round and replaced such that its cilia beat away from the uterus, the conceptus moves only up to this point of oviducal reversal and then stops.

On reaching the uterus, the conceptus engages in an elaborate interaction with the mother in which several messages are transmitted in both directions. This interaction, which may be likened to a conversation, has two important components. First, the conceptus establishes physical and nutritional contact with the maternal endometrium at *implantation*; failure to do so properly would deprive the conceptus of essential nutritional substrates and arrest its growth. The conversation operating during implantation involves short-range signals, and is the major subject of this chapter. Second, the conceptus *signals its presence* to the maternal pituitary–ovarian axis; failure to do so would result in the normal mechanisms of luteal regression coming into operation, causing progesterone levels to fall and loss of the conceptus. Somehow the conceptus must convert the whole of the female's reproductive system from a cyclic pattern, in which oscillating dominance of oestrogens and progestagens occurs, to a non-cyclic, pregnant pattern, in which progestagens dominate throughout. The conversation operating during this *maternal recognition of pregnancy* involves long-range signals, and is the major subject of Chapter 10. Thus, pregnancy is not initiated with fertilization but only when the conceptus has signalled its presence successfully to the mother.

GROWTH OF THE PRE-IMPLANTATION CONCEPTUS

The control of development

During its period in the oviduct, the conceptus continues its cellular division at a rate characteristic for each species (Table 9.1). Each cell or *blastomere* undergoes a series of divisions, during which the total size of the conceptus remains much the same (Fig. 9.1). In consequence, with each of these so-called *cleavage divisions*, the size of the individual blastomeres is progressively reduced, restoring to normal the exaggerated cytoplasmic : nuclear ratio of the oocyte. The large volume of oocyte cytoplasm, distributed to blastomeres during cleavage, contains materials essential for this cleavage process. These include: ribosomes and the full protein biosynthetic apparatus for making proteins (including a vast diversity of mRNA species); mitochondria and an ATP generating system, which is based initially on the use of pyruvate and then on glucose as a metabolic substrate; a Golgi system for production and modification of glycoproteins; and a cytoskeletal system essential for cyto- and karyokinesis. As the spermatozoon has little

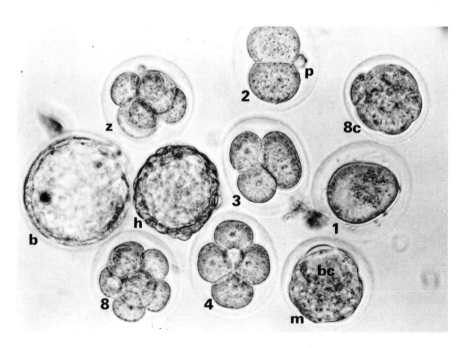

Fig. 9.1 Photographs of various pre-implantion mouse stages (all except 'h' surrounded by zona ellucida). (1) Unfertilized oocyte surrounded by spermatozoa; (2) two-cell with polar body (p); (3) three-cell; (4) four-cell; and (8) eight-cell cleavage stages; (8c) eight-cell compacted morula; (m) 32-cell late morula/early blastocyst, with beginning of blastocoelic cavity (bc); (b) fully expanded blastocyst; (h) blastocyst that has lost its zona pellucida (z).

cytoplasm when compared with the oocyte, the bulk of these materials is derived from the ooplasm, giving rise to the notion of a *maternal cytoplasmic inheritance*, distinct from, and in addition to, the *genetic inheritance* provided in equal measure via the chromosomes from each parent. Indeed, it is clear that the earliest stages of development occur in the complete absence of any mRNA synthesis by the conceptus's own chromosomes, which are transcriptionally inactive. Thus, these early stages are, in effect, controlled by the products of oogenesis, and thus by the maternal genome. Moreover, the developing conceptus (as well as the later fetus and neonate) possesses only maternal mitochondria. These carry their own DNA and are autonomously replicating. Any mitochondria carried into the oocyte by the spermatozoon do not replicate and are lost. This almost total dependence of the early conceptus on its maternal cytoplasmic inheritance means that any deficiency in oocyte maturation will result in impaired or failed early development and possible pregnancy loss.

The one non-chromosomal component that does seem to be contributed by the spermatozoon, at least in the human and large farm animals, is a centriole. It is clear that the centriole that enters with the sperm head is active in nucleating microtubules in the newly fertilized oocyte and that elements of it can contribute to the construction of mitotic spindles during cleavage. Whether it is essential, however, is doubtful, as cleavage can occur in oocytes activated parthenogenetically in the absence of a spermatozoon! The male contribution appears to be redundant during early development.

However, eventually the reservoir of maternally inherited cytoplasmic information ceases to be the sole controller of development. At a developmental stage that is characteristic for the species, transcription is activated and the genes of the conceptus itself start to contribute to development (e.g. mouse two-cell, pig four-cell, human four/eight cell, cow/sheep eight/16 cell). At the same time, all the maternally derived mRNA is destroyed. However, despite the fact that over a period of a few hours all *newly synthesized proteins* are switched from being maternally encoded to being embryonically encoded, proteins that were synthesized earlier on maternal templates persist until the blastocyst stage and beyond.

Shortly after gene activation, the conceptus shows a marked quantitative increase in its biosynthetic capacity. Net synthesis of RNA and protein increases, transport of amino acids and nucleotides into the cells rises, and changes occur in the synthetic patterns of phospholipids and cholesterol. Maturational changes occur in mitochondria, Golgi and the endocytic and secretory systems of the conceptus. From this time onwards, the growth

and metabolic activity of the pre-implantation conceptus *in vitro* has been shown to be stimulated by a number of growth factors, including the insulin-like growth factors (IGF1 & 2), transforming growth factors α (TGFα) and β1 & 2 (TGFβ1 & 2), epidermal growth factor (EGF), and platelet-derived growth factor A (PDGFA). Receptors for these factors have been identified on the early conceptus. Moreover, synthesis of many of these growth factors has been detected either in the conceptus itself (IGF1 & 2, PDGFA, TGFα) or in maternal tissues, as a result of which they appear in uterine fluids (insulin, IGF1, TGFα, EGF). It is therefore reasonable to conclude that these factors act as *autocrine or paracrine* agents to promote early development *in vivo*.

Early differentiative events

At around the eight- to 16-cell stage in most species studied, the cleaving conceptus changes its morphology by undergoing the process of *compaction* to yield a *morula* (Figs 9.1–3). This process involves maximizing intercellular contacts and also transforming cell phenotype from radially symmetric to highly polarized or epithelioid (Fig. 9.3). This polar phenotype is important developmentally, as when an eight-cell blastomere divides to the 16-cell stage, one of their offspring may inherit their basal domain while the other inherits their apical domain. This differential inheritance then prompts these two cell types to develop into inner cell mass and trophectoderm cells, respectively, in the *blastocyst*, which forms shortly thereafter, at about the 32- to 64-cell stage in most species (Figs 9.1 & 9.2). The precise form of the blastocyst varies in different species, but in all species the blastocyst contains two distinctive types of cell: (a) an outer rim of *trophectoderm* cells surrounding a *blastocoelic cavity* containing *blastocoelic fluid*; and (b) an *inner cell mass* (ICM), which is eccentrically placed within the *blastocoelic cavity* against, or embedded within, one wall of the trophectoderm. The expansion of the blastocoele is stimulated by growth factors EGF and TGFα. The trophectoderm cells constitute the first so-called *extra-embryonic* tissue, as they do not contribute to the embryo or fetus proper. Instead, they give rise to part of an accessory fetal membrane, the *trophoblast* of the *chorion*, which is concerned with the nutrition and support of the fetus. Indeed, the first 14–18 days of human development are concerned, in large part, with the further elaboration within the conceptus of various extra-embryonic tissues (Table 9.2 & Fig. 9.7), and only after this time can distinct and separate tissues be defined that will give rise exclusively either to a single fetus or to extra-embryonic supporting tissues. Thus, only at this time is it appropriate to refer to an *embryo* (i.e.

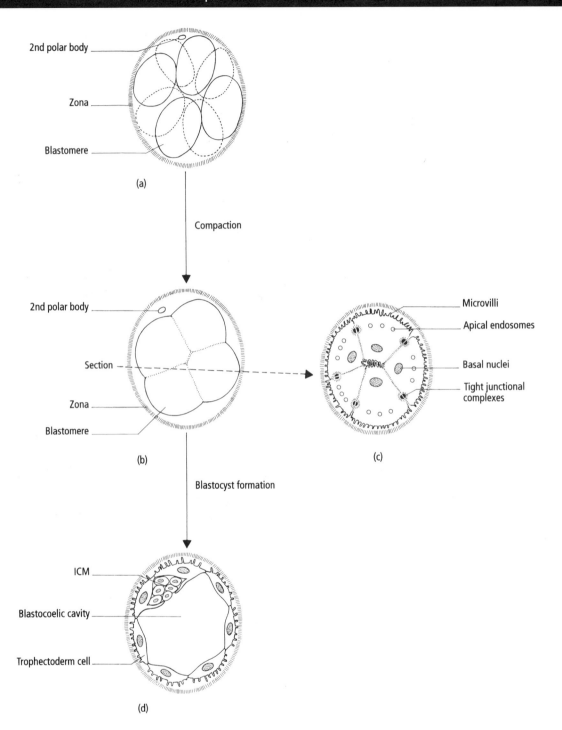

Fig. 9.2 (a–c) Compaction. Spherical cells become wedge-shaped and, by apposing adjacent surfaces, maximize cell contact. Tight junctional complexes develop between outer membranes of adjacent cells; these are punctate at first, but later become zonular, forming a barrier to intercellular diffusion between the inside and outside of the conceptus. Each cell also becomes polarized: the nucleus occupying a more basal position, endosomes and other organelles being apical, and microvilli being restricted to the exposed surface and points of contact with other cells basally. (d) Section through a 34- to 64-cell blastocyst; fluid accumulation within the blastocoelic cavity is possible, as the tight junctional complexes between adjacent trophectoderm cells prevent its escape (ICM: inner cell mass; see also Fig. 9.3).

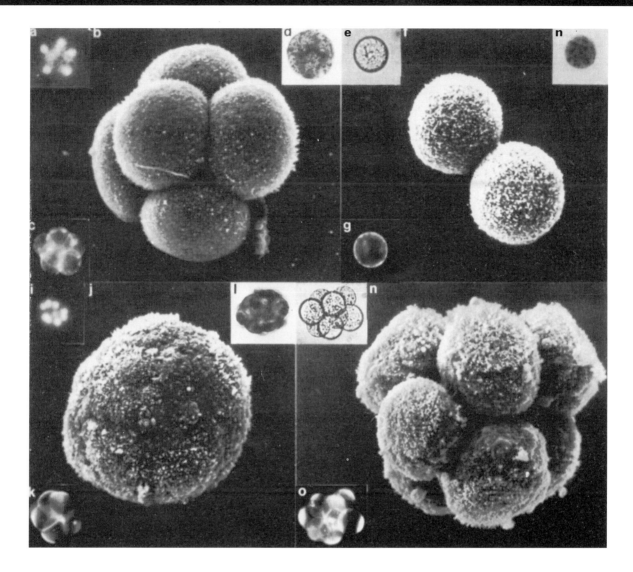

Fig. 9.3 The eight-cell mouse conceptus. (a–d) Intact, early eight-cell mouse stages. Note that individual blastomeres are spherical and distinctly separated from each other whether observed: (a) after staining of nuclei; (b) with the scanning electron microscope; (c) after labelling with a fluorescent surface marker; or (d) after they have been allowed to ingest a black dye (speckles in cytoplasm). The spherical form and the symmetry of the cells is shown further when isolated blastomeres are viewed in the same ways (e–h). In contrast, a late, eight-cell conceptus has 'compacted' (i–l); this involves flattening of cells on each other so that clear cell outlines cannot be resolved (j). Moreover, the nuclei cluster centrally (i), and the ingested black dye is not accumulated throughout the cytoplasm but is concentrated just above the nucleus and just below the outward-facing membrane (l). If this surface membrane is examined more closely, it is shown to concentrate binding of the fluorescent surface marker (k). These features are observed more readily if the flattening component of compaction is reversed by briefly incubating the conceptus in medium that is low in Ca^{2+}; the cells pull apart (m–o), revealing that they are no longer radially symmetrical but are highly polarized, with a tuft of microvilli facing outwards that stains for fluorescence (n–o). Thus, at compaction, the cells not only flatten on each other but also become epithelial-like, as judged by surface morphology, basal or central nuclei, and peripheral concentration of ingested dye.

that which gives rise to a fetus and then a baby), and prior to this stage the total product of fertilization is properly called the conceptus (also pre-embryo or pro-embryo).

Throughout development, from fertilization to the blastocyst, the conceptus remains enclosed within the zona pellucida. The zona probably has two functions. First, it prevents the blastomeres of the conceptus from falling apart during early cleavage prior to compaction. If the conceptus does become divided into two distinct groups of cells at this stage, *identical twins* result (*monozygotic*: derived from one zygote and therefore genetically identical, compared with *dizygotic* twins, which are derived from the independent fertilization of two oocytes from the

Table 9.2 Derivatives of a typical blastocyst

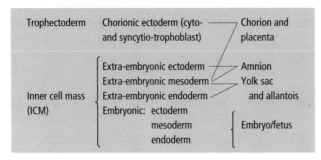

same ovulatory episode). Second, and conversely, the zona prevents two genetically distinct conceptuses from sticking together to make a single *chimaeric* conceptus composed of two sets of cells each of distinct genotype.

It is during the transition from a morula to a blastocyst that the conceptus enters the uterus (Table 9.1), and it is, therefore, the blastocyst that engages in the conversations vital to its survival and future development. Next we will consider how the blastocyst and the uterine endometrium interact locally with one another to effect implantation.

IMPLANTATION

The free-living blastocyst is bathed in the uterine secretions from which it draws the oxygen and metabolic substrates required for continued growth and survival. It actively accumulates organic molecules and ions by specific, transport mechanisms, while the exchange of oxygen and carbon dioxide is diffusional. There is a limit to the size that a free-living conceptus can attain before such exchanges become inadequate. Before this critical stage is reached, the growing conceptus develops its own blood vascular system that exchanges essential metabolites at its extra-embryonic surface and distributes them throughout its tissues. Conceptuses also develop one or more anatomically distinct and highly vascularized regions of their extra-embryonic surface through which the interchange of materials with maternal tissues is particularly facilitated. This special region draws on two major maternal sources of nutrition. During the initial *pre-, peri- and early post-implantation* stages of pregnancy, it continues to utilize almost exclusively maternal 'tissue juices' of various sorts (such as secretions or cell debris, see later for details). This sort of nutrition is called *histiotrophic*. However, these sources become progressively inadequate in most species, and so with the successful conclusion of *implantation* there develops in the maternal endometrial tissue a corresponding specialized and vascularized region. This zone of adjacent and highly vascularized contact between

mother and conceptus is called the *placenta*. In the placenta, the two discrete circulations lie sufficiently close that rapid and efficient transfer of materials can occur between them. This sort of nutrition is called *haemotrophic*. The placenta, then, is the ultimate outcome of the primary interactions between mother and conceptus that occur at implantation, and so *the form that the mature placenta takes is determined largely by the pattern and timing of the events of implantation*.

On entering the uterus, the conceptus is positioned for implanting at a site (or sites) within the uterus that is characteristic for each species. Uterine muscular activity may be important in this process, as its inhibition leads to abnormal sites of implantation. The appropriate location and, in *polytocous species* having several conceptuses and bicornuate uteri, spacing of implantation sites is important in minimizing physical and nutritional competition between conceptuses.

The first stage of implantation involves close apposition and adherence of the trophectodermal cells of the blastocyst to the luminal epithelial cells of the endometrium. This process is called *attachment*. However, before attachment can occur, the zona pellucida interposed between these cells must be removed. The proteolytic enzymes required can come from either the trophectodermal cells themselves or the uterine secretions, depending on the species. The main consequence of attachment is the induction of changes in the underlying endometrial stromal tissue, initiating its development as the maternal component of the placenta. How this is achieved varies with the species. In some, implantation is *invasive*, the conceptus breaking through the surface epithelium to invade the underlying stroma. In other species, implantation is *non-invasive*, epithelial integrity being retained (or at least only breached locally, transiently or much later in gestation: see later) and the epithelium becomes incorporated within the placenta.

Invasive implantation (dog, cat, mouse, rabbit, human and all primates except lemurs and lorises)

The attachment phase

The blastocysts of these species are first nurtured by uterine secretions but, compared with species in which implantation is non-invasive, this free-living phase is relatively short-lived (Table 9.1, column 4). In consequence, invasive conceptuses tend to be smaller at attachment and only a few trophectodermal cells are involved in making contact with the maternal epithelium. Yet, within a few hours, an increased vascular permeability in the area of stromal tissue underlying the conceptus is observed. This is associated with an oedema, localized changes in the intercellular matrix composition and stromal cell morphology, and a progressive sprouting and ingrowth of capillaries (Fig. 9.4). This *stromal reaction* is particularly marked in primates and rodents, where it is called the *primary decidualization reaction*. After 2–3 days, the decidualization spreads to give a larger secondary decidua, as the major endometrial component of the placenta is prepared. In other invasively implanting species the decidual reaction may be less marked but is functionally equivalent. Thus, a primary, highly localized signal from the small conceptus has been effectively transduced through the epithelium to the stroma where it is amplified rapidly.

The invasive phase

Within a few hours of attachment, the surface epithelium underlying the conceptus becomes eroded (Fig. 9.4).

tissue thus functions as a large 'yolk reservoir' equivalent to the yolk of a bird's egg. As the mammalian conceptus does not leave the female tract to develop, it does not carry its yolk with it and the mother only makes the yolk-equivalent (decidualization) if a pregnancy has occurred.

The depth to which the conceptus invades maternal tissues varies with species and also in certain pathological conditions. Those species in which the conceptus invades the stroma so deeply that the surface epithelium becomes restored over it are said to implant *interstitially* (e.g. the human, chimpanzee and guinea-pig; Table 9.3). In other species, the stroma is only partially invaded and the conceptus continues to project to varying degrees into the uterine lumen: so called *eccentric* implantation (e.g. the rhesus monkey, dog, cat and rat). Indeed, in these species, secondary contact and attachment by the conceptus to the uterine epithelium

Species	Depth of invasion	Extent and shape of attachment (Fig. 9.5)	Maternal tissue in contact with conceptus (Fig. 9.6)	No. of layers of chorionic trophoblast (Fig. 9.6)	Histological type (Fig. 9.6)
Invasive					
Human	Interstitial	Discoid	Blood	1	Haemomonochorial
Rabbit	Eccentric	Discoid	Blood	2	Haemodichorial
Rat/mouse	Eccentric	Discoid	Blood	3	Haemotrichorial
Rhesus monkey	Eccentric	Bi-discoid	Blood	1	Haemomonochorial
Dog/cat	Eccentric	Zonary	Capillary endothelium	1	Endotheliochorial
Non-invasive					
Pig/mare	Central	Diffuse	Epithelium	1	Epitheliochorial
Ewe/cow	Central	Cotyledonary	Syncytium of epithelium and binucleate cells	1	Synepitheliochorial

Table 9.3 Classification of implantation and placental forms in several species

Trophectodermal processes seem to 'flow' between adjacent epithelial cells, isolating and then dissolving and digesting them. Some trophectodermal cells fuse together and form a syncytium (*syncytiotrophoblast*), while others retain their cellularity (*cytotrophoblast*) and serve as a proliferative source for generating more trophoblastic cells. The uterine glandular tissue and the decidual tissue most adjacent to the invading trophoblastic front of the conceptus is destroyed. It releases massive quantities of primary metabolic substrates (lipids, carbohydrates, nucleic acids and proteins), which are taken up by the conceptus for use by the growing embryo. The decidual

on the opposite side of the uterine lumen may result in two sites of placental development, a *bi-discoid* placenta (e.g. the rhesus monkey, see Table 9.3), or a complete 'belt' of placenta round the 'waist' of the conceptus, a *zonary* placenta (e.g. the dog and cat; see Table 9.3 & Fig. 9.5b).

The invasiveness of the conceptus is influenced by the effectiveness of the decidual response. Thus, where the decidual response is inadequate, either pathologically in the uterus, or after implantation at non-uterine *ectopic* sites, invasion is much more aggressive and penetrating. This observation has been taken by some to suggest a

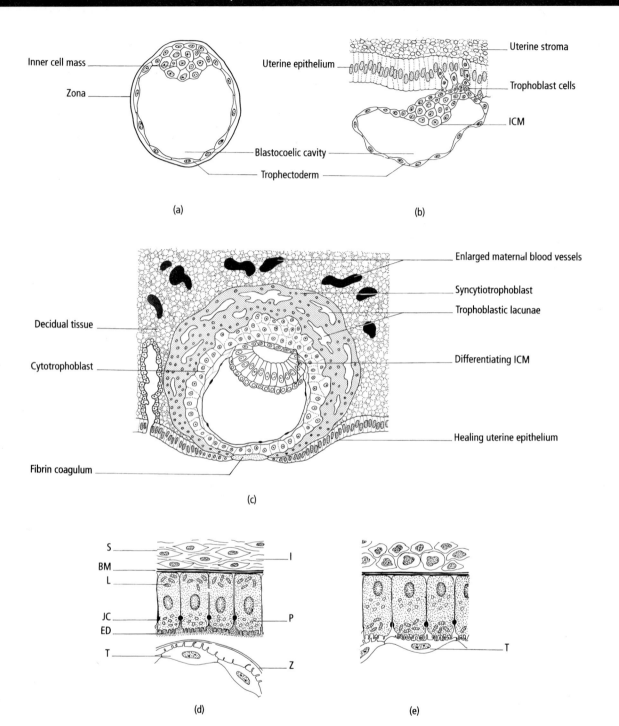

Inner cell mass

Zona

Blastocoelic cavity

Trophectoderm

(a)

Uterine epithelium

Uterine stroma

Trophoblast cells

ICM

(b)

Decidual tissue

Cytotrophoblast

Fibrin coagulum

Enlarged maternal blood vessels

Syncytiotrophoblast

Trophoblastic lacunae

Differentiating ICM

Healing uterine epithelium

(c)

S

BM

L

JC

ED

T

I

P

Z

(d)

T

(e)

'restraining' influence of the decidual tissue over the conceptus, but it could reflect an attempt by the invading trophoblast to overcome the inadequate nutritional support of a defective decidual response.

There are also species differences in the degree of proximity ultimately established between the circulations of the invasive conceptus and the mother, as well as in the physical depth of penetration. In some species, a relatively high degree of integrity of maternal tissue is retained, but in others, such as the human, the maternal blood vessels are eroded to bathe the trophoblast cells in maternal blood (summarized in Table 9.3 & Fig. 9.6a–d). There is little evidence that the number of layers of tissue that ultimately lie between the circulation of mother and conceptus can in any way be related to the *efficiency of placental transfer*. However, it may well influence the *types of transport mechanism* used (see Chapter 11 for more details).

With the formation and invasion of the decidua, implantation is completed, a physical hold and a nutritional source of decidual 'yolk' is established, and the basis of placental development, leading to adjacent circulations and exchange of nutrients, is initiated.

Non-invasive implantation (pig, sheep, cow and horse)

Attachment of non-invasive conceptuses is, in general, initiated relatively later than that of invasive conceptuses (Table 9.1). During the prolonged interval prior to attachment, the free-living conceptuses continue to draw on richer and more copious secretions from uterine glands: the non-invasive equivalent of the decidual 'yolk'. This *uterine 'milk'* is a rich source of ions, sugars, amino acids and lipid precursors, and the conceptus utilizes these resources and grows to a much greater pre-implantation size than the human or rodent invasive conceptus. The growth may be prodigious and rapid. The pig conceptus, for example, elongates from a spherical blastocyst 2 mm in diameter on Day 6 of pregnancy to a membranous, highly convoluted thread some 1000 mm long by Day 12. This growth is almost exclusively confined to the extra-embryonic tissues of the

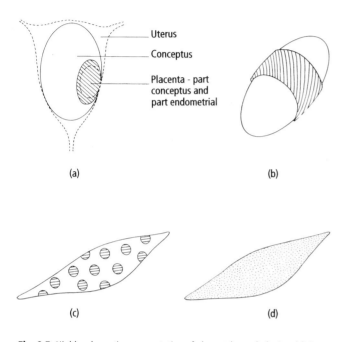

(a)   (b)

(c)   (d)

Fig. 9.5 Highly schematic representation of placental morphologies. (a) A discoid human placenta. The egg-shaped sac is the conceptus, which has a specialized disc on its surface that forms the placental component. A corresponding area on the endometrium also contributes to the placenta. Discoid placentae are also found in the mouse, rat, rabbit, bat and insectivores. Other types of placentae (uterine part is not shown) are: (b) zonary (e.g. the cat, dog, bear, mink, seal and elephant); (c) cotyledonary (e.g. the cow, sheep, giraffe, deer and goat); and (d) diffuse (e.g. the horse, pig, camel, lemur, mole, whale, dolphin and kangaroo).

Fig. 9.4 (*opposite*) Scheme of conceptus–uterine relations during invasive implantation (human). (a) 4–5 days: free-living blastocyst within zona pellucida; (b) 6 days: zona is lost, and trophectoderm (overlying ICM) transforms to trophoblastic tissue, attaches to epithelial cells, and, inducing changes in these and underlying stromal tissue, then starts to penetrate epithelium; (c) 9 days: decidualization in underlying stromal tissue spreads out rapidly from attachment site; trophoblast erodes surface epithelium, invades and destroys adjacent decidual tissue, and becomes embedded. Surface epithelium heals over; (d) and (e) show details of early phase. (d) Part of blastocyst in zona (Z), with microvillous trophectoderm (T), lies free in uterine lumen adjacent to epithelial cells linked by zonular junctional complexes (JC), with surface microvilli interdigitating with a thick surface secretion, which is electron dense (ED). Within epithelial cells are pinocytotic vesicles (P), central nuclei, and basal lipid droplets and mitochondria (L). The cells rest on a basement membrane (BM) underlain by condensed fibrous connective tissue. Stromal cells (S) are spindle-shaped, and lie in an extensive extra-cellular matrix (I) abundant in collagen fibres. (e) At attachment, zona has been lost, microvilli become flat and interdigitate; trophoblast and epithelial cells become very closely apposed. Epithelial nuclei assume basal position and mitochondria are apical, with dispersed fat droplets in between. In stroma, oedema and increased vascular permeability are accompanied by loss of collagen fibres and swelling of stromal cells, which subsequently develop extensive endoplasmic reticulum, polysomes, enlarged nucleoli, lysosomes, glycogen granules and lipid droplets, and become 'decidual cells'. In the rat, extensive intercellular gap junctions are also seen. Nuclei frequently become polyploid. Peripherally, sprouting and ingrowth of maternal blood vessels occurs. *Note:* in the human endometrium, some decidual-like changes may occur in stromal cells in the absence of a conceptus during the late luteal phase. These changes are often called 'pre-decidualization'.

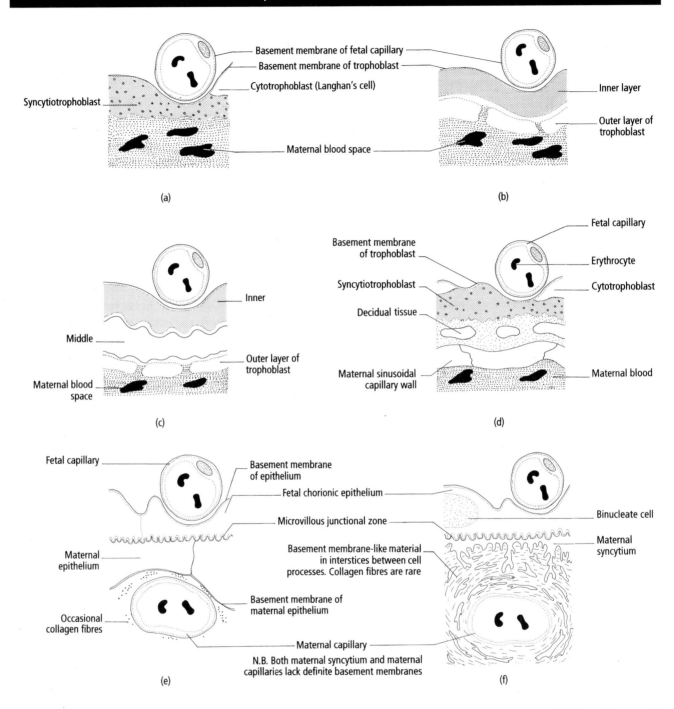

Fig. 9.6 Schematic view of microstructure at the mature placental interface of various species. (a) Human: haemomonochorial; (b) rabbit: haemodichorial; (c) rat and mouse: haemotrichorial; (d) dog (also bears, cats and mink): endotheliochorial; (e) mare and pig (also whales, lemurs, dolphins, deer and giraffe): epitheliochorial; (f) sheep and cow: synepitheliochorial: the maternal uterine epithelial cells fuse with binucleate cells from the fetal side to form a syncytium; the binucleate cells may also invade maternal tissues later in pregnancy.

conceptus, and establishes a vast surface area over which exchange of metabolites with the uterine milk can occur.

This large surface area of extra-embryonic chorionic trophoblast also permits an extensive subsequent attachment to uterine epithelial cells. In the horse and pig, attachment occurs at multiple sites over most of the external surface of the conceptus, but, in ruminants, attachment is limited to uterine caruncles, distinct areas of projecting aglandular uterine mucosa, which in sheep and goats may be up to 90 in number. As a result, the ultimate form of the placentae in these species differs from that of invasive conceptuses, and is said to be *diffuse* (e.g. in the mare and pig) or *cotyledonary* (e.g. in the cow and sheep) (see Fig. 9.5c,d). Between the sites of attachment, uterine glands continue to secrete a nutrient 'milk' for the conceptus, especially in the pig and horse. Although penetration and erosion of the epithelium does not occur at implantation in these species (and therefore the conceptus is said to implant *centrally*), junctional complexes are observed between trophoblast and the adjacent epithelial cells (e.g. in the pig and sheep), and invasion of conceptus cells into the endometrium may occur either transitorily (e.g. in the horse) or much later in development (e.g. in the sheep). No proper decidual response is induced in the underlying stroma (the 'yolk' is provided by secretion of uterine milk and the epithelium acts to control invasion), but vascularity increases and distinct changes in cellular morphology occur, providing evidence of the recognition of an embryonic presence. These changes constitute the beginnings of the formation of the placenta and mark the end of the protracted implantation seen in these species.

Uterine–blastocyst conversations and the control of implantation

The endocrine dependence of implantation

The blastocyst provokes a response of considerable magnitude in the uterus. However, it can only do so over a very narrow window of time during a normal oestrous or menstrual cycle. During this time, the endometrial epithelial surface is *receptive* to the embryo. Prior to this time, the endometrium is said to be *pre-receptive* and is characterized by long microvilli, a thick glycocalyx, which includes transmembrane mucin glycoproteins integral to the endometrial epithelial cells, and high surface charge: all features likely to impair attachment. Accordingly, the transition to receptivity involves loss of negative charge, thinning of the mucin coat and shortening of microvilli. This receptive phase is then followed by a *refractory* phase. Uteri in both the pre-receptive and refractory phases not only resist attachment by the blastocyst, but are also, in many species, embryotoxic to any conceptus that leaves the oviduct prematurely or arrives too late (e.g. in rodents, sheep and rabbits, but not in humans or monkeys). Thus, the uterus can be thought of as a primarily hostile environment able to control carefully a potentially dangerous invasive trophoblastic tissue. Clearly, for the conceptus to survive, its early development and transport must be coordinated precisely with the changing receptivity of the uterus. This coordination is achieved by the mediation of the steroid hormones.

Progestagenic domination is required if the uterus and implanting blastocyst are to engage effectively. We saw in Chapter 7 that the luteal phase of the cycle in many species was characterized by distinctive patterns of uterine secretions rich in sugars and specific proteins. Likewise, the changes in epithelial surface properties are driven by hormonal changes. However, while progesterone is critical, the release of specific secretions and the changes in receptivity seem to require an additional endocrine input: namely, the superimposition of oestrogen during the luteal phase. Thus, female rats or mice can be deprived of this oestrogen by ovariectomy during the first 2–3 days of pregnancy while maintaining progesterone levels by daily injections. Under such conditions the conceptus enters the uterus as normal, but instead of proceeding to implant, remains free in the uterine lumen. The blastocyst may spend many days in the uterus in this quiescent state, its metabolism 'ticking over'. If a single injection of oestrogen is then given, blastocyst metabolism increases, attachment occurs rapidly, and decidualization is initiated.

The oestrogen affects the epithelial cells of the uterus in two ways. First, it stimulates the release of a glandular secretion of a characteristic composition including specific macromolecules, which may include growth factors (see below), as well as glucose, amino acids and ions. This secretion stimulates activation of the blastocyst. Second, the oestrogen also acts on the epithelial cells to make them responsive or sensitive to a blastocyst signal, so that they can transmit evidence of the blastocyst's presence at attachment to the underlying stromal cells, and thereby initiate decidualization. This facility of the rat and mouse to suspend the blastocyst *in utero* for several days is called *delayed implantation*, and it provides a clear example of the uterine control over the development of the blastocyst.

Delayed implantation should not be thought of as purely an artificial or experimentally induced phenomenon. It occurs in the rat, due to the natural suppression of endogenous oestrogen secretion in females, that are suckling young of the previous litter (see Chapter 13 for details). This type of natural delay is often called *facultative delayed implantation*, as it only occurs under conditions of suckling. It is shown by mice, gerbils, some marsupials and the bank vole, and is clearly useful as the mother delays the growth of her uterine litter for as long as she is suckling the previous litter. If this delay did not occur, the helpless newborn of the second litter would have to compete for milk with older and bigger siblings.

In addition, there are many species in which an *obligatory delayed implantation* is an essential and normal part of their pregnancy (e.g. in the roe-deer, badger, elephant seal, fruit bat, mink, stoat, wolverine and brown bear). In these species, a prolonged period of weeks or months may be spent with blastocysts in delay (or *diapause*, as an obligatory delay is also called). Diapause confers distinct biological advantages. The roe-deer, for example, mates in July and August when the adults are well fed and in their prime, ensuring effective competition for mates. The conceptus remains as a blastocyst until January when it reactivates for delivery of young in May and

June when the nutritional conditions will have become optimal for the new offspring.

In many species examined, including humans, secretion of oestrogen is imposed on the progesterone background of the luteal phase, and it is therefore possible, but not yet proved, that, in all species, the oestrogen acts to stimulate endometrial release of messages crucial to activation of the conceptus. It has been established that, in at least some of these species, all or part of the oestrogen is *not of ovarian origin*, but instead is synthesized by *the conceptus itself*. In the pig, for example, the luteal phase lacks a significant rise in oestrogen, but if a conceptus is present a clear oestrogen peak is detectable. Pre-implantation stages of the pig, rabbit, sheep, cow and hamster have all been shown to be capable of steroid synthesis and metabolism *in vitro*. Thus, the interactions between the conceptus and the uterus in these species may prove to be even more complex than those in the rat and mouse. The conceptus may first produce oestrogens to stimulate the progesterone-primed uterus; the uterine epithelium may then respond by secreting embryotrophic factors and showing increased sensitivity to blastocyst attachment. Finally, the blastocyst may co-respond by attachment, and so initiate implantation.

The molecular basis of implantation

What is the nature of the messages passing between conceptus and uterus involved in these signalling processes? Both cytokines and prostaglandins have been implicated.

The earliest observed, localized uterine response to a mouse blastocyst is the appearance in the endometrial luminal epithelium of mRNA encoding *heparin binding EGF-like growth factor* (HB-EGF; see Table 2.6). This response occurs only in cells adjacent to the blastocyst, and *precedes* dissolution of the zona pellucida, attachment, and any increase in stromal vascularity. Significantly, both *EGF receptors* and *heparan sulphate proteoglycans* (HSPG) are expressed on trophectoderm cells, thereby providing double binding sites. Moreover, when HB-EGF binds to EGF receptors on the blastocyst, it induces receptor phosphorylation and, presumptively, activation of the intracellular second messenger cascade (Fig. 2.10). This presumption is confirmed by the observation that local dissolution of the zona pellucida and invasive behaviour by trophoblast cells follows. Not surprisingly, blastocysts genetically deficient in EGF receptors die at implantation, and exposure of blastocyts to inhibitors of HSPG renders them incompetent to attach. Moreover, blastocysts in the uteri of mice in delayed implantation due to oestrogen deprivation do *not* induce HB-EGF, unless oestrogen is injected into the mothers. These observations enable us to start to build a picture of the molecular language used in

the conversations occurring between uterus and conceptus. The conceptus is clearly aware in some way of the endocrine state of the uterus. If oestrogen is present, the conceptus then somehow signals its presence to the local endometrium, induces HB-EGF locally, and then, in turn, responds to the HB-EGF by shedding its zona, attaching and invading.

What might tell the conceptus that oestrogen is present? A second growth factor, leukaemia inhibitory factor (LIF; Table 2.6) is a possible candidate. LIF is produced by the cells of the endometrial glands, but only when oestrogen is imposed on the background of luteal progesterone. Moreover, the uteri of mice genetically deficient in LIF cannot support implantation. Receptors for LIF exist on trophectoderm. It remains to be established whether LIF prompts the blastocyst to send an HB-EGF-inducing signal to the endometrium. The nature of this inducing signal is also not established, and neither is the precise sequence of molecular events that follows the initial attachment and invasive response to HB-EGF. However, a number of other factors are implicated.

The invasiveness of the trophoblast is clearly dependent upon a variety of different proteolytic enzymes, as well as other molecules that regulate their proteolytic activities. These are essential if controlled passage of the blastocyst through the epithelial layer, the underlying basement membrane and the stromal intercellular matrix is to occur. Also implicated is granulocyte/macrophage colony stimulating factor (GM-CSF; Table 2.6), which is produced by both the trophectoderm and the decidual cells, and receptors for which are present on the uterine epithelium. The attachment of the blastocyst to the endometrium seems to be promoted by GM-CSF both *in vitro* and *in vivo*. In a complementary way, macrophage colony stimulating factor 1 (M-CSF1; Table 2.6) becomes elevated in the endometrial epithelium at implantation, and receptors that recognize it have been detected on both trophectoderm and decidual cells. Mice lacking the gene for M-CSF1 show severely impaired implantation.

Both histamine and prostaglandins are implicated in the stromal response, as inhibitors to them reduce decidualization, and their injection, locally, induces a decidual response. Within the decidual tissue itself, transforming growth factor β (TGFβ; Table 2.6) seems to promote the implantation process, while tissue necrosis factor α (TNFα; Table 2.6) promotes resorption. One important feature of the decidual response is the increase in vascular permeability and the growth of new capillaries that occurs (*angiogenesis*). TGFβ promotes these processes, as do the platelet-derived growth factors (PDGFα & β) and the recently discovered, and related, placental growth factor (PlGF; Table 2.6), all of which are present in the decidua/trophoblastic tissues. Angiogenesis is an enduring need for the growing placenta and fetus (see later), and so the detection of these cytokines well into pregnancy is unsurprising.

The cellular events of decidual remodelling and trophoblastic invasion are complex, and a number of extracellular matrix molecules and cell surface adhesion molecules have also been implicated in the process. However, the precise patterns of interplay between all these various factors, where in the sequence of interactions between blastocyst, uterine epithelium, stroma and decidua they might act, and precisely how the endocrine milieu might influence their roles, remain to be elucidated and are the subjects of intensive research. It is clear, however, that the conversation occurring between the conceptus and the uterine tissues is complex and employs a rich molecular language.

Summary

We have seen that the earliest physical attachment of the conceptus to the uterine endometrium shows considerable interspecies variability in timing, extent and degree of invasiveness. This variation at the earliest stage of pregnancy results, subsequently, in a correspondingly diverse pattern of placental forms that frequently confuses students of reproduction. Certain simple rules, however, help to clarify this confusion. The invasive conceptus attaches early when still small, in order to yield the prize of uterine decidual nourishment. Its placenta therefore tends to be compact and to have fewer interposed layers between the fetal and maternal circulations. The non-invasive conceptus attaches late when larger, being nourished initially by uterine secretions. Its placenta therefore tends to be extensive and to be epitheliochorial. In both cases, the maternal endocrine condition influences the effectiveness with which mother and conceptus communicate during implantation.

So far we have considered only the anchoring of the conceptus to the mother. We must now consider precisely how the histiotrophic nourishment of the embryo by uterine secretory and decidual products is superseded by the more direct haemotrophic support provided by the intervascular exchange route of the mature feto-placental unit.

THE CHANGE FROM HISTIOTROPHIC TO HAEMOTROPHIC SUPPORT

The pre-implantation conceptus derives its nutrition from endogenous reserves and via the secretions of the genital tract. During invasive implantation these exocrine secretions are supplemented by decidual material released by the invading trophoblast. For how long does this histiotrophic route for nutritional and excretory exchange persist and what replaces it? In order to answer this question, we need first to examine briefly the structure of the developing conceptus and the principal routes of metabolic exchange that become available.

The fetus and fetal membranes

The development of the fetal membranes, and the interspecific variety of their organization, provides one of the most enduring confusions for students of embryology. A simplified scheme to explain the origin of the various fetal membranes is shown in Fig. 9.7, and this figure and its legend should be studied carefully before you read further. During the histiotrophic phase of growth, the conceptus takes up materials from, and excretes waste products into, the surrounding endometrial fluids. These materials pass through the thin 'shell' of trophoblast and distribute by simple diffusion through the cavities and tissues of the conceptus itself. However, as the mesoderm of the conceptus becomes more extensive, blood vessels invade it, and these link up to form an extensive vascular network. Blood formation occurs in the yolk sac mesoderm, and a primitive heart forms in the cardiac mesoderm (Fig. 9.7f) within the developing embryo itself. Blood can thus be pumped throughout the extensive mesodermal tissue of the whole conceptus, through both its embryonic and its extra-embryonic parts. This blood passes throughout the chorionic mesoderm where equilibration with maternal secretory and decidual fluids occurs.

With further development, the vascularity in the conceptus becomes particularly marked in the yolk sac mesoderm and where the yolk sac and chorionic mesoderm fuse together (arrowed in Fig. 9.7e), and a corresponding vascularity develops within the endometrium adjacent to this site of fusion. Together, the two adjacent, highly vascular sites form the *yolk sac (or choriovitelline) placenta*. The yolk sac, and the yolk sac placenta to which it contributes, is a structure homologous to the yolk-containing sac found in the eggs of reptiles, birds and monotremes (e.g. the platypus). This comparatively primitive origin is reflected in the transitory existence and function of the yolk sac placenta in most mammals, a second exchange site called the *chorio-allantoic placenta* taking over (Fig. 9.7f arrowed). However, in some species (e.g. marsupials), the yolk sac placenta functions alone throughout pregnancy and, in others (e.g. the rabbit, rat and mouse), it persists and remains func-

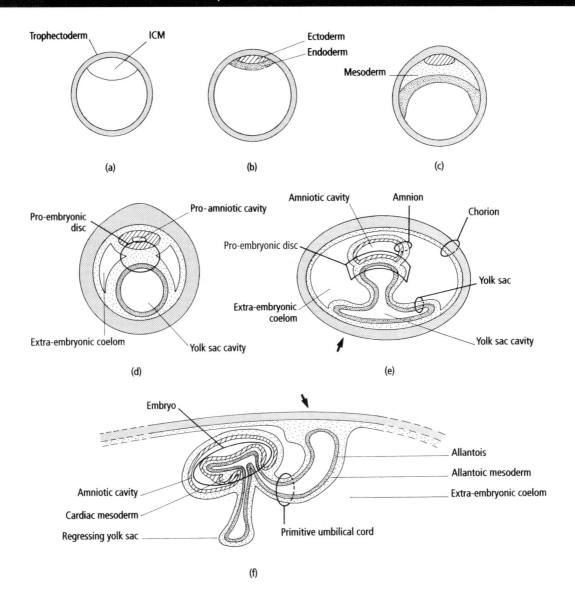

Fig. 9.7 (a) Blastocyst with trophectoderm (shaded) and inner cell mass (white). (b) Development has occurred within the inner cell mass: two layers of cells have formed, namely ectoderm (striped) also called epiblast) and endoderm (heavy stipple) also called hypoblast). (c) A third mesodermal tissue develops (light stipple). The endoderm spreads out over the underside of the trophectoderm, which is now transforming to mature trophoblast. (d) The endoderm edges meet to form a hollow spherical cavity: the yolk sac cavity. Mesoderm extends between the endoderm and trophoblast and the ectoderm and trophoblast. Cavities develop within the mesodermal tissue (the extra-embryonic coelom) and ectodermal tissue (pro-amniotic cavity). (e) The extra-embryonic coelom, pro-amniotic cavity and yolk sac cavity enlarge and change shape. Between the amniotic and yolk sac cavities, three cell layers persist, forming a sandwich, the ectoderm and endoderm forming the 'bread' around a mesodermal 'filling'. It is from this 'trilaminar or pro-embryonic disc' that the definitive embryo will arise (see below), and all the other tissues are extra-embryonic and form the fetal membranes. Thus, the amnion consists of a layer of extra-embryonic ectoderm fronting the amniotic cavity and a layer of extra-embryonic mesoderm outside it. The yolk sac consists of a layer of extra-embryonic endoderm fronting the yolk sac cavity and a layer of extra-embryonic mesoderm outside it. The chorion consists of a layer of trophoblast fronting the uterine tissue and a layer of extra-embryonic mesoderm within it. *Note:* (1) Blood vessels develop throughout the embryonic and extra-embryonic mesoderm, and the embryonic blood starts to flow through them. (2) The point at which the yolk sac mesoderm and chorionic mesoderm fuse is arrowed; this is the site of formation of the yolk sac (or chorio-vitelline) placenta. (f) The trilaminar embryonic disc curls up to give the definitive embryo with its outer ectoderm surrounded by amniotic fluid in the amniotic cavity; the derivatives of this ectoderm include the outside 'skin' of the fetus. Within this curled up disc, there is an endoderm-lined cavity (continuous with the yolk sac cavity) that is the primitive gut. The interposed filling of embryonic mesoderm (which will form many of the fetal tissues between the gut and the skin and their derivatives) is highly vascular and includes, at the anterior or head end of the embryo, the primitive heart (cardiac mesoderm) that pumps blood through the network of embryonic and extra-embryonic vessels. A diverticulum of the endo-derm, the allantoic endoderm, develops at the posterior or tail end of the embryo and it grows out surrounded by mesoderm to form the allantois. The allantoic and chorionic mesoderm (both rich in blood vessels) fuse (arrowed) and this is the site of formation of the chorio-allantoic placenta. The connection that the allantois makes between the embryo and the chorio-allantoic placenta will become the umbilical cord (see Fig. 9.8).

tional, subserving a specific subset of transport functions even after the definitive chorio-allantoic placenta takes over the major role. In higher primates (e.g. the human), the yolk sac placenta is never functional at all.

The chorio-allantoic placenta forms as a result of the outgrowth of an endodermal diverticulum (*the allantois*) from the hindgut region of the developing embryo (Fig. 9.7f). Together with its ensheathing mesoderm, containing the pro-umbilical vessels, it contacts the chorionic mesoderm and thereby determines the site (or sites) of chorio-allantoic placentation. As we saw earlier, the consequence of allantoic outgrowth in species with

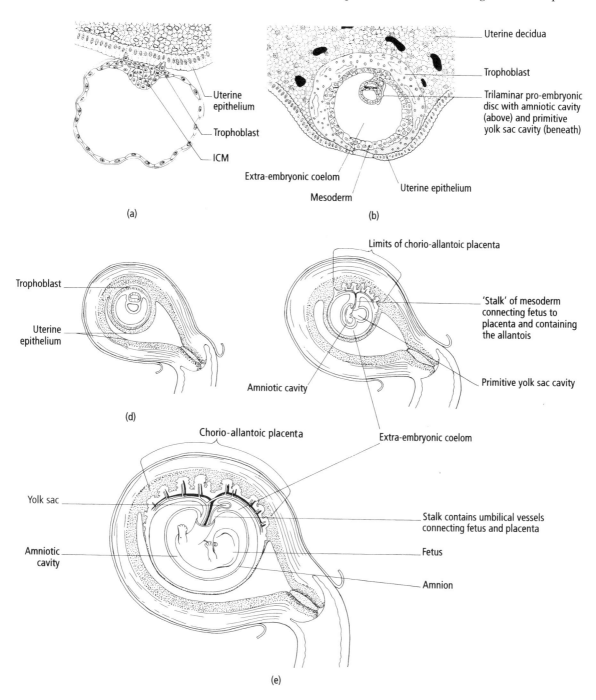

Fig. 9.8 Highly schematic views (not to scale) of a human conceptus at: (a) attachment; (b) invasion; and (c)–(e) progressively later stages of pregnancy. The general relationship of fetus, extra-embryonic membranes, placenta and maternal vessels is shown. Note how the fetus itself ends up floating in amniotic fluid and linked to the placenta by an umbilical stalk.

early invading conceptuses is a restricted discoidal, bi-discoidal or zonary placenta, whereas late-attaching conceptuses acquire a more extensive and diffuse chorio-allantoic placenta.

It is important to re-emphasize that the functional placenta incorporates not only a contribution from the conceptus, as shown developing in Fig. 9.7, but also from endometrial tissue. This is shown schematically in Fig. 9.8 for the developing human conceptus, which has a discoid, chorio-allantoic placenta. Having considered the general disposition of membranes, fetus and placenta, we will now take a closer look at the organization of the placental interface itself.

The placental interface

The functional chorio-allantoic placenta is characterized by: (a) extensive proliferation of the chorionic tissue to give a large surface area for exchange; (b) highly developed vascularity of both fetal and maternal components; and (c) intimately juxtaposed, but physically separate, fetal and maternal blood flows. The precise arrangement of vessels in the functional placenta varies considerably, and here we examine the human (discoidal, haemomonochorial) and sheep (cotyledonary, synepithelio-chorial) placentae as illustrative representatives.

The human placenta

In the mature *haemomonochorial* placenta of the human, tongues or *villi* of chorionic syncytiotrophoblast-containing cores of mesodermal tissue in which fetal blood vessels run, penetrate deeply into the maternal tissue to form an extensive network (Fig. 9.9). Each villous blood vessel, by progressive branching of the main divisions of the umbilical vessels, is separated from the surface by a thin syncytiotrophoblastic layer (Fig. 9.9c). At the tips of the terminal villi, the capillaries are dilated and form tortuous loops (Fig. 9.10). Thus, fetal blood flow through the tips will be slow, allowing for exchange of metabolites with maternal blood. The branches of the villi are arranged, with a somewhat variable degree of regularity, to form 'fenestrated bowls' (rather in the shape of the bowl of a brandy glass). Their terminal villi project inwards into the central space of the bowl, between the adjacent villi that form the fenestrated wall of the bowl, and outwards into the space peripheral to the bowl. Each 'bowl' unit is sometimes called a *fetal lobule*. It is suggested that, in the human, the maternal spiral artery at the decidual base of the placenta ejects its blood into the space that forms the bowl of this lobule, filling the brandy glass as it were. The pressure of the blood causes its circulation through the fenestrated wall

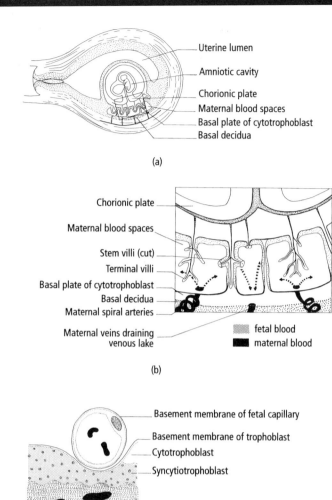

(a)

Uterine lumen
Amniotic cavity
Chorionic plate
Maternal blood spaces
Basal plate of cytotrophoblast
Basal decidua

(b)

Chorionic plate
Maternal blood spaces
Stem villi (cut)
Terminal villi
Basal plate of cytotrophoblast
Basal decidua
Maternal spiral arteries
Maternal veins draining venous lake

fetal blood
maternal blood

(c)

Basement membrane of fetal capillary
Basement membrane of trophoblast
Cytotrophoblast
Syncytiotrophoblast
Maternal blood space

Fig. 9.9 Schematic representation of structures of human placental interface. (a) Stem villi connect chorionic plate to basal plate forming a labyrinthine series of spaces. (b) From the stem villi and the chorionic plate, smaller villi ramify into the intervillous space forming a network of fine filamentous terminal villi, which are the principal sites of metabolic exchange. (c) At these sites, only a thin layer of chorionic syncytiotrophoblast separates the fetal blood vessels from maternal blood. Villi are classified as *primary*, when comprised of solid trophoblast, *secondary*, when mesoderm involves the villous core, and *tertiary* when blood vessels penetrate the mesoderm. As a villus grows and extends, it goes through each of those stages.

of the lobule over the fine terminal villi. The blood is then thought to drain back via the basal venous openings into maternal veins (Fig. 9.9b). Up to 200 such lobular units form in the mature human placenta, which is a 3 cm-thick 'pancake' of 15–20 cm in diameter. Several lobular units are grouped together to form a *lobe*, the boundaries of which may be seen grossly on the placenta, defined by fibrous septa.

Fig. 9.10 Low-power view of a cast made from the microvascular capillaries in the terminal villi on the fetal side of the placental circulation. Note how the terminal vessels form convoluted knots compared with the larger calibre vessels adjacent. The straight capillaries supplying the knots can be seen (arrow head), as can the terminal dilatations (arrowed) in which blood flow is slower and at which exchange of metabolites between fetal and maternal blood takes place.

This anatomical description is probably somewhat idealized, and the consistency with which such a regular arrangement of maternal vessels and villi occurs in the human placenta is perhaps questionable. The important point to understand is that maternal blood circulates across the fine terminal villi containing the fetal capillaries. The general anatomical organization of the placental interface is achieved by 3–4 weeks of pregnancy, and until recently, it was believed that the circulation of the maternal blood within the placenta was established at the same time. However, recent measurements on maternal blood flow and oxygenation at the placental interface suggest that, despite the anatomical competence, a fully mature *maternal* blood flow may not develop until 10–12 weeks. Thus, for the first third of human pregnancy the embryo is developing in an environment of relatively low oxygen compared with that achieved at later fetal stages.

The ovine placenta

The sheep placenta is *synepithelio-chorial* and *cotyledonary*, having some 80–90 independent sites of close vascular proximity. Attachment occurs at each uterine epithelial caruncle, the fetal chorion at the point of attachment showing specialization as a cotyledon. The fetal cotyledon and maternal caruncle together constitute the functional unit of the placenta known as a *placentome* (Fig. 9.11). In the mature cotyledon, an ingrowth of chorionic villi indents and compresses the epithelium of the maternal caruncles. Within the villi, a core of vascularized fetal mesoderm develops. Each cotyledon receives one to three branches of the umbilical vessels. The vessels divide and ramify within the villi where they come to lie under the surface of the trophoblast at the tips of the villi (Fig. 9.11). On the maternal side, the surface epithelium of the caruncle becomes syncytial, and the underlying stroma becomes acellular and vascular. Binucleate cells form in the fetal cotyledon, and some of these contribute to this maternal syncytium (an equivalent penetration into maternal tissues may occur in other large farm animals that are none the less appropriately classified as non-invasive at implantation itself). Tortuous, coiling maternal arteries supply each caruncle and split into capillaries between the penetrating terminal villi of the fetal cotyledon. The capillaries then run back along the long axis of the terminal villus towards the tip, where they drain into maternal veins (Fig. 9.11). This organization of fetal and maternal vessels means that, in the

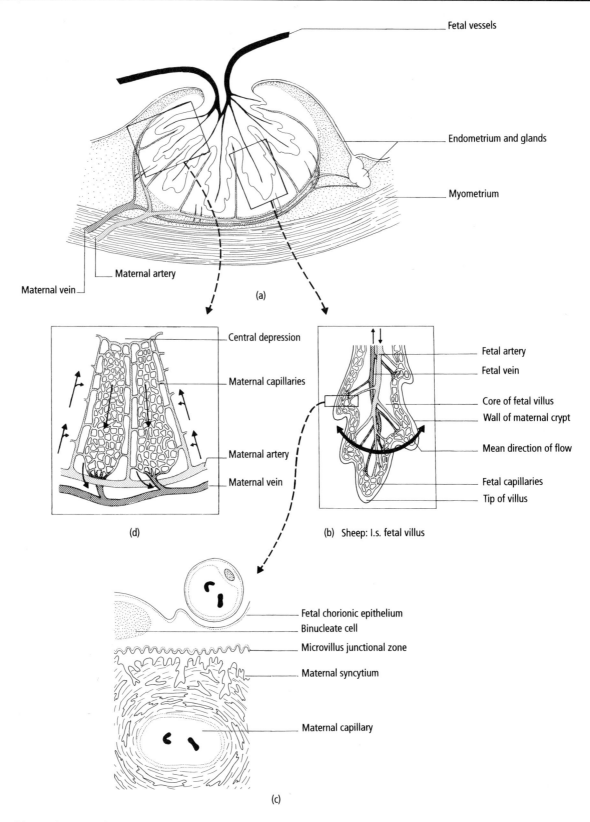

Fig. 9.11 (a) General structure of a sheep placentome. Fetal villi (shown in longitudinal section in (b)) interdigitate with maternal tissue giving close apposition of circulations (c). Maternal blood flows around the fetal villus, as indicated in (d). A similar interdigitation of fetal villi with maternal tissue also occurs in the horse and the cow.

capillary beds, the blood could flow in opposite directions (Fig. 9.11), which would maximize opportunities for metabolic exchange.

Blood flow in the placenta

Haemotrophic support is more efficient than histiotrophic support at establishing metabolic gradients to drive diffusional and carrier-mediated exchange of metabolic substrates and excretory products. The effectiveness with which gradients are established depends largely on how effectively rates of blood flow through the two sides of the placental circulation are regulated, as well as on the diffusional barriers and special transport mechanisms that might exist between them. The latter are considered in detail in Chapter 11. Here we are concerned with rates of blood flow and the factors affecting them.

Cardiac output is increased by up to 25% in pregnancy in response to the extra peripheral load. Maternal blood reaches the placenta via uterine and ovarian vessels and constitutes about 10% of the total cardiac output by the end of pregnancy. In the human, the uterine arteries course along the lateral walls of the uterus giving off nine to 14 branches, each of which penetrates the outer third of the myometrial tissue. At this level, anastomosis of these arteries with the ovarian arteries may occur, and from the anastomosis a series of arcuate arteries run within the anterior and posterior myometrial walls of the uterus, thereby encircling it. From this enveloping vascular network, radial arteries penetrate through the remaining myometrium into the basal endometrial tissue. Here the so-called basal arteries distribute spiral arteries to supply the endometrial decidua. This convoluted or spiral nature of terminal endometrial arteries, together with a tendency to dilate terminally and a lack of responsiveness to vasoconstrictor transmitters and drugs, is a feature common to many species. In addition, in higher primates, trophoblastic cells actively migrate into the eroded ends of the terminal branches of the spiral arteries (so-called *endovascular trophoblast*), partially obstructing them. Each of these features tends to diminish the arrival velocity of the maternal blood, which is further diminished to 0.1–10 ml/second in the primate by the extensive volume of the intervillous blood spaces.

This sluggish flow may protect the early conceptus from being dislodged by 'spurts' of blood, and also gives ample time for the exchange of metabolites at the placental interface (estimated mean transit time of 15 seconds in the full-term monkey placenta). A similar slowing of flow occurs on the fetal side of the circulation, where the total cross-sectional area of blood vasculature increases due to both the profusion of vascular branching and the terminal capillary dilatations (Fig. 9.10) at which exchange occurs. This massive expansion of the fetal blood vasculature means that the fetal circulation operates at low pressure. The protection of the fetal blood vessels from collapse is assured by a corresponding low maternal perfusion pressure (4–10 mmHg) within the intervillous spaces.

The maternal arteries have a sympathetic innervation restricted to their myometrial course, which enables them to constrict in response to sympathetic nerve stimulation or sympathomimetic drugs. Reduced placental perfusion can, therefore, result either from local vasoconstriction or from lowered systemic pressure. Transient reductions in perfusion pressure do not seem to have adverse effects on placental exchange or fetal growth, but chronic reductions do, particularly later in pregnancy. Thus, chronic anxiety, heavy smoking and stress during pregnancy result in smaller term babies, probably caused in part via effects on placental perfusion. Similarly, administration of drugs to relieve maternal hypotension, as, for example, in asthma, will cause an increase in visceral vasoconstriction, which will already be elevated reflexly, and thereby further reduce placental perfusion. During maternal exercise, some reduction in blood flow to the uterus occurs, but the conceptus itself seems to be relatively protected, unless exercise is severe and prolonged. In addition to the adverse effects of impaired flow towards the placenta, occlusion or slowed blood flow at the placental interface will also reduce the efficiency of metabolite exchange. Such an effect occurs in conditions of increased maternal blood viscosity, for example sickle cell anaemia, or after the expansion of placental villous structures, with a consequent reduction in the intervillous space available for circulation. This occurs in conditions of increased umbilical vein pressure, such as erythroblastosis or vascular occlusion of the fetal liver, both of which cause distension of the villi. Smoking in pregnancy also exerts direct effects on the fetal vasculature in the placenta, there being fewer, narrower and less convoluted capillaries in the terminal villi. Finally, occlusion of the maternal venous drainage, as can occur during compression of the inferior vena cava when lying supine or during uterine contractions at parturition, will reduce flow through the placental interface. Engorgement of the intervillous space can then, via pressure effects, reduce fetal blood circulation and the efficiency of placental exchange.

SUMMARY

The growing conceptus faces a hazardous few days of early life. Not only must it pass from oviduct to uterus at exactly the right time, but it must also establish adequate nutritional support for its growth and development. The very earliest days are marked by little growth at all, but the rising progesterone stimulates the secretory phase of the uterus to provide a nutritious fluid support. The superimposition of an oestrogenic stimulus allows attachment and implantation to occur. The conceptus can then secure a physical hold on the mother, establish its own circulation and stimulate, at adjacent endometrial sites, the establishment of special maternal circulatory changes. The transition from a histiotrophic to a more efficient haemotrophic route of metabolic exchange has been achieved via this placental structure.

However, a successful pregnancy requires more than this. The uterus will only provide a congenial environment as long as appropriate endocrine conditions persist. While the conceptus is establishing a secure physical and nutritional anchorage within the uterus, the luteal phase of the cycle is progressing. In some species, such as the dog, the luteal phase is of the same duration as pregnancy (Table 9.1). In others, such as the rat, rabbit and mouse, coitus is equated with possible pregnancy, and so the coital stimulus neurogenically activates a pituitary secretion of luteotrophic prolactin that prolongs the normal brief luteal phase of the oestrous cycle into a pseudopregnancy, which is about half the length of pregnancy (Table 9.1; Chapters 4 & 5). In primates and the large farm animals, however, pregnancy greatly exceeds the normal life of the corpus luteum (Table 9.1), and even in the rat, rabbit and mouse, the prolonged luteal phase is not as long as pregnancy itself. As we saw earlier, a uterus dominated by oestrogen and lacking progesterone is hostile to the conceptus. The progesterone-dominated state must therefore be extended in some way. The conceptus must somehow prevent or neutralize corpus luteum regression. This achievement is all the more remarkable when considering that, in species such as the pig and cow, attachment has not even occurred by the expected time of luteal regression (Table

9.1). How does the conceptus achieve this task? How does it signal its presence to the maternal organism? This is the subject of the next chapter.

FURTHER READING

Barnea ER, Hustin J & Jauniaux E (Eds) (1992) *The First Twelve Weeks of Gestation.* Springer Verlag, Berlin.

Clark DA (1993) Cytokines, decidua and early pregnancy. *Oxford Reviews in Reproductive Biology* **15,** 83–111.

Cupps PT (Ed.) (1991) *Reproduction in Domestic Animals,* 4th edition. Academic Press, London.

Das SK *et al.* (1994) Heparin-binding EGF-like growth factor gene is induced in the mouse uterus temporally by the blastocyst solely at the site of its apposition: a possible ligand for interaction with blastocyst EGF-receptor in implantation. *Development* **120,** 1071–1083.

Denker H-W & Aplin JD (Eds) (1990) *Trophoblast Invasion and Endometrial Receptivity.* Plenum Press, New York.

Flint APF, Renfree MB & Weir BJ (Eds) (1983) *Embryonic Diapause in Mammals.* Supplement 29. Journals of Reproduction and Fertility Ltd.

Johnson MH *et al.* (1986) A role for cytoplasmic determinants in the development of the mouse early embryo? *Journal of Embryology and Experimental Morphology* (now Development) (Suppl.) **97,** 97–121.

Lotgering FK, Gilbert RD & Longo LD (1983) Maternal and fetal responses to exercise during pregnancy. *Physiological Reviews* **65,** 1–36.

Pampfer S, Arceci RJ & Pollard JW (1991) Role of colony stimulating factor-1 and other lympho-haematopoietic factors in mouse preimplantation development. *BioEssays* **13,** 535–540.

Schultz GA & Heyner S (1993) Growth factors in preimplantation mammalian development. *Oxford Reviews in Reproductive Biology* **15,** 43–81.

Soares MJ *et al.* (1993) *Trophoblast Cells.* Springer Verlag, Berlin.

Steven DH (Ed.) (1975) *Comparative Placentation.* Academic Press.

Strickland S & Richards WG (1992) Invasion of the trophoblasts. *Cell* **71,** 355–357.

Wegner CC & Carson DD (1994) Cell adhesion processes in implantation. *Oxford Reviews in Reproductive Biology* **16,** 87–137.

Wilkinson AW (Ed.) (1976) *Early Nutrition and Later Development.* Pitman Medical, London.

Maternal Recognition and Support of Pregnancy

C O N T E N T S

If the implanting conceptus is to survive, it must signal its presence to the mother and prevent the withdrawal of progestagenic support that would normally occur with luteolysis. Moreover, in many species, that support must be prolonged not just briefly, but sustained over weeks or months. The mechanisms by which the endocrine support of pregnancy is first established and then sustained are the subjects of this chapter.

MATERNAL RECOGNITION OF PREGNANCY

We saw in Chapter 4 that normal luteal survival depended upon the balance between a positive luteotrophic complex and negative luteolytic factors which, in species other than primates, seem to be uterine prostaglandins. If luteal life is to be prolonged, then clearly either normal luteolytic factors must be neutralized or normally dwindling luteotrophic factors must be stimulated or supplemented. In practice, both mechanisms are probably at work.

The primate

The primate blastocyst prolongs the life of the corpus luteum by production of a luteotrophic factor of its own, which takes over from the inadequate support provided by the very low levels of pituitary luteinizing hormone (LH) present during the luteal phase (see Chapter 4). If blood samples from pregnant women 8–12 days post-fertilization are compared with those taken from women in the comparable period of a non-pregnant cycle, they are found to contain rising levels of the glycoprotein human chorionic gonadotrophin (hCG; see Table 2.4). hCG is synthesized in the syncytiotrophoblast of the implanting blastocyst from as early as 6–7 days after fertilization and is released to pass into the maternal circulation (Fig. 10.1a,b). It is carried to the ovary where it binds to LH receptors on the luteal cells to exert a luteotrophic action. This luteotrophic stimulus overcomes any tendency to luteolysis. The evidence for this effect of hCG comes from two types of experiment.

If non-pregnant women are given daily injections of hCG, starting in the mid-luteal phase, regression of their corpora lutea is prevented or delayed and progestagen levels remain elevated. Conversely, antibodies specific to the terminal amino acid sequence unique to the hCG β-chain (or indeed to CG derived from baboons, rhesus monkeys and marmosets, which make bCG, rhCG and mCG, respectively) will bind to CG but not to LH. Injection of this antiserum throughout the luteal phase of a pregnant human or monkey cycle, neutralizes CG production and luteal regression occurs on time (Fig. 10.1c). Female monkeys *actively* immunized to their

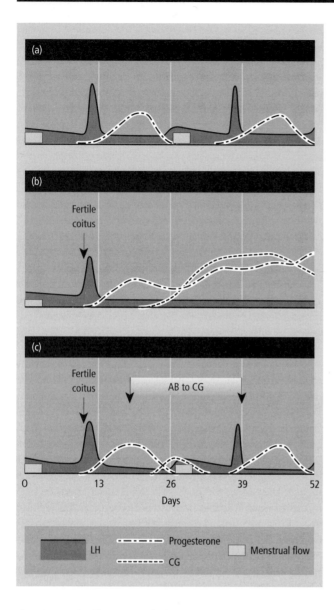

Fig. 10.1 Levels of hormones in the blood during two cycles of a higher primate. (a) Non-pregnant cycles. (b) Cycles in which fertile mating occurs. (c) Similar to (b) but passive administration of a highly specific anti-CG antibody is given (arrows); depression of CG and loss of pregnancy occurs but there is no effect on LH levels or cyclicity.

own CG, have normal LH levels and menstrual cycles, and fertile matings, but no pregnancies. These lines of evidence clearly implicate CG as necessary and sufficient for the prevention of luteal regression during early pregnancy in the primate.

Large domestic animals

In the large farm animals, luteolysis is normally induced

by uterine prostaglandins (see Chapter 4). We saw that in the absence of these prostaglandins, for example after hysterectomy, luteal life is prolonged or lasts indefinitely. In these species, the luteotrophic support by the pituitary is presumably therefore quite adequate. Not surprisingly, therefore, CGs have not been detected in the blood of pigs, cows and sheep bearing pre-implantation conceptuses. The pig conceptus does produce oestrogens from about Day 12 onwards, and oestrogens are a major luteotrophic agent in this species (see Table 4.4). It is therefore possible that oestrogens help prolong luteal life in the pig in the way that CGs do in primates. However, in the large farm animals the conceptuses appear primarily to *suppress luteolytic activity* by neutralizing the action of prostaglandin $F_{2\alpha}$ ($PGF_{2\alpha}$).

In pregnant animals, the levels of $PGF_{2\alpha}$ and its metabolites in the blood draining the uterus are reduced compared with the non-pregnant animal. As was seen in Chapter 4, luteal oxytocin provides the stimulus for $PGF_{2\alpha}$ release, but the endometrial oxytocin receptors only appeared towards the end of the luteal phase. In the presence of a conceptus, this luteal oxytocin receptor expression in uterine cells is found to be depressed, thereby preventing oxytocin stimulation. The receptor depression is caused by the action of *trophoblast interferon* (IFNτ; also called *trophoblastin* and *trophoblast protein 1* or *TP-1*), a peptide belonging to the type-1 interferon family of cytokines (Table 2.6). IFNτ secretion from mononuclear trophoblast cells is restricted to the 8–20-day-old bovine and ovine conceptus, precisely the period over which suppression of $PGF_{2\alpha}$ activity is required. In the pig, oestrogens from the conceptus act on the uterus to suppress $PGF_{2\alpha}$ release, as well as exerting a possible direct luteotrophic effect on the corpus luteum.

MATERNAL ENDOCRINE SUPPORT OF PREGNANCY

The successful establishment of pregnancy creates a new and extraordinary parabiotic liaison between mother and conceptus, which may last for a period of months in some species. In the human, in which pregnancy approximates to 9 months, its course is often described as lasting three *trimesters* (a trimester being 3 months). Over this period, the conceptus is totally dependent upon the mother for protection and nutrition. It subverts many of her metabolic and physiological activities to its own ends, so that there is adequate mobilization of oxygen, salts and organic precursors to supply its needs. One way in which it achieves this is by inducing and taking part in the formation of the vascularized placenta where the bulk of these substances is exchanged for its own metabolic waste. During pregnancy, the mother is also prepared for the future requirements of the fetus and neonate. Thus, hypertrophy of the uterine musculature, which partici-

pates in fetal expulsion at parturition, and the development and maintenance of mammary glands for postpartum lactational nutrition are both stimulated during pregnancy. This take-over of the mother's metabolism is controlled by the *pregnancy hormones*. In

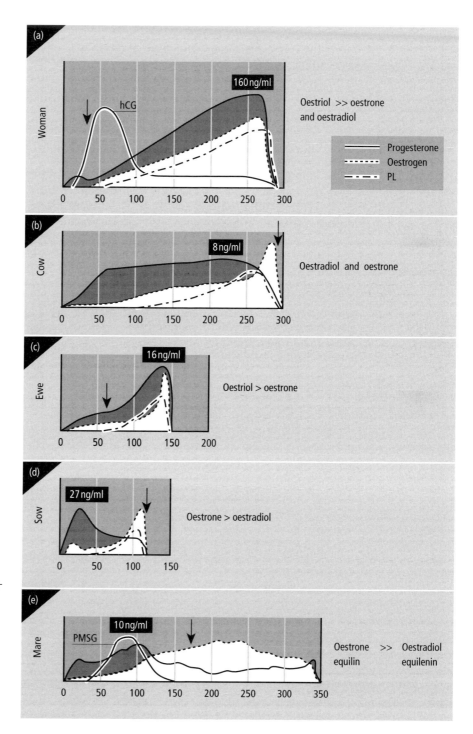

Fig. 10.2 Patterns of plasma hormones during pregnancy in various species. The maximal progesterone level is indicated in the black box and the principal oestrogens are recorded to the right of each panel. PL: placental lactogen; PMSG: pregnant mare serum gonadotrophin (which is also equine chorionic gonadotrophin). Pregnancy is timed in days. The vertical arrow indicates the time at which avariectiomy no longer terminates pregnancy. hCG, human chrionic gonadotrophin.

the remainder of this chapter, we will consider the nature and origin of these hormones. In Chapters 11–13, we look at how, where and when they achieve their effects.

As we have already seen, the extended secretion of progesterone is absolutely critical for the initiation of pregnancy. This absolute requirement for progesterone persists throughout pregnancy, and, indeed, in many species, progesterone levels in maternal blood rise continuously as pregnancy proceeds (Fig. 10.2). Moreover, as pregnancy advances, the level of oestrogens in the maternal blood also rises (Fig. 10.2). These steroid hormones appear to be important for the maintenance of pregnancy. Indeed, both oestrogens and progesterone may reach plasma levels many times greater than those seen in a normal luteal phase. In many species, the ovarian corpora lutea, which are under the control of the maternal pituitary, continue to secrete an essential proportion of these steroids. Removal of either the ovary or pituitary therefore results in abortion (see Table 10.1, group C). In other species (Table 10.1, groups A and B), pregnancy clearly does not depend totally on steroids

secreted by pituitary–ovarian interactions, as, at varying periods during pregnancy, one or both glands may be removed without inducing abortion. Where does the endocrine support come from in these species? Some of the clearest answers to this question have come from studies on human pregnancy, which shows the least dependence on the maternal ovarian–pituitary axis of any species. We will therefore examine the endocrinology of human pregnancy in some detail and discuss the evidence from other species in relation to the human pattern.

Human pregnancy

We saw in the first section of this chapter that within 2 weeks of fertilization the human conceptus synthesizes and releases the hormone hCG. This hormone then maintains the progestagenic activity of the corpus luteum. Within a further 2–3 weeks, the conceptus is also synthesizing all the steroidal hormones required for pregnancy, and although the maternal corpus luteum

Species	Duration of pregnancy (days)	Duration of non-pregnant luteal phase (days)	Day of pregnancy when hypophysectomy without effect	Day of pregnancy when ovariectomy without effect
Group A				
Human	260-270	12-14	?	40
Monkey (M. mulatta)	168	12-14	29	21
Sheep	147-150	16-18	50	55
Guinea-pig	60	16	3	28
Group B				
Rat	22	10-12	12	Term
Mouse	20-21	10-12	11	Term
Cat	63	30-60	Term?	50
Horse	330-340	20-21	?	150-200
Group C*				
Cow	280-290	28-20	?	Term
Dog	61	61	Term?	Term?
Pig	115	16-18	Term	Term
Rabbit	31	12	Term	Term
Goat	150	16-18	Term	Term

* NB These results do not mean that these species are solely dependent on pituitary and ovarian function. It is established that some are partially dependent on support from placental sources which are inadequate when acting alone (see Table 10.3).

Table 10.1 Dependence of pregnancy on maternal ovarian and pituitary function in various species

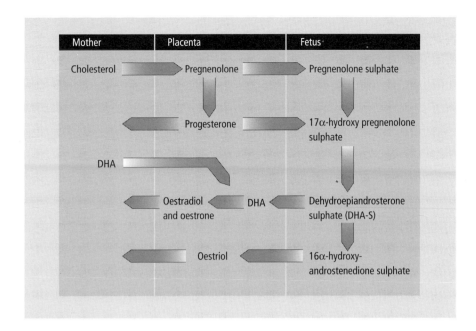

Fig. 10.3 Summary of principal routes by which the human feto-placental unit synthesizes oestrogens.

remains active for the whole of pregnancy, it can be dispensed with after only 4–5 weeks and plays only a trivial role in total progesterone output at later stages. *The human conceptus thus shows a remarkable endocrine emancipation.* It asserts complete control over the mother.

Our understanding of the role of the conceptus in steroidogenesis has come from several sources. For example, much has been learnt from observations on steroid output and interconversions in abnormal pregnancies, and from isolation of tissues from different parts of the normal conceptus and assessment of their steroid biosynthetic capacity *in vitro*. However, by far the most informative and important data have come from *in vivo* studies using the fetuses of induced pregnancy terminations. Direct sampling of umbilical and maternal blood, and the infusions of minute quantities of radio-labelled steroid precursors into the maternal, fetal or placental circulations, with subsequent analysis of their interconversions, has provided information of great clinical value about the sites of steroid synthesis and interconversion in the conceptus (Fig. 10.3).

Progesterone

The output of progesterone in late human pregnancy exceeds 200 mg/day and blood levels are high. Part of this rise in levels is accounted for by a three-fold increase in transcortin, which increases the proportion of bound progesterone in the blood (Table 2.9). The transcortin rise is, in turn, stimulated by a direct effect of oestrogens on the liver. Another consequence of the

transcortin rise is, of course, elevated cortisol levels during pregnancy. As the corpus luteum is not required for the production of this progesterone, from where does it come?

A clue comes from the observation that, in pathological pregnancies, in which an embryo fails to develop, progesterone output exclusively by the extra-embryonic components of conception is often only marginally less than in normal pregnancy. For example, *choriocarcinomata* or *hydatidiform moles* (malignant and benign tumours of the chorion, respectively) secrete progesterone in the absence of any embryonic tissue. Such observations suggest that the placental trophoblast itself is the principal source of progesterone. *In vitro* studies on the biosynthesis of progesterone by cultures of syncytiotrophoblast confirm this conclusion, and also indicate that the trophoblast can only use cholesterol (not acetate) as a substrate. The cholesterol is usually derived from the maternal rather than the fetal circulation. The rising output of progesterone through pregnancy appears to be completely autonomous; no external controlling mechanism has yet been discovered.

Clinical observations on patients whose ovaries have been removed at various times in early pregnancy indicate that the placenta is capable of synthesizing an adequate, supportive level of progesterone by 5–6 weeks. In the normal pregnant woman, there is a plateau, or even a slight fall, in the concentration of blood progesterone between 6 and 9 weeks. The plateau coincides with a marked fall in 17α-hydroxyprogesterone (an ovarian progestagen), and it is probable, therefore, that it is over this period of 6–9 weeks that the placenta normally takes over the major progestagenic support. Both pregnenolone and progesterone pass from the placenta into both the fetal and maternal circulations, concentrations in umbilical venous blood being high and infused radiolabelled progestins being distributed widely within the fetus. A similar universal distribution of progesterone to maternal tissues is also seen. The principal excreted metabolite of progesterone is *pregnanediol*, but it has been found that assays of urinary pregnanediol (some 15% of the progesterone produced) are not a useful indicator of fetal well-being and only a crude indicator of placental function. The failure to correlate progesterone output with fetal well-being is not surprising, as the fetus itself plays no part in progesterone synthesis. The poor correlation with placental function is mainly a result of the wide variation in levels of progesterone in plasma and of its urinary metabolites observed in normal pregnancies that makes its use predictively of little value.

Oestrogens

The conceptus is not only responsible for the output of progesterone during pregnancy; it is also the source of the high level of oestrogens (Fig. 10.2). The principal oestrogen in human pregnancy is not oestradiol 17β but the less potent oestriol (Table 10.2). Oestrogen output differs from that of progesterone, however, in that it is severely reduced in cases of choriocarcinoma and hydatidiform mole, in which a fetus is lacking. Moreover, although the incubation of placental tisssue with radiolabelled cholesterol or pregnenolone yields labelled progesterone, no labelled oestrogens are produced. Clearly, the placenta alone is inadequate for synthesis of oestrogens. From where then do they come? Could it be from the fetus itself?

When radiolabelled pregnenolone was infused into the circulation of isolated perfused human fetuses, little or no labelled oestrogens were produced. Therefore, the fetus alone cannot be the site of oestrogen synthesis. However, when labelled precursors were injected into an intact feto-placental unit, complete synthesis of oestrogens occurred. We can conclude from these observations that, in the human, the fetus and placenta cooperate to produce oestrogens. A series of observations on the result of infusing various radiolabelled steroid substrates has shown that, while the placenta is capable of synthesizing oestrogens from C19 androgens, notably DHA, it is *not* capable of synthesizing its own androgens from progestagens. The *fetal adrenal*, in contrast, *can* synthesize the C19 androgens (DHA and some androstenedione) but cannot aromatize them. These C19 steroids, which are made in a *special fetal zone* of the adrenal, must therefore pass to the placenta, which can then convert them to oestrogens. The principal pathways involved are summarized in Fig. 10.3. This cooperative organization is reminiscent of that observed between luteal and granulosa cells in the production of ovarian oestrogens (see Chapter 4).

Two very important points should be noted about this cooperative event. First, the initial step in the *fetal handling* of *all steroids* arriving from placenta or mother is *conjugation to sulphates*. This conjugation occurs mainly in the fetal liver but also in the adrenals. The conjugated steroid is then converted biosynthetically by the fetal adrenal to other steroid derivatives via sulphated intermediates. Conversely, the placenta takes conjugated steroids arriving from the fetus and deconjugates them, releasing free steroids into the maternal (and fetal) circulation. Thus, the *fetus sulphates*, and the *placenta desulphates*. This apparently bizarre manoeuvre is in fact extremely important, as conjugated steroids are both water soluble and biologically inactive. The conjugation may protect the fetus against untoward steroidal penetration into, and activity within, fetal tissues (e.g. masculinization of females by weak androgens, see Chapter 1). Thus, this strategy permits the high ambient levels of steroid required by the *maternal organism* for the maintenance of pregnancy to coexist with the low level of active steroids necessary to protect the *fetal organism* from unwanted side-effects. An X-linked placental sulphatase deficiency exists and is associated with deficiencies in the aromatization of C19 steroids to oestrogens, with a consequent maternal oestrogen deficiency.

Second, the placental synthesis of oestrone and oestradiol 17β can occur from DHA sulphate derived

Table 10.2 Plasma levels of various steroids in women during late pregnancy

Steroid	Values in pregnancy (ng/ml)	Values in luteal phase of cycle (ng/ml)
Progesterone	125–200	11
Oestriol	7	–
Oestriol conjugates	106 } 113	
Oestrone	7	0.2
Oestrone conjugates	46 } 53	
Oestradiol 17β	10	0.2
Oestradiol conjugates	5 } 15	

either from the fetal adrenal (40% but rising towards term) *or* from the maternal adrenal. Oestriol synthesis also occurs in the placenta but the 16α-hydroxylated substrate required comes *solely* from the *fetal* liver. It follows therefore that 16α-hydroxylated steroids, such as oestriol, should provide an indicator not only of *placental function* but also of *fetal well-being*. In pregnancies at risk, a consistent decline over a period of days of 16α-hydroxylated steroids correlates with fetal distress and may indicate that premature delivery should be induced. A parallel decline in oestradiol 17β is not seen because 50–60% of its synthesis by the placenta utilizes maternal not fetal DHA sulphate. In anencephalic fetuses, in which fetal adrenal function is impaired, the ratio of oestradiol and oestrone : oestriol is greatly increased, as would be expected. Thus, 16α-hydroxylated steroids can have a diagnostic value.

Corticosteroids

Blood levels of cortisol rise in pregnancy, and this rise can be mimicked by oestrogen injection into non-pregnant women. The elevated levels are partly due to a decrease in the metabolism of free cortisol, but predominantly due to an oestrogen-stimulated synthesis of transcortin from 3.5 mg% to 10 mg%. The transcortin binds both cortisol and progesterone, and this explains the very high level of total progesterone observed in human pregnancy.

Chorionic gonadotrophin (Table 2.4)

We saw earlier in this chapter that hCG was critical to the initiation of pregnancy by extending luteal life from 2 to 6–7 weeks when the placental steroids take over. The rising blood levels and critical role of hCG over this period make it a useful hormone to measure when undertaking pregnancy tests. The action of hCG may be limited to this short-term luteal support, as its blood levels only remain elevated for the first 8 weeks of pregnancy (Fig. 10.2). The fall in blood hCG concentrations correlates with the falling levels of 17α-hydroxyprogesterone (a luteal product) and the emancipation of the feto-placental unit from ovarian dependence. Low levels of hCG remain detectable for the remainder of pregnancy (although in the rhesus monkey rhCG levels are trivial from 40 days onwards).

The somatomammotrophins (Table 2.5)

As hCG levels decline, syncytiotrophoblastic secretion of two other protein hormones rises steadily. Human placental lactogen (hPL) and the placental variant of human growth hormone (hGH-V; Table 2.5) are produced from the end of the first trimester onwards, although blood levels of these hormones only rise during the last trimester. As plasma levels of hGH-V rise, those of pituitary hGH fall, suggesting that the placental variant is exerting a negative feedback control. It is not entirely clear what controls the production of either of these hormones.

In addition, release of pituitary prolactin is stimulated by oestrogens (see Chapter 5), and maternal plasma concentrations reach over 200 ng/ml by the last trimester. Prolactin is also synthesized by the decidual tissue. Some synthesis is detectable by the spontaneous decidual cells that appear towards the end of each luteal phase (see Chapter 9), but much larger amounts are secreted between 10 and 30 weeks of pregnancy. Progesterone is implicated in the control of its secretion. Little decidual prolactin appears in either the fetal or maternal circulation. Most of it seems to pass into the amniotic fluid where it may have a role in maintaining the water and electrolyte balance of the fetus (see Chapter 11).

Pregnancy in other species

Steroids

Elevated steroid levels are a feature of pregnancy in most mammals studied, although it is clear, despite incomplete data, that the patterns and absolute levels of steroids vary considerably from species to species (Fig. 10.2). As we saw in Table 10.1, the degree of independence from the pituitary–ovarian axis varies, but even in species, such as the cow and the pig, that require both pituitary and ovary to be present throughout pregnancy, a fetal steroid contribution also occurs (Table 10.3). Conceptuses of both the horse and the sheep become independent of the ovary by about one-third of the way through pregnancy. Both are comparable to the human with a fetal source of DHA sulphate being deconjugated and aromatized by the placenta. In the horse, the principal oestrogens formed are oestrone, together with two oestrogens found only in equids: *equilin* and *equilenin*. The fetal DHA sulphate used for aromatization, in the horse, is derived not from the fetal adrenal but from the interstitial tissues of the fetal gonads. The spectacular increase in weight of the equine fetal gonads between 100 and 300 days of gestation is due mainly to hypertrophy of interstitial glands. By birth, the gonads have regressed. The sheep fetus, like the human fetus, synthesizes sulphated DHA in the adrenal and maternal sources of DHA may also be available. In addition, the sheep placenta, unlike the human placenta, can undertake conversion of progesterone to oestrogens, and this property is of great importance at parturition (see Chapter 12).

Species		Progestagens	Oestrogens	Gonadotrophins
Human		Placenta	Fetal adrenal and placenta	Placenta (hCG and hPL)
Horse:	early	Corpus luteum	Ovarian follicles	Placenta (eCG)
	mid–late	Placenta	Fetal gonad and placenta	–
Sheep:	early	Ovary and placenta	Placenta (oestradiol and oestrone sulphate)	–
	mid–late	Placenta	Ovary (oestrone), placenta (oestrone sulphate)	Placenta (oPL)
Cow		Ovary (and placenta)	Placenta (and ovary)	Pituitary and placenta later (for bPL)
Pig:	early	Ovary	Placenta	Pituitary
	late	Ovary (and placenta)	Placenta	Pituitary

Table 10.3 Major sites of hormone synthesis during established pregnancy

Protein hormones

Protein hormones of pregnancy have also been detected in the sheep, cow, goat and horse. The latter half of sheep, cow and goat pregnancies is characterized by the production of ovine, bovine and caprine placental lactogens from binucleate cells of the chorion. However, these hormones are much more closely related to prolactin than hPL is to human prolactin, and so have less GH-like activity than does hPL.

Between days 40 and 120 of pregnancy, in the mare, a gonadotrophin with weak follicle stimulating hormone- and strong LH-like activity, called originally *pregnant mares' serum gonadotrophin* (PMSG), but now *equine chorionic gonadotrophin* (eCG) on the basis of its structural similarity to other CGs, is secreted from the trophoblastic cells of the endometrial cups. How its secretion is controlled is not known, although the quantity of its secretion is determined by the genetic constitution of the mare. One effect of the eCG is to promote follicular growth and even secondary ovulation in the maternal ovaries between days 40 to 120 of pregnancy; as a result, secondary corpora lutea may develop. These remain an active source of progesterone until the decline of eCG and the take-over by placental progesterone at around 140–150 days.

In many species, but notably in the guinea-pig and and pig, a cytokine called relaxin (a member of the insulin family; Table 2.6) has been detected in blood at low levels during pregnancy, rising just prior to parturition. This hormone appears to be produced by the corpus luteum of pregnancy, and has also been detected in humans. Its actions will be discussed in the context of parturition (see Chapter 12).

Summary

Three comments of general relevance need to be made when comparing the available data on pregnancy hormones. First, the tendency for the feto-placental unit to take over endocrine control from the mother in whole or in part is seen in most species. Second, while the levels of plasma oestrogens and progesterone rise, there is great species variation in the absolute level achieved (compare 160 ng progesterone/ml in the human with 8 ng/ml in the cow). Third, the temporal patterns of plasma steroids recorded through pregnancy differ markedly among various species. The large variation, both qualitative and quantitative, in plasma hormone levels is difficult to explain. This difficulty is compounded by the fact that we do not yet have a complete understanding of the functions of many of these hormones during pregnancy. These issues are addressed in Chapters 11–13.

FURTHER READING

Barnea ER, Hustin J & Jauniaux E (Eds) (1992) *The First Twelve Weeks of Gestation*. Springer Verlag, Berlin.

Ciba Foundation Symposium 62 (1979) *Maternal Recognition of Pregnancy*. Excerpta Medica (New Series).

Cupps PT (Ed.) (1991) *Reproduction in Domestic Animals*, 4th edition. Academic Press, New York.

Flint APF, Stewart HJ, Lamming GE & Payne JH (1992) Role of the oxytocin receptor in the choice between cyclicity and gestation in ruminants. *Journal of Reproduction and Fertility* Suppl. **45**, 3–58.

Fuchs F & Klopper A (Eds) (1977) *Endocrinology of Pregnancy*, 2nd edition. Harper Row, London.

Roberts RM, Farin CE & Cross JC (1990) Trophoblast proteins and maternal recognition of pregnancy. *Oxford Reviews in Reproductive Biology* **12,** 147–180.

Soares MJ *et al.* (1993) *Trophoblast Cells.* Springer Verlag, Berlin.

Steven DH (Ed.) (1977) *Comparative Placentation.* Academic Press, New York.

Webley GE & Hearn JP (1994) Embryo-maternal interactions during the establishment of pregnancy in primates. *Oxford Reviews in Reproductive Biology* **16,** 1–32.

CHAPTER 11

The Fetus and its Preparations for Birth

The fetus is not a quiescent, passively-growing product of conception tucked neatly away in its protected uterine environment. While undoubtedly dependent on the mother's nutrient supplies for its growth and survival, it none the less enjoys considerable independence in the regulation of its own development. Indeed, the fetus exerts profound effects on maternal physiology via hormones secreted by the fetal part of the placenta into the maternal circulation. These, in part, determine the mother's ability to meet the metabolic requirements of pregnancy. In addition, the fetus and mother must develop physiological mechanisms that anticipate the transition from a uterine to an external, independent existence at parturition. This extraordinary change of circumstance must be achieved while maintaining the internal environment of the neonate.

FETAL GROWTH

The pattern of fetal growth is determined primarily by the *genome of the fetus*, but additional fetal and maternal factors modulate the effects of its expression. *Insulin-like growth factors* (IGFs or *somatomedins*) produced by a large range of fetal cell types, but particularly by the fetal liver, provide a major endocrine stimulus to fetal growth. Their production is stimulated by placental lactogen. Fetal thyroid hormones also stimulate growth in the latter part of pregnancy, but growth hormone, in contrast to its postnatal role, and although produced by the fetus, is not effective in stimulating fetal growth.

Maternal nutrition and health are clearly of great significance, the effects of severe malnutrition on fetal well-being and neonatal survival being well known. Additional factors, such as parity (primiparous mothers have smaller babies than multiparous mothers), maternal size, multiple pregnancy (more than one fetus carried simultaneously), and self-inflicted damage, for example smoking (see Chapter 9), may all affect birth weight. It is reasonable to assume that growth is an indicator of fetal well-being. Babies of *low birth weight* (less than 2500 g) may be so either because they are born prematurely or, if born at full term, because of some retarding influence on fetal growth. As many as one-third of low birth weight babies come into the latter category and are said to be '*small for dates*' (defined as being of a weight that is two standard deviations below the weight expected of a baby of a particular gestational age).

The rate of fetal growth is relatively slow up to the 20th week of pregnancy but accelerates to reach a maximum around weeks 30–36, declining thereafter until birth (Fig. 11.1a). A postnatal peak in growth velocity occurs during week 8. Protein accumulation occurs early in fetal development to reach its maximum, about 300 g, by week 35, and precedes fat deposition, most of which is subcutaneous, and which only comes to exceed the weight of protein by week 38. By term, some three times as much energy is stored as fat than is stored as protein.

The relative amount of water in the fetus (95% in the young fetus) also decreases during development, as the proportion of solids increases. Amniotic fluid also increases in volume until week 34 of pregnancy, after which time it declines. The placenta increases in size slowly and steadily until birth. However, the rapid growth of the fetus before birth means that the ratio of placental weight to fetal weight falls significantly during the later stages of pregnancy (Fig. 11.1b). This observation has led to the suggestion that the placenta may limit the transport of nutrients to the fetus late in pregnancy and thereby underlie part of the fall in growth velocity after week 36.

The main ingredient of the fetal diet is carbohydrate, and about half the calories needed for growth and metabolism come from glucose, the remainder coming equally from amino acids and from lactate formed from glucose in the placenta. The fetus must also, if normal growth and development are to occur, be provided with the basic building materials: essential amino acids, fatty acids, vitamins and minerals. These are transported selectively from mother to fetus, mostly across the placenta, which provides a large and specialized surface area for transport, but transfer of some materials across the amniotic membranes also occurs. Thus, a considera-

tion of metabolic activity in the fetus requires an understanding of transport across the maternal–fetal interface.

MATERNAL METABOLISM, FETAL METABOLISM AND PLACENTAL TRANSPORT

The discrete nature of the maternal and fetal circulations, separated by cellular and acellular layers, confers an important barrier property on the placenta. Simple diffusional exchange between the circulations will only therefore be significant in the case of low molecular weight molecules, such as blood gases, Na^+, water and urea, or in the case of non-polar molecules, such as fatty acids and non-conjugated steroids. In contrast, hexose sugars, conjugated steroids, amino acids, nucleotides, water-soluble vitamins, plasma proteins, maternal cells, potentially infective agents, such as viruses and bacteria, and molecules, such as cholesterol (which are complexed, and transferred, in large lipoprotein particles), will not gain access to the fetal circulation unless either special transport mechanisms exist or the integrity of the barrier is breached. 'Bleeds' across the placenta do occur, but are rare (except at parturition) and probably occur mainly in a feto-maternal direction. However, recent evidence suggests that less traumatic erosions of the integrity of the trophoblastic layer at the materno-fetal interface may not be uncommon.

The identification of transfer routes between mother and fetus is difficult. The system is highly complex: both maternal and fetal components are dynamic and their physiology changes with the duration of pregnancy; the fetal, placental and maternal components may each utilize (or produce) the substance under study and so complicate quantitative measurements; there are several potential routes of transfer (amnion, chorio-allantoic placenta, yolk sac placenta), each of which may also have some heterogeneity in its transport systems. Studies have used radiolabelled markers and serial sampling in: an intact materno-feto-placental unit; a perfused feto-placental unit; isolated placentae, cultured placental fragments or trophoblastic cells *in vitro*.

The factors that will influence exchange between mother and fetus are: the thickness and organization of the tissues interposed between them; the maternal and fetal blood flows (already considered in Chapter 9); the fetal and maternal concentrations of the substances to be transported; and the types of transport mechanism available.

We saw in Chapter 9 that there is considerable interspecies variability in the microstructure of the layers separating maternal and fetal circulations (see Table 9.3 & Fig. 9.6). In addition, within a species, the facility with which diffusional exchange occurs varies during pregnancy. For example, in early pregnancy, terminal

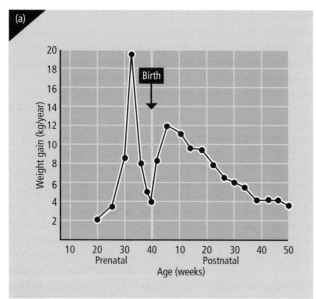

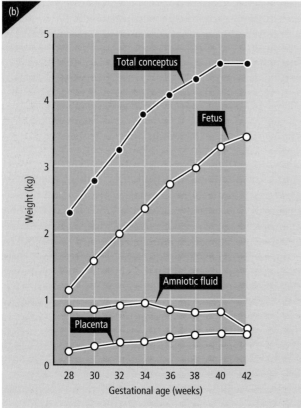

Fig. 11.1 (a) Change in velocity of growth in weight of singleton fetuses and children. (b) Weight changes of conceptus, fetus, placenta and amniotic fluid during pregnancy.

villi in the human placenta are large, with a diameter of 150– 200 μm, and the fetal vessel is centrally located beneath a 10-μm layer of syncytiotrophoblast. Thus, metabolites must diffuse a considerable distance between the two circulations. Furthermore, as the syncytiotrophoblast is itself metabolically active, for example synthesizing chorionic gonadotrophin, it will intercept and utilize some of the maternal metabolites, such as oxygen. As *pregnancy progresses*, the *villi thin* to 40 μm in diameter, and the *fetal vessel occupies a more eccentric position*, indenting the overlying syncytiotrophoblast to leave only a 1–2-μm layer separating it from the maternal blood in the intervillous space. This altered anatomical relationship not only increases the capacity for diffusional exchange but also reduces the consumption by the trophoblast of oxygen during its diffusional passage. A similar thinning of diffusional barriers also occurs in the synepitheliochorial placenta of sheep.

Although the interspecies variation in the numbers of layers separating the two circulations may influence the relative efficiency of diffusional exchange (e.g. diffusional movement of Na^+ across the placenta is considerably faster in the haemochorial primate than in the synepitheliochorial ungulate). The more freely diffusible molecules, such as O_2, are much more affected by blood flow than any consideration of placental barrier thickness. The important point to appreciate about the microstructure of the placental interface is that it either permits adequate diffusional exchange (i.e. has a large safety factor) or employs special transport systems to promote selective transport of less freely diffusible, but essential, substances, such as glucose, fructose, amino acids and various proteins. Both methods of placental transport will be affected critically by circulatory factors that were discussed in Chapter 9. Here we will now consider the transport mechanisms themselves, and the changes in maternal and fetal metabolism and physiology that determine the relative blood levels of metabolites, and thereby the gradients across the placental interface.

Oxygen and carbon dioxide

During pregnancy, maternal oxygen (O_2) consumption at rest and during exercise is increased compared with the non-pregnant female proportionate to the growing tissue mass of the conceptus. Physical working capacity and the efficiency with which work is performed are not significantly affected by pregnancy. Cardiac output does increase during the first third of pregnancy by

	Maternal blood Arterial	Fetal blood Venous	Umbilical artery	Umbilical vein
1 Oxygen tension (P_{O_2}) mmHg	90	35	15	30
2 Saturation O_2%	95	70	25	65
3 Oxygen content vol. %	14	10	5	13
4 CO_2 tension (P_{CO_2}) mmHg	30	35	53	40
5 pH	7.43	7.40	7.26	7.35

Table 11.1 O_2 and CO_2 composition of human maternal and fetal blood

about 30% but little more thereafter. Increases in both stroke volume and rate account for this increase. Mean arterial systemic blood pressure increases slightly, but the increase in output is mainly accommodated by a reduction of peripheral resistance, by as much as 30%, associated with the increasing demands of the conceptus (see also Chapter 9, 'Blood flow in the placenta'). Blood volume rises by up to 50% near term in humans, partly due to a 20–30% increase in erythrocytes and partly due to increasing plasma volume (up 30–60%).

Maternal *pulmonary ventilation increases by 40%* during pregnancy, possibly due to a direct effect of progesterone on respiratory mechanisms in the brainstem. A decrease of about 25% in maternal P_{CO_2} results, with a corresponding fall in bicarbonate concentration and a slight increase in pH. The reduction in bicarbonate buffering capacity leads to greater changes in pH with exercise than are observed in non-pregnant females. Oxygen is needed in relatively continuous supply by the fetus because fetal stores of the gas are very small: a 3-kg fetus near term requires 18 ml O_2/min, but stores are only 36 ml or 2-minutes worth. Oxygen, being a non-polar molecule, readily diffuses across the placental interface, as does the carbon dioxide (CO_2) generated by fetal metabolism, with its diffusion constant 20 times higher than O_2.

The gradients of the gases at the transplacental interface may be estimated from the figures shown in Table 11.1. Clearly the tension of O_2 in the fetal blood is low and that of CO_2 is high relative to the tension of the gases in maternal blood. Gradients to drive diffusional exchange therefore exist. However, inspection of Table 11.1 reveals that although the P_{O_2} in the oxygenated fetal venous blood leaving the placenta in the umbilical vein is relatively low, the O_2 saturation and content is not much less than maternal arterial blood. Thus, a much lower O_2 tension leads to a very similar O_2 content, indicating that the 'O_2 trapping' properties of fetal blood must be more effective than those of maternal blood. This higher affinity of fetal blood is shown graphically in Fig. 11.2. How is it achieved?

The earliest site of *erythropoiesis* in implanting mammalian embryos is the yolk sac mesoderm (Fig. 9.7). The primitive *embryonic erythrocytes* formed are nucleated, and are replaced later during embryogenesis by *fetal erythrocytes* (also nucleated) that are made in the liver. Finally, the spleen and bone marrow take over erythropoiesis as parturition approaches. This latter transition is regulated by the rising levels of fetal cortisol occurring towards term (see later). Embryonic and fetal erythrocytes, like those of the adult, contain haemoglobin as the O_2-carrying molecule. The embryonic and fetal haemoglobins consist of four globin chains coupled to a haem group; however, they differ from the adult in that the constituent globin chains are not two α and two β

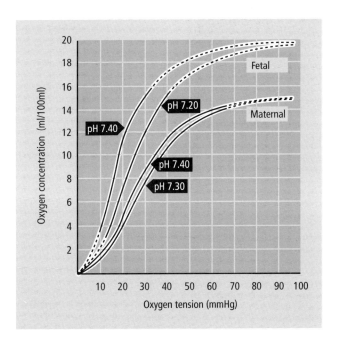

Fig. 11.2 The effect of varying oxygen tension on the oxygen content of human maternal and fetal blood under conditions likely to be encountered in the placenta. Note the greater oxygen concentration in fetal blood at any given P_{O_2}. Note also that the pH shifts occurring in the placenta (falling pH in maternal blood and rising pH in fetal blood) will further facilitate unloading of oxygen from maternal blood and its uptake by fetal blood.

chains. Rather, embryonic haemoglobin contains two ϵ and two ξ chains, while fetal haemoglobin contains two α and two γ chains. Each globin chain is coded for by a different gene, and during development a programme of gene switching occurs to yield the embryonic, the fetal and the adult sequences. The switch from fetal to adult haemoglobins, like the site of erythropoiesis, is also regulated by fetal cortisol. It is this difference in globin chain composition that results in the different oxygen binding curves (Fig. 11.2).

In adults, the highly-charged molecule 2,3-diphospho-glycerate (DPG) can bind to sites on the β chain that are exposed only on deoxygenation. This binding stabilizes the deoxyhaemoglobin, thereby reducing the opportunity for O_2 binding. Thus, a higher Po_2 is needed to load haemoglobin with O_2. The β chain is lacking in embryonic and fetal haemoglobins, and its equivalents (the γ and ϵ chains) are relatively insensitive to DPG. Therefore, at any given Po_2, fetal and embryonic haemo-globins will bind more O_2 than will maternal haemoglo-bin. In some species, the fetal DPG levels may also be lower, which will further assist the O_2 loading process.

In addition, as Fig. 11.2 shows, a *double Bohr effect* occurs in the placenta to facilitate transfer of about 10% of the O_2 in a fetal direction. Thus, the fall in pH of maternal blood due to uptake of fetal CO_2 drives release of maternal O_2, and the rise in fetal pH due to the removal of its CO_2 facilitates uptake of O_2. Under non-pathological conditions the maternal Po_2 and placental perfusion are unlikely to be limiting, and thus the rate of transfer of O_2 across the placenta will vary simply with the Po_2 of the fetal blood in the umbilical circulation. Thus, *the level of fetal oxygenation will be regulated simply by the fetal requirement for O_2*.

Glucose and carbohydrate

The fetus has little capacity for gluconeogenesis largely because the necessary enzymes, although present, are inactive at low arterial Po_2. At birth, when arterial Po_2 rises, gluconeogenesis is initiated. The fetus must there-fore obtain its glucose from maternal blood, in which glucose levels depend on both maternal nutritional status and the integrated endocrine control mechanisms, which maintain free plasma glucose levels within narrow limits. Thus, the secretion of insulin from the maternal pancreas prevents glucose levels from rising too high by increasing glucose utilization for glycogen and fat synthesis and storage. Conversely, absorption of glucose from the gut and gluconeogenesis (by the util-ization of glycogen stores under the action of cate-cholamines, corticosteroids, glucagon and growth

hormone) help prevent maternal glucose levels from falling.

Early on in gestation, progesterone increases maternal appetite and stimulates the deposition of glucose in fat stores. Later in pregnancy, human placental lactogen, via its growth hormone-like activity, assists in the mobiliza-tion of fatty acids from these fat 'depots' for use by the mother as an energy source. This source of fatty acids can be important for maternal metabolism because, as pregnancy proceeds, maternal tissues become progress-ively less sensitive to insulin. The cause underlying the *insulin insensitivity* is uncertain, but placental lactogen activity has been implicated. The consequences of the insensitivity are two-fold.

First, *latent diabetes mellitus* in women may manifest itself overtly for the first time from mid-pregnancy onwards. Second, maternal blood glucose, being taken up less by maternal tissues, is more available for placen-tal transfer to the growing fetus. This transfer occurs by facilitated diffusion, in which concentration gradients drive the transport of D-glucose via specific carrier mechanisms, thereby enhancing the rate of transfer (maximum rate of flux is 0.6 mmol/min/g placenta). The levels of glucose in the fetus will be related simply and directly to those in the mother, as the carrier system saturates only at supra-physiological, maternal serum concentrations (*c.* 20 mmol/l). The placenta also utilizes a considerable amount of the 'in transit' glucose meta-bolically, and generates lactate, some of which is distrib-uted to the fetal circulation at about one-third the equivalent flux of glucose. Thus, it contributes significantly to fetal metabolism. Under normal condi-tions there is little *direct* flux of lactate from mother to fetus.

The rate at which glucose is utilized by growing fetal tissues is probably determined largely by the actions of fetal insulin secreted by its own pancreas in response to glucose load. If this secretion is high, as can occur, for example, in maternal diabetes mellitus, fetal growth and fat storage are promoted and an overweight neonate results. It would seem, then, that the rate of glucose util-ization, rather than glucose availability, is the primary determinant of fetal growth under normal nutritional conditions.

The storage of glucose as glycogen, particularly in the fetal liver, is important if the metabolic needs of the neonate are to be provided until feeding begins. A pro-gressive increase in the activity of the fetal adrenal cortex near term (see below) is especially important in promoting the deposition of liver glycogen. Indeed, fetal adrenal hypoactivity is associated with major reductions in liver glycogen stores, a situation reversed by cortico-steroid replacement, while exogenous adrenocorticotrophic hormone (ACTH) or corticosteroids enhance liver glycogen content in normal fetuses.

The high concentration of glycogen in fetal cardiac muscle probably explains why the heart can maintain its contractile activity in the face of severe hypoxia. By contrast, the brain has no glycogen stores and therefore relies totally on a supply of glucose from the circulation. Thus, we see why hypoglycaemia can have such deleterious effects on the brain, especially if accompanied by hypoxia, the risk being higher in a fetus with low cardiac glycogen reserves (as occurs through glucoprivation accompanying placental insufficiency). Postnatal hypoglycaemia may also be seen in babies born to a diabetic mother, as the reduction in the maternal supply of glucose at parturition is not immediately accompanied by a corresponding cut in the fetal hypersecretion of insulin. A sharp drop in neonatal blood glucose results. Fortunately, although neonatal hypoglycaemia is potentially a major cause of brain damage it is highly treatable.

The storage of glucose as fat in the fetus is regulated primarily by insulin. Thus, when glucose levels are maintained optimally, such as occurs when maternal nutrition is good, the glucose available after the requirements for growth are fully met is diverted by fetal insulin into fat stores. In conditions where glucose supplies to the fetus are reduced, the needs of growth have priority over storage, glucose is released from fetal stores, and the newborn consequently has an emaciated appearance.

In addition to these storage depots of white fat, brown adipose tissue or 'brown fat' is also found in the fetus, newborn and infant. It is deposited in five sites: (1) between the scapulae in a thin diamond shape; (2) small masses around blood vessels in the neck; (3) in the axillae; (4) in the mediastinum between the oesophagus and trachea, as well as around the internal mammary vessels; and (5) a large mass around the kidneys and adrenal glands. It is different in structure to white fat, the lipid being distributed multilocularly with, in close apposition, large numbers of mitochondria bearing prominent cristae. These deposits of brown fat are of immense importance in temperature regulation, having the ability to generate large quantities of heat, independent of other mechanisms, such as increased muscular movement and shivering. Indeed, this form of heat production is termed *non-shivering thermogenesis*. As neonatal development proceeds, brown fat becomes of less importance thermogenically, and regresses.

Amino acids and urea

Amino acids in the adult are derived directly from digestion of both dietary and endogenous protein, and in the case of the non-essential amino acids, by interconversion from other amino acids. Deamination of amino acids during catabolism results in release of ammonia, levels of which are kept low by hepatic conversion to urea, which constitutes the major source of urinary nitrogen excretion. Traditionally, protein supplements have been considered an important and desirable feature of pregnancy diets, to cope with the increased protein synthetic demands of the growing conceptus. However, except in cases of extreme malnourishment, there is little or no evidence to support this view, and protein supplements appear largely to be used for conversion as energy sources.

As the fetus clearly does grow and thereby increases the total protein content of the pregnant mother, where do the amino acids required come from? There is no evidence for improved maternal digestion of dietary protein (already exceeding 95% in the non-pregnant state). However, the efficiency of the intermediary metabolism of amino acids does appear to increase. Thus, urea excretion falls markedly in pregnancy, suggesting that the same intake of dietary amino acids is being utilized more efficiently. A reduced capacity of the maternal liver to deaminate amino acids can be detected during pregnancy, and results largely from the anabolic action of progesterone. Therefore, the human conceptus, via its production of progesterone, regulates maternal metabolism of amino acids such that no extra dietary protein intake is required to support fetal growth. The 'extra' amino acids retained in the mother are transported actively to the fetal circulation, and fetal urea, produced by the limited catabolism of fetal amino acids, diffuses passively into maternal blood with its already lowered endogenous urea levels.

Water and electrolytes

During pregnancy, the elevated progesterone binds competitively to the maternal renal aldosterone receptor but does not activate it, leading to natriuresis. A compensatory ten-fold increase in aldosterone is stimulated, an exaggerated version of the luteal aldosterone rise observed in the normal menstrual cycle (see Chapter 7). In addition, oestrogens stimulate angiotensinogen output by the liver four- to six-fold, further stimulating aldosterone output. The net effect is an increase in maternal sodium and water retention.

Exchange of water between mother and fetus occurs at two main sites: the placenta, and the remainder of the non-placental chorion where it abuts the amnion internally (Figs 9.7–9). The relative quantitative contributions of each route, particularly early in pregnancy, is unclear, although the placenta is suspected to be the main site of exchange. Most studies in humans and in animals indicate that both amnion and chorion are freely permeable to water molecules. There is no evidence to suggest that water crosses by active transport or is secreted by the membranes themselves; so either diffusional or hydrostatic fluxes must account for its transfer. However, a large hydrostatic pressure difference between maternal and fetal circulations within the placenta cannot occur, as, if large in the materno-fetal direction, the fetal vessels would collapse and fetal–placental exchange would be impaired, while, if large in a feto-maternal direction, the fetus would dehydrate. Most likely, then, small or intermittent hydrostatic gradients are responsible for moving the large amounts of water necessary for the fetus.

There is a large measurable traffic of sodium and other electrolytes across the placenta in both directions. Much of this flux is in association with energy dependent co-transport of other molecules via a transcellular route, but the evidence that this route is used primarily for ion accumulation, or even contributes significantly to net flux, is not secure, at least for humans. As ions can also diffuse, and so equilibrate, via a paracellular route, it is probable that the net flux is mainly diffusionally driven.

Significant quantities of water, sodium and other electrolytes may also cross the amniotic membranes. There is some evidence that the high concentration of decidual prolactin in the amniotic fluid of late pregnancy may assist this exchange.

Iron

Iron is present in both fetal and maternal blood in both an unbound form and bound to the protein transferrin. However, fetal blood contains iron at two to three times the concentration of maternal blood, and as the iron-binding capacity of fetal and maternal blood are similar, the higher fetal serum iron concentration is due to an increased concentration of unbound iron, which accumulates through active transport across the placenta. Trophoblast cells contain intracellular iron as a ferritin complex, which is involved in iron transport. In pregnancy, there is a high incidence (40–90%) of maternal iron deficiency based on serum iron levels, although anaemia is less frequent. Additional iron in pregnancy is required to replace an average loss of 300 mg to the fetus, 50 mg to the placenta and 200 mg in blood loss after labour. In addition, about 500 mg is required to increase the maternal haemoglobin mass, but this amount is not ultimately lost. The net need, therefore, is about 550 mg. However, absorption is enhanced from 10% in the first trimester to 30% or more in the third. A

food intake of 12 mg/day and 10% absorption would provide an estimated 335 mg, and such a diet would be sufficient for the majority of pregnant women, but, in general, a supplement of 100–150 mg/day iron, especially in the second half of pregnancy, is recommended.

Calcium

The fetus places a considerable demand for calcium on the mother, largely during ossification in the last trimester. However, the average daily maternal intake of calcium appears greatly to exceed adequacy; there is a positive calcium balance throughout pregnancy partly due to the more efficient absorption of dietary calcium from the intestine. This efficiency increases further during lactation and is a result of higher levels of parathormone stimulating conversion by the kidney of vitamin D to the active derivative 1α, 25-dihydroxy-vitamin D_3. Lack of vitamin D during pregnancy is a more likely cause of maternal osteomalacia than is dietary deficiency of calcium.

Calcium (and phosphate) is transferred to the fetal circulation against a concentration gradient by a saturable active transport mechanism that is sensitive to metabolic and competitive inhibitors.

Vitamins

Folic acid and vitamin B_{12} are two essential compounds obtained from the diet that have very fundamental metabolic actions, making them vital for normal fetal development. They are provided for the fetus at the expense of maternal stores, so making fetal deficiency unlikely. However, vitamin deficiencies in the mother may affect the fetus indirectly via the resultant maternal metabolic disorders. It is perhaps pertinent to remember here that folic acid and its co-enzyme forms are involved in 1-carbon transfers, and thereby in nucleoprotein synthesis and amino acid metabolism: processes that are vital for normal fetal development.

Vitamin B_{12} is a co-factor in folate metabolism and in the metabolism of some fatty acids and branched amino acids. In pregnancy, serum folate levels decrease progressively towards term, and, in megaloblastic anaemias of pregnancy, they are generally lower than in the folate-deficient anaemias of non-pregnant women. Red cell folate activity accounts for 95% of the blood folate, levels falling progressively from the first to third trimester. The incidence of folate deficiency varies considerably according to the type of population studied, but it is of the order of 2%. The incidence of megaloblastic anaemia is much lower. The precise consequences of folate deficiency in the fetus have not been established definitively, but appear to be associated with prematurity and abortion. Folic acid supplements are generally considered to be desirable in pregnancy as a means of preventing the development of maternal anaemia.

Vitamin B_{12} is absorbed relatively slowly across the mucosa of the terminal ileum to be transported in blood mainly by a protein, transcobalamin II. Maternal serum levels of vitamin B_{12} decrease during pregnancy, falling to a minimum at 16–20 weeks, although this does not indicate a true deficiency, which is rare in pregnancy; indeed, vitamin B_{12}-deficient women are unlikely to become pregnant in the first place.

Bilirubin

Bilirubin is a lipid-soluble product of haemoglobin catabolism and is present in fetal, neonatal and adult plasma in both a free form and bound to serum albumin. Bilirubin passes readily across the placenta in either direction by diffusion. In the mother, bilirubin is transported in the blood to the liver where the enzyme uridine diphosphate (UDP) glucuronyl-transferase converts it to bilirubin glucuronide, and this polar conjugate cannot cross the placenta but is excreted in the maternal bile (Fig. 11.3). The fetal liver lacks the conjugating

enzyme, which is first synthesized late in pregnancy as one of the last clusters of fetal liver enzymes, so preparing the neonate for its own bilirubin excretion. Thus, the diffusion of the non-polar bilirubin from fetus to mother, and her hepatic conjugation activities preventing its return to the fetus, ensure adequate elimination of fetal bilirubin (Fig. 11.3).

Bilirubin transport and metabolism is important clinically. *Hyperbilirubinaemia* is common in the newborn and in its mild form (so-called 'physiological hyperbilirubinaemia') is known as *jaundice* because of the associated yellow coloration of the skin and mucous membranes. A number of factors may be associated with its occurrence, such as accelerated red blood cell breakdown in cases where an infant has a large blood volume, as can arise if the umbilical cord is clamped too late, when some 30% of babies may show jaundice. Or the effectiveness of bilirubin conjugation by the neonate may be depressed as a result of dehydration, low caloric intake or even the actions of steroid hormones, particularly progestagens, which actually inhibit conjugation.

Severe, or pathological hyperbilirubinaemia is particularly dangerous as it can result in *encephalopathy* (also known as *kernicterus*). This severe hyperbilirubinaemia may arise as a result of much increased fetal red cell breakdown, such as occurs if the mother develops an immune response to fetal erythrocytes (see later in this chapter). Impairment of maternal bilirubin conjugation

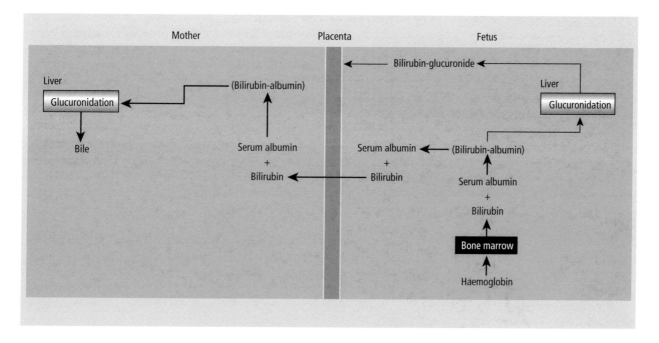

Fig. 11.3 The metabolism of bilirubin. The heavy arrows represent the principal route of metabolism during pregnancy. Fetal glucuronidation does not occur significantly until after birth.

can also give rise to bilirubin toxicity. Damage to the fetal liver as a result of injection of drugs, maternal diabetes, congenital defects and prematurity may all be associated with poor conjugation of bilirubin. The hyperbilirubinaemia (or its consequences) may be prevented by a number of approaches: exposure to ultraviolet light (phototherapy) causes the breakdown of bilirubin to a non-toxic product; early feeding prevents hypoglycaemia and dehydration; early clamping of the umbilical cord decreases red cell volume and hence bilirubin plasma levels; if the above fail, or in severe cases, exchange transfusion is used.

AMNIOTIC FLUID

The composition and turnover of amniotic fluid has become a subject of renewed interest with the advent of the sampling procedure, *amniocentesis*, in the diagnosis of fetal abnormalities. The volume of amniotic fluid increases during pregnancy from about 15 ml at 8 weeks post-conception to 450 ml at week 20, after which time net production declines to reach zero by week 34. The composition of amniotic fluid (Table 11.2) suggests that it is a dialysate of maternal and/or fetal fluids, save that the concentration of protein is only 5% that of serum. During the last third of pregnancy, total solute concentration in amniotic fluid falls, while the concentrations of urea, uric acid and creatinine increase. Amniotic fluid is in a dynamic state, complete exchange of its water component occurring every 3 hours or so.

There are several routes by which water and solutes enter and leave the amniotic cavity. Exchange with the fetus occurs via its gastrointestinal, urinary and respiratory tracts, and, until week 20 or so, the skin of the fetus is not keratinized and so amniotic fluid (osmolarity 260–280 mosmol/l) exchanges freely with fetal extracellular fluid. In addition, the amniotic epithelium, which during the later stages of pregnancy expands to fill the extra-embryonic coelom and becomes apposed to both chorion and umbilical cord (see Fig. 9.8), is also a route of exchange. Indeed, the finding that removal of the fetus does not prevent formation of amniotic fluid in the rhesus monkey emphasizes the capability of the amnion in this context.

It is clear that, during early life, the fetus swallows from 7 ml of amniotic fluid per hour at week 16 to around 120 ml per hour at week 28. Moreover, radio-opaque dye injected into amniotic fluid becomes concentrated in the fetal stomach, presumably because after swallowing, most of the water is absorbed by the fetus. Thus, the fetal gastrointestinal tract is a pathway for removal of amniotic fluid from the amniotic cavity.

The fetal lungs produce a fluid that fills the alveoli and also contributes to amniotic fluid, as surfactant compounds (see below) are found within it during late pregnancy. Micturition also occurs into the amniotic cavity and, undoubtedly, the fetal urinary system plays an important role in maintaining amniotic fluid volume. It has been estimated, using ultrasound techniques, that, at 25 weeks, some 3–5 ml/hour of hypotonic urine is formed, rising to 26 ml/hour (500–600 ml/day) by week

Table 11.2 Compositions of amniotic fluid in early and late pregnancy and the full-term maternal and fetal serum

Fluid	Total osmotic pressure (mosmols)	Na (mM)	Cl (mM)	K (mM)	Non-protein nitrogen (mg/ 100 ml)	Urea (mg/ 100 ml)	Uric acid (mg/ 100 ml)	Creatinine (mg/ 100 ml)	Protein (gm/ 100 ml)	Water content (%)
Amniotic fluid First and second trimester	283	134	110	4.2	24	25	3.2	1.23	0.28	98.7
Amniotic fluid Third trimester	262	126	105	4.0	27	34	5.6	2.17	0.26	98.8
Maternal serum Full term	289	137	105	3.6	22	21	–	1.55	6.5	91.6
Fetal serum Full term	290	140	106	4.5	23	25	3.6	1.02	5.5	–

40, after which time it drops rapidly. The relative importance of these pathways for exchange at various times of gestation has not, however, been determined, although the fetal kidney would seem to be a principal source of amniotic fluid later in pregnancy. Renal agenesis causes *oligohydramnios* (insufficient amniotic fluid, 'Potter's syndrome'), while excessive accumulation of amniotic fluid, *hydramnios*, is associated with impaired or no swallowing, for example in anencephaly or oesophageal atresia.

The diagnostic value of amniotic fluid can be considerable. For example, the glycoprotein α-*fetoprotein* is normally found in very low concentrations in amniotic fluid. However, if the fetus has a defect in neural tube formation, as occurs in *spina bifida* or *anencephaly*, the concentration of α-fetoprotein is elevated markedly. The recovery of fetal cells in amniotic fluid makes it possible to scan the karyotype of the fetus for gross chromosomal abnormalities (e.g. Down's syndrome) and assessment of fetal sex. It is also possible to identify various genetic disorders by use of recombinant cDNA probes that recognize DNA sequences restricted to chromosomes carrying mutant genes (e.g. those associated with Duchenne muscular dystrophy, phenylketonuria or haemophilia). The recovery of amniotic fluid at around 16–20 weeks for these diagnostic tests is not without risk of fetal damage, abortion or rhesus sensitization. However, where an appreciable risk of fetal abnormality exists, as, for example, in mothers over 39 years of age, a parent with a balanced translocation, exposure of the mother to potential teratogens, or prior evidence of familial risk, then an offer of amniocentesis with the possibility of therapeutic abortion is becoming routine.

DEVELOPMENT OF FETAL SYSTEMS AND THEIR MATURATION FOR POSTNATAL LIFE

The fetus must exist both for the present and for the future. The physiological basis of life within the maternal environment is very different from that to be experienced after parturition. The fetus must therefore develop mechanisms that anticipate this change and gear its own metabolism to adapt more or less instantaneously. As has already been indicated, the rising fetal output of corticosteroids towards term plays an important part in orchestrating the maturation of fetal physiology as well as its timing. Other critical anatomical and functional changes also occur. In this section, these various adaptations to uterine and then extra-uterine life are considered.

The cardiovascular system

The fetal circulation differs from that in the adult because the placenta, not the lung, is the organ of gaseous exchange. The adaptations that achieve this end are ingenious: the two fetal ventricles pump in parallel, not in series as in the adult, and there are several vascular shunts that divert the fetal circulation away from the lungs and towards the placenta. These adaptations are exquisitely designed such that conversion to the adult form of circulation is initiated instantaneously at the first breath taken by the newborn. The fetal circulation is shown diagrammatically in Fig. 11.4(a). Oxygenated blood returns from the placenta and is carried into two channels. The larger of these is the *ductus venosus*, a fetal shunt that bypasses the hepatic circulation and delivers blood directly into the inferior vena cava. The smaller channel perfuses the liver and enters the inferior vena cava through the hepatic veins. The inferior vena cava carries blood to the right atrium where it is split into two streams by the *crista dividens*, the free edge of the interatrial septum, which projects from the foramen ovale. The larger stream passes through the foramen ovale, another fetal shunt, into the left atrium, thereby avoiding the pulmonary circulation. The smaller stream continues through the right atrium, as does blood returning from the head region via the superior vena cava and also returning coronary blood. This blood flows into the right ventricle and out through the pulmonary artery. Thereafter, it also is split into two channels: the largest passing through yet a third fetal shunt, the *ductus arteriosus*, which carries blood to the aorta, while the smaller channel conveys blood to the fetal lungs. The small amount of poorly oxygenated blood passing through the pulmonary circulation returns to the left side of the heart.

The combined cardiac output consists of about one-third from the left and two-thirds from the right ventricle. The three fetal vascular shunts, the ductus venosus, foramen ovale and ductus arteriosus, combine in function to ensure the optimal distribution of oxygenated blood to the head and body. It is instructive to examine the quantitative aspects of this process. Blood leaving the placenta in the umbilical vein is 90% saturated with O_2. Most of it is shunted past the liver to join with blood only poorly oxygenated (20%) in the inferior vena cava, the resultant mix reaching the heart being 67% saturated. The blood shunted through the foramen ovale to the left atrium is joined by O_2-poor blood returning from the lungs. The result is blood 62% saturated, which leaves the heart, via the brachiocephalic

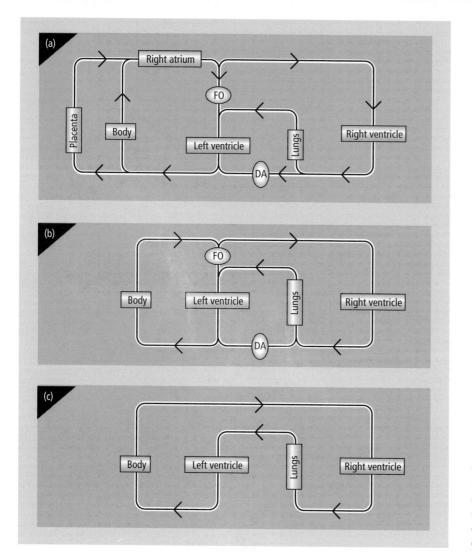

Fig. 11.4 Highly schematic representation of the circulations of: (a) the fetus; (b) the neonate; and (c) the adult. The transitory form in (b) occurs through the occasional opening of fetal shunts prior to their complete, anatomical closure. DA: ductus arteriosus; FO: foramen ovale.

artery, to supply, in large part, the head region. The remainder of the O_2-rich blood from the inferior vena cava enters the right atrium to mix with poorly oxygenated blood (31% saturated) returning from the head region via the superior vena cava. The blood thereby entering the pulmonary artery via the right ventricle is 52% saturated with O_2, and the largest proportion of it is shunted through the ductus arteriosus to join O_2-rich blood in the aorta to yield an O_2 saturation of 58% in the descending aorta. The effectiveness of the ductus arteriosus shunt results largely from the high pulmonary vascular resistance due to constriction of pulmonary arterioles in response to the low fetal O_2 tension (20–25 mmHg, compared with 80–100 mmHg in the adult).

The changes in the fetal circulation at birth involve closure of the three fetal shunts, thereby replacing the placental circulation with a pulmonary circulation (Fig. 11.4b,c). With the obliteration of the umbilical circulation, the ductus venosus ceases to carry blood to the heart. At the same time, there is a dramatic fall in pulmonary vascular resistance due to inflation of the lungs with the first breath and the rise in pulmonary Po_2. Thus, overall, there is a net drop in pressure on the right side of the heart (with a loss of umbilical input and rise in pulmonary outflow) and a rise in pressure on the left side (with a return of pulmonary venous blood). This pressure imbalance leads to a brief reversal of blood flow through the ductus arteriosus, the muscular wall of which responds to the elevated Po_2 of neonatal blood by contracting. The foramen ovale has a flap valve over it in the left atrial chamber. In the fetus, this is maintained open by the stream of blood from the right atrium, but with the reversal of interatrial pressure, the flap is pressed against the interatrial wall, thereby separating the two sides of the heart to yield two pumps working in series. There is a recognizable, transitory form of circulation in the newborn that is the result of functional, rather than anatomical, closure of the foramen ovale and ductus, as these shunts are able to reopen from time to time. The ductus venosus is closed permanently in most individuals within 3 months of birth, the ductus arteriosus by 1 year, and the foramen ovale obliterates very slowly and not in all individuals; in 10% of adults, a probe may be passed through it. The rising cortisol levels observed as parturition approaches (see later) seem to underlie increases in fetal cardiac output, peripheral resistance and blood pressure towards term.

The respiratory system

It is clear that the fetus spends at least 1–4 hours each day making rapid respiratory movements, which are irregular in amplitude and frequency and generate negative pressures of 25 mmHg or more in the chest. Interestingly, these movements occur in episodes of up to 30 minutes during rapid eye movement sleep (see 'The nervous system' below) but not during wakefulness or slow wave sleep. The breathing movements are purely diaphragmatic and move amniotic fluid in and out of the lungs (see 'Amniotic fluid' above). The functions of these breathing movements probably include an element of 'practice' of the reflex neuromuscular activities, to be initiated in breathing at birth in order to fill the lungs with air, and also the promotion of the growth that follows lung distension. Prevention of fetal breathing, as occurs in congenital disorders of the nervous system or diaphragm, retards development of the lungs to the extent that they may be incapable of supporting extra-uterine life.

The fetal lungs undergo major structural changes during pregnancy, especially as parturition approaches. Primitive air sacs are apparent in the lung mesenchyme from about week 20 and blood vessels appear soon afterwards at around week 28. The pressure required to expand the fetal lung decreases as the time of birth approaches and, in large measure, this is the result of the appearance of a surface active agent, or *surfactant*, which reduces the surface tension of pulmonary fluid, thereby aiding lung expansion. The surfactant is a phospholipid (a disaturated lecithin, mainly dipalmitoyl lecithin) that is attached to an apoprotein. The enzymes required for its synthesis are found in the human fetal lung from weeks 18 to 20 onwards, but increase enormously in concentration in the 2 months before birth. One of the most important findings in fetal physiology is that synthesis of

surfactant is promoted by fetal corticosteroids, particularly during their pre-term rise (Fig. 11.5). The corticosteroids stimulate conversion of noradrenaline to adrenaline by activation of phenylethanolamine N methyl transferase (PNMT) in both the adrenal medulla and, locally, in the lungs. In addition, catecholamine receptor number in the lungs increases. Within the lungs, both water resorbtion and surfactant production rises. Failure to produce sufficient surfactant has serious consequences for lung expansion, as is seen in so-called *idiopathic respiratory distress syndrome* (hyaline membrane disease). Lung maturation can be accelerated by injecting ACTH to stimulate fetal adrenal activity, or by administering potent corticosteroids to the mother. Such treatment is highly effective for infants about to be born prematurely.

To initiate normal, continuous breathing at birth, the first breath must overcome the viscosity and surface tension of fluid in the airways and also the resistance of lung tissues. Removal of fluids in respiratory pathways occurs through the mouth during vaginal delivery due to the rise in intrathoracic pressure, but this does not occur in babies born by Caesarean section. Remaining fluid is resorbed through the agency of pulmonary lymphatics and capillaries, helped by the cortisol/adrenaline stimulation. Prenatal episodic breathing is replaced rapidly with normal, continuous postnatal breathing, and the gaseous exchange function of the lungs is established quite rapidly, within 15 minutes or so of birth.

The mechanisms bringing about the pronounced inspiratory effort at birth are varied. Cold exposure, tactile, gravitational, auditory and noxious stimuli may all enhance respiration at birth, but are not absolutely necessary for the initiation of continuous postnatal breathing. Whichever factors are important, they operate on a newborn in which the neuromuscular activities of respiration and swallowing (see below), including the ejection of foreign bodies from the pharynx and trachea, have been rehearsed, and medullary respiratory rhythmicity has been established. Pulmonary stretch receptors and arterial and central chemoreceptor mechanisms are also functional by this time, and, again, corticosteroids seem to be involved in their activation.

The gastrointestinal system

By using intra-amniotic injections of inert substances, which may subsequently be identified in maternal urine, or of opaque media, which are visualized by subsequent X-ray, it has been demonstrated that, near term, the human fetus swallows about 500 ml of amniotic fluid daily. The opaque medium is transferred rapidly through the stomach to the large bowel where it remains

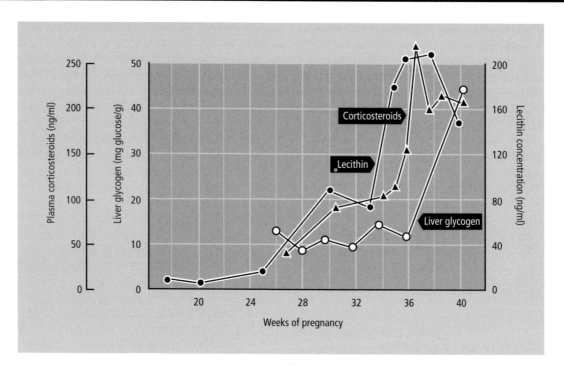

Fig. 11.5 Time course of concentrations of corticosteroids in plasma of umbilical cord, lecithin in amniotic fluid, and glycogen in fetal liver. Lecithin may be used as an indicator of surfactant production in the fetal lung. The rise of lecithin content of human amniotic fluid just before birth is reflected in the lecithin : sphingomyelin ratio test for monitoring fetal development and well-being. (Sphingomyelin is a phospholipid, the concentration of which does not change near term and serves, therefore, as a baseline against which to measure lecithin.) A ratio > 2 indicates surfactant production is normal, and < 2 indicates the fetus may have hyaline membrane disease (see text). The close temporal relationship between corticosteroid concentration in blood and amniotic lecithin levels is suggestive of a causal relationship between the two (see text). The increase in liver glycogen near term is also the result of the rise in corticosteroids, which induces the late cluster of fetal liver enzymes, including those required for glycogen synthesis.

for the duration of pregnancy. Water, which forms the major part of the swallowed fluid, is absorbed readily through the small bowel as are electrolytes and other small molecules, such as glucose. Debris from fetal skin found in amniotic fluid, as well as larger molecules, accumulate in the large bowel together with sloughed cells from the small intestine and bile pigments, to form a green faecal mass, *meconium*. Defecation *in utero* does not normally occur.

Shortly before birth, there is a major and rapid increase in glycogen concentration in the liver, associated with the development of a cluster of liver enzymes (the *neonatal cluster*) apparently induced by corticosteroids. Enzymes involved in gluconeogenesis are also induced at this time. Fetal corticosteroids also appear to regulate maturation of the pancreatic islets of Langerhans. We have discussed already that the concentration of glucose in fetal blood is fairly constant coming, as it does, from a relatively constant maternal pool. The β-cell response to increases in circulating levels of glucose is very weak, or even absent, for much of pregnancy unless hyperglycaemia is sustained, as in the case of maternal diabetes. Thus, enhanced secretion of fetal adrenal corticosteroids near term has a major, orchestrating role in the maturation of pancreatic and liver functions essential for postnatal survival.

The renal system

Although the placenta is the main organ of excretion, the fetal kidneys are functional during pregnancy and produce substantial quantities of hypotonic urine, the tubules being inefficient at Na⁺ reabsorption. Fetal urine contributes to total amniotic fluid volume to the extent of half a litre or so per day. Renal agenesis (Potter's syndrome) is associated with a marked reduction of amniotic fluid volume, but such fetuses survive to term and are born alive although with various growth and developmental defects. At parturition, renal function must undergo a quite radical alteration, as the constant supply of water, sodium and other electrolytes through the placenta will be lost. Soon after birth, urine flow occurs at a high rate while sodium reabsorption is rather low, but over the following hours, urine flow is reduced rapidly, rising again at the end of the first week. Glomerular filtration rate is relatively low in the newborn, perhaps only one-third of what would be estimated in relation to body size. Indeed, a mature rate of glomerular filtration is not achieved until 1.5–2 years of age. The newborn is in danger of hyponatraemia as the ability to retain sodium is poor. Prematurity represents an important risk in this context, as the kidney leaks up to three times more sodium than that seen in babies born at term. The improvement seen in term babies reflects the action of rising fetal cortisol on Na^+, K^+, ATPase activity in cortical tubules, so improving Na^+ reabsorbtion. The developmental changes in the urogenital tract during early pregnancy have been discussed in Chapter 1, with special reference to the role of sex steroids in the differentiation of internal and external genitalia.

The nervous system

Fetal hormones, particularly sex steroids and thyroxine (see Chapter 1 and below), have major effects on neural development, but, in general, rather little is known about the development of function in the fetal brain. The fetus is clearly capable of responding to extraneous stimuli. Loud noises and intense light, noxious stimulation of the skin and rapid decreases in the temperature of its fluid environment will result not only in movement but also in autonomic responses, such as acceleration of heart rate. Presumably the level of stimulation in these sensory modalities is normally rather low and unvarying in the fluid-cushioned, constant temperature, light-and-sound-attenuated chamber in which the fetus grows.

Fetal movements occur early in pregnancy and these can be felt readily by the mother by week 14. The function of the movements is uncertain, but 'exercise,' which will both contribute to muscle growth and limb development, is likely to be among them. During the long, human gestation period, the innervation of muscles by motor nerves and partial maturation of both ascending sensory and descending motor systems in the central nervous system (CNS) means that some well-coordinated movements become possible late in pregnancy. A number of simple postural and other stereotyped reflexes is also apparent in the fetus from a relatively young gestational age. However, *complete myelination* of the long motor pathways, for example the corticospinal tract, does not occur until after birth, which is why fine movements of the fingers, for example apposition of fingers and thumb, are not possible until several months after birth.

The fetus shows periods of slow wave (SW) and rapid eye movement (REM) sleep between periods of wakefulness. The neonate sleeps for about 16 hours each day, and a much greater portion of this time is spent in REM sleep when compared with the adult. This sleeping time progressively decreases over the first 2 years of life to about 12 hours each day, and the proportion of REM sleep decreases to about one-quarter compared with SW sleep.

Of tremendous clinical and sociological significance is the impact of drugs from the maternal circulation on the developing brain of the fetus. Many are lipid soluble and have no barrier to their free diffusion both across the placenta and, therefore, into the brain as well. The immature status of the blood–brain barrier also ensures that other chemical agents may gain access to the fetal brain in a way not seen in the adult. Abused drugs, such as the opiates (heroin, morphine), taken by pregnant women produce withdrawal symptoms, and, indeed, dependence in their babies. Pain-killing drugs, which are sometimes given during labour, may depress the behavioural repertoire of the newborn; for example, sucking reflexes may be impaired. This may have adverse consequences both for lactation and mother–infant interaction (see Chapter 13).

Summary

In this section, we have examined the ways in which selected fetal systems operate during intra-uterine life and the developmental changes, which occur towards the time of parturition, that are essential for extra-uterine survival. The mechanisms involved vary from the instantaneous (in the cardiovascular system) to the gradual (after birth in the urinary system), with some occurring late in pregnancy (gastrointestinal and respiratory systems) and many regulated, like parturition itself (see Chapter 12), by endocrine events involving the fetal adrenal cortex.

FETAL AND NEONATAL ENDOCRINOLOGY

It has already been emphasized that the fetal endocrine system exerts important influences on maternal physiology, especially through the hormones of the feto-placental unit (see Chapter 10). In contrast, the maternal endocrine system does not influence the fetus directly, except in pathological circumstances, and few maternal hormones, other than unconjugated steroids, cross the placenta. In general terms, the fetal endocrine organs function by the end of the first quarter of pregnancy, and, from this time, the fetus is autonomous in its endocrine requirements and most of its hormones do not cross the placenta to the maternal circulation. Furthermore, the fetal endocrine system has a number of unique functions that are not apparent in the adult, for example in the differentiation of the reproductive tract, lungs, gastrointestinal system and even the brain itself.

Anterior pituitary hormones

The anterior pituitary is functional throughout most of fetal life, as evidenced, for example, by early-occurring thyroid, gonadal and adrenal activity. Thyroid hormone production under thyroid stimulating hormone (TSH) and hypothalamic thyrotrophin releasing hormone (TRH) control is discussed under 'The thyroid gland' below. The gonadotrophins, follicle stimulating hormone (FSH) and luteinizing hormone (LH), are essential for the androgenic output of the testis, and their secretion is regulated by a combination of pulsatile gonadotrophin releasing hormone (GnRH) secretion and steroidal feedback. The importance of corticotrophin releasing hormone-dependent ACTH secretion is discussed under 'The adrenal gland' below. Growth hormone is produced by the fetal pituitary, but its role, if any, in the fetus is not clear. Prolactin is present in amniotic fluid in concentrations 100-fold greater than those in either maternal or fetal circulations, but most of this prolactin is of decidual origin. After week 30 or so, the low levels of prolactin in fetal plasma rise markedly until term, presumably due to fetal pituitary activity, but decline in the neonate after a brief postpartum rise. The functions of fetal pituitary prolactin are unknown, although, as pointed out earlier, decidual prolactin may have a role in regulating the permeability of the chorion and amnion to water. It may also be an important co-hormone in facilitating the effects of corticosteroids on the production of lung surfactant.

The thyroid gland

Thyroxine (T_4) is essential for the normal development of the fetus. Fetal hypothyroidism is associated with a bone age far behind chronological age, deficiency in body hair and, most important, behavioural retardation. Thyroxine is essential for the normal differentiation of the CNS. Maternal T_4 does have limited access to the fetal circulation as it crosses the placenta, particularly during the second half of pregnancy, although not in amounts sufficient to meet fetal needs. The secretion of T_4 from the fetal thyroid increases, under the influence of TSH, from about week 20 of gestation such that its circulating levels may even exceed those in the maternal circulation at term.

Infusing TRH into fetal lambs early in gestation increases circulating levels of tri-iodothyronine (T_3) and prolactin, and these hormones are able greatly to facilitate the effects of relatively low-circulating levels of corticosteroids that on their own are insufficient to affect lung maturation. These data have been interpreted to suggest that T_3, as well as prolactin, are important synergists of the corticosteroid effects on lung maturation near term. As TRH readily crosses the placenta, maternal treatment with glucocorticosteroids and TRH has the potential to accelerate fetal lung maturation at a gestational age when the response to corticosteroids alone is attenuated.

The adrenal gland

We have already encountered the importance of adrenal cortical activity for the fetus (in this chapter) and the endocrine function of the placenta (see Chapter 10), and will do so again in Chapter 12 in the context of parturition. There is some evidence that ACTH from the anterior pituitary has an inductive role in the growth and development of the adrenal cortex. Fetal serum ACTH concentrations are quite high during weeks 12–19 of gestation and gradually decline by around week 40. Very high levels are found in the fetus at term, probably reflecting a stress response to parturition. Fetal pituitary ACTH secretion is under the control of fetal hypothalamic corticotrophin releasing factor (CRF) and changes in the content of CRF mRNA precede alterations in the circulating levels of ACTH and cortisol.

Cortisol itself is found in fetal blood by week 10 and levels increase as gestation proceeds, to become especially high during labour. After birth, the adrenal shows major structural changes. The fetal zone, which is responsible for the synthesis of DHA (substrate for placental aromatization to oestrogen; Chapter 10), regresses by the end of the first neonatal month, its job having been done, while the cortex proper differentiates into the distinctive three zones characteristic of the adult gland. Table 11.3 summarizes the actions of fetal corticosteroids as parturition approaches. Aldosterone levels rise late in pregnancy, but appear not to become responsive to a reduction in blood volume or nephrectomy until after birth. Atrial natriuretic factor (ANF) is also produced prenatally and is able to regulate Na^+ excretion; its production may be regulated by adrenal corticosteroids.

Function	Mechanism and/or relevant section
1 Lung maturation	Induction of enzymes necessary for surfactant synthesis, stimulation of alveolar water resorbtion and central respiratory mechanisms (see 'The respiratory system')
2 Parturition	Induction of placental oestrogen-synthesizing enzymes. Increase oestradiol precursor (DHEA) concentrations (function of fetal zone) (Chapter 12)
3 Glucose storage and gluconeogenesis	Induction of enzyme systems in liver and myocardium (see 'Glucose and carbohydrate metabolism'; 'The gastrointestinal stystem')
4 Insulin secretion	Regulates maturation of fetal islets (see 'Glucose and carbohydrate metabolism')
5 Lactogenesis	Ductal-lobule-alveolar growth in pregnancy (Chapter 13)
6 Synthesis of adrenaline	Induction of phenylethanolamine-*N*-methyl transferase in adrenal medulla
7 Production of thyroxine	May promote conversion of T3 to T4
8 Haemoglobin formation	May promote 'switch' in production of fetal to adult haemoglobin and shift of haematopoiesis to bone marrow
9 Maturation of salt:water	Activation of ANF? stimulation of GFR regulation and reabsorbtion of Na+

Table 11.3 Functions of glucocorticosteroids secreted by the fetal adrenal cortex

The parathyroid glands and calcium-regulating hormones

The parathyroid glands are capable of secreting parathormone by about week 12 of gestation, although plasma concentrations are low until 2 or 3 days after parturition. This relative suppression of parathormone secretion is largely a reflection of the rather high circulating levels of maternally-derived calcium (see 'Calcium' above). Thyrocalcitonin secreted from thyroid parafollicular cells, on the other hand, is at relatively high concentrations in fetal plasma and declines slowly after birth. This hormone is able to exert its hypocalcaemic effects, and abnormally high circulating levels may cause a severe hypocalcaemia in the fetus.

Glucagon and insulin

The endocrine pancreas is active early in pregnancy, glucagon (α cells) and somatostatin (δ cells) being the predominant hormones at first, followed by insulin (β cells), which is clearly present by 10 weeks in humans. Development of β-cell function appears to depend upon anterior pituitary growth hormone and ACTH activity. Initially, each type of cell is clustered separately, and only later do β cells become surrounded by α and δ cells. Abnormalities in early pancreatic development tend to affect α and δ cells preferentially, leading to relatively uncontrolled β-cell activity, hyperinsulinaemia and hypoglycaemia (*nesidioblastosis syndrome*). Insulin secretion from mid-pregnancy onwards responds positively to amino acids, glucose and short-chain fatty acids, and negatively to catecholamines.

Details of how insulin acts in the fetus are given above under 'Glucose and carbohydrate metabolism'.

IMMUNOLOGICAL ASPECTS OF PREGNANCY

On the first page of this book, we discussed the particular value of sexual reproduction in generating genetic diversity. The whole edifice of sexual differentiation, reproductive cyclicity and pregnancy, with their social ramifications, is constructed upon the biological advantages conferred by producing a genetically distinct individual. Yet we have now come full circle. For the genetically distinct individual will also be phenotypically unique. A component of this unique phenotype is the array of cell-surface glycoproteins that constitute the system of histocompatibility antigens. Thus, the biological advantages of genetic variation appear to confront and conflict with those of viviparity. The problem may be illustrated dramatically. If the skin of a newborn child is grafted to its mother, she rejects it. Why then does she not reject the whole fetus, as it is, in a sense, grafted onto the maternal uterus? Several mechanisms have been proposed to explain the survival of the fetus *in utero*. None is in itself adequate.

Fetal antigenicity

The developing fetus does not lack target antigens. The major histocompatibility antigens appear on embryonic cells shortly after implantation and, although present in smaller amounts than in the adult, are detectable throughout pregnancy. Thus, grafting of fetal tissues to another individual (or to the mother) results in their rejection.

Maternal immune responsiveness

The pregnant mother is competent to respond immunologically to the fetus. Indeed, examination of maternal blood in late pregnancy indicates that a regular feature of pregnancy is an immunological reaction against the conceptus. Fetal cells can be detected in the maternal circulation where they can presumably induce this antibody and lymphocyte response to their antigens. Even more strikingly, the active pre-sensitization of the mother, by injecting or grafting paternal tissues, does not prejudice the establishment and maintenance of subsequent pregnancies by the same father. Thus, neither a generalized nor a specific depression of maternal immune responsiveness can adequately explain fetal survival. There is some evidence that during pregnancy the quality of the maternal immune response is different, perhaps modified by the high levels of pregnancy hormones. Helper T cells decline relative to suppressor cells, and the classes of immunoglobulin produced change their balance. It is possible that these qualitative changes contribute to the survival of the fetal 'graft', but they alone cannot easily provide a complete explanation.

Immunological filter

The fetus and its circulating blood are separated from the mother by the investments of fetal membranes (Fig. 9.8). These membranes, the outermost of which is usually chorionic trophoblast, are part of the conceptus and therefore are also genetically alien. However, if the chorionic trophoblast was itself able to resist maternal rejection, and also effectively precluded damaging maternal antibodies and lymphocytes from entering the fetal circulation, then protection for the fetus might be achieved. A number of studies on the antigenicity of syncytio- and cytotrophoblast have revealed convincing evidence to suggest that it lacks conventional Class I and II histocompatibility antigens on its surface. Some trophoblastic cells, particularly those in the human that invade the termini of the spiral arteries, do express a non-polymorphic (monomorphic) Class I histocompatibility antigen. This antigen is not a target for cytotoxic action. Thus, it appears that the trophoblast layer is presented to the mother as being effectively 'antigenically neutral'. In addition, the placenta is bathed in fluids containing high levels of progesterone, corticosteroids and chorionic gonadotrophin. There is evidence that these

hormones may act as immunosuppressants to reduce the cytotoxic effectiveness of immune cells locally.

We saw earlier that the discrete nature of the fetal and maternal circulations prevented appreciable passage of maternal cells to the fetus. In many species, including the pig, sheep, cow and horse, antibody is also excluded. However, in humans and monkeys, immunoglobulin G (IgG) antibodies normally pass across the chorio-allantoic placenta into the fetal circulation via a special transport mechanism. The maternal antibodies will include some directed against prevalent bacteria and viruses and thus will confer passive immunity on the fetus and afford the neonate temporary protection for a few weeks postnatally. (In rats, mice and rabbits, the yolk sac placenta performs this function, while, in farm animals, a similar protection is afforded by antibodies transmitted postnatally in the milk; see Chapter 13.) However, in addition to antibodies reactive to bacteria, IgG antibodies directed against fetal antigens presumably could also be transferred. That such transfer does indeed occur is seen in cases of sensitization to the major blood group antigen called rhesus. A woman may lack the rhesus antigen (rhesus negative) and if she carries a fetus possessing it (rhesus positive), she may mount an IgG immune response to the antigen, particularly at parturition when extensive fetal bleeding into the mother may occur. In subsequent pregnancies, the IgG antibody is transferred across the placenta and destroys the fetal erythrocytes. Here, then, is a clear example of the mother rejecting her fetus immunologically. But the rhesus antigen is only one of many antigens by which mother and fetus may differ. Can we gain any clues from the rhesus example as to why fetuses are not normally rejected?

The rhesus antigen differs in two ways from most of the other cellular antigens expressed on fetal cells. First, the rhesus antigen is present only on red blood cells. Most of the other important histocompatibility and blood group antigens are also present on several other types of fetal cell and are thus widely distributed among the tissues of the fetus. Second, the rhesus antigen exists only as a structural component of the cell membrane. Other antigens, such as ABO blood group and major histocompatibility antigens, seem to be present not only as structural membrane components but also in solution in the fluids of the fetus, such as the blood and amniotic fluid. Thus, if an antibody, or indeed the odd lymphocyte, directed against these other antigens should enter the fetal circulation, it could first be 'mopped up' harmlessly by free soluble antigen. Any remaining antibody will be distributed among a wide range of cell types and so would effectively be 'diluted-out'. Any given single cell will be unlikely to bind a large number of antibody molecules, and, as only the binding of a large number of antibodies will debilitate the cell, gross tissue damage is avoided. In the case of the rhesus antigen, 'mopping up' and 'diluting-out' cannot occur and so the chance of tissue damage increases considerably with obvious pathological consequences.

Thus, the protection of the fetus from the immune response of the mother appears to depend upon: (a) an antigenically inert trophoblast forming the front-line defences, possibly in association with local, endocrinologically-mediated depression of immune reactivity; (b) a complete (or, in humans, highly selective) barrier to the transmission of immune cells or antibodies from mother to fetus; (c) properties of fetal antigens such that aggressive immune cells or antibodies that do get across the placenta are mopped up and diluted out before they can cause extensive tissue damage.

SUMMARY

Viviparity provides the developing embryos with the optimal environment for growth. The uterus may be viewed as the ultimate 'nest': temperature-controlled, a continuous supply of food and protection from predators. This environment is created at the mother's expense and her metabolism is brought, to varying degrees, under the control of the fetus, which to a large extent functions autonomously within its protected environment. It should, however, be clear from this account that we still have much to learn about fetal and maternal function in pregnancy and its control. Central to the maintenance of pregnancy is the trophoblast of the placenta. This remarkable tissue elaborates and secretes the steroid and protein hormones; it is involved in the transplacental passage of a variety of essential substances and also the products of metabolism, both by diffusion and by active transport; it acts as a selective barrier between the two circulations and presents an antigenically inert front to the mother's immune system. The role of the trophoblast ends with delivery of the new infant but the mother still has a major and crucial role to play postnatally. This subject is discussed in the following chapters.

FURTHER READING

Biggers JJ (1979) Fetal and neonatal physiology. In: *Medical Physiology* (Ed. Mountcastle V), pp. 1947–1982. Mosby, St Louis.

Ciba Foundation Symposium 86 (1981) *The Fetus and Independent Life.*

Dancis J (1975) Feto-maternal interaction. In: *Neonatology* (Ed. Avery GB). J.B. Lipincott, Philadelphia.

Garel JM (1987) Hormonal control of calcium metabolism during the reproductive cycle in mammals. *Physiological Reviews* **67,** 1–66.

Gluckman PD (1986) The role of the pituitary hormones, growth factors and insulin in the regulation of fetal growth. *Oxford Reviews of Reproductive Biology* **8,** 1–60.

Jones CT (Ed.) (1988) *Research in Perinatal Medicine (VII): Fetal and Neonatal Development.* Perinatology Press, Ithaca, N.Y.

Jones CT & Rolph TP (1986) Metabolism during fetal life: a functional assessment of metabolic development. *Physiological Reviews* **65,** 357–430.

Knobil E & Neill JD (Eds) (1994) *The Physiology of Reproduction* Volume 2, 2nd edition. Raven Press, New York. (Particularly chapters by: Morris FH, Boyd RDH & Mahendran, M, pp. 813–862; Stock MK & Metcalfe JM, pp. 947–984.)

Künzel W & Jensen A (Eds) (1988) *The Endocrine Control of the Fetus: Physiologic and Pathophysiologic Aspects.* Springer-Verlag, Berlin.

Liggins GC (1994) The role of cortisol in preparing the fetus for birth. *Reproduction, Fertility and Development* **6,** 141–150.

Loke YW (1986) Female reproductive immunology. In: *Scientific Foundations of Obstetrics and Gynaecology* (Eds Philip EE, Barnes J & Newton M), pp. 66–77. Heinemann, Oxford.

Philip EE, Barnes I & Newton M (Eds) (1986) *Scientific Foundations of Obstetrics and Gynaecology.* Heinemann, Oxford.

Shearman RP (Ed.) (1979) *Human Reproductive Physiology.* Blackwell Scientific Publications, Oxford.

C H A P T E R 12

Parturition

C O N T E N T S

In the previous 11 chapters, we have described the events leading to the development of a mature fetus. In this chapter, we consider *parturition*. Factors affecting the onset of parturition in women are poorly understood and we must therefore rely on experimental data largely from the ewe and the goat. Although these species differ from the human in both the source and profile of their pregnancy hormones (see Chapter 10), they nevertheless share with humans at least some features controlling the mechanism and timing of onset of parturition. Indeed, ultimately the endocrine process underlying parturition is probably identical in goats, sheep and humans, and possibly in all species, even if the means by which this process is initiated shows interspecies variation.

Prior to the onset of parturition, the fetus lies within its fetal membranes in the uterus and is retained there by the cervix (Fig. 12.1). Expulsion of the fetus requires coordinated contractions of the myometrium (assisted later in labour by contractions of striated muscle in the abdominal wall and elsewhere). However, the large increases in intra-uterine pressure that accompany these contractions and the expulsion of the neonate will be resisted by the cervix, unless its structure changes to become soft and compliant. Thus, expulsion of the fetus requires both *myometrial contractions* and *cervical softening*. Before examining the processes underlying each of these changes, we will first consider briefly the nature of each tissue involved.

THE TISSUES

The myometrium

The myometrium consists of bundles of non-striated muscle fibres, intermixed with areolar tissue, nerves, blood and lymph vessels. During pregnancy, myometrial bulk increases enormously, due primarily to an increase in muscle cell size from about 50 to 500 μm (hypertrophy) and increased glycogen deposition, both stimulated by oestrogens. Functionally, this system of muscle cells can behave as a syncytium, being electrically coupled via specialized regions of contact, the so-called gap junctions or *nexuses*, which allow the flow of electrical current and thereby coordination of the spread of contraction through the myometrium. However, the extent to which this nexus functions to achieve contrac-

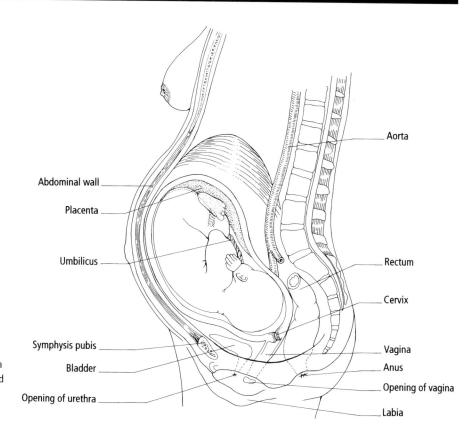

Fig. 12.1 Diagram of a sagittal section through a pregnant woman. Notice, particularly, the size and position of the uterus, and how much of the abdomen it occupies.

Labels: Abdominal wall, Placenta, Umbilicus, Symphysis pubis, Bladder, Opening of urethra, Aorta, Rectum, Cervix, Vagina, Anus, Opening of vagina, Labia

tion is under tight endocrine control. In order to understand this process, we will first consider briefly the physiology of myometrial contractility.

The contraction of myometrial cells depends on a rise in intracellular calcium ion concentration, both by liberation from intracellular stores and by entry into the muscle cells from the extra-cellular fluid. The calcium then binds to regulatory sites on the contractile proteins, actin and myosin, to allow expression of ATPase activity, and hence contraction. This release of calcium is stimulated by the presence of action potentials within the muscle cell. In the myometrium, spontaneous depolarizing pacemaker potentials occur. If the magnitude of such potentials exceeds a critical threshold, a burst of action potentials is superimposed on the pacemakers, a sharp increase in intracellular Ca^{2+} occurs and a contraction follows. Ca^{2+} is then pumped back into intracellular stores and out of the cell, and the muscle relaxes. Contractility could therefore be modulated by changing the pacemaker potentials, the relationship between these potentials and the threshold for spiking, and/or the effect of spiking on calcium release. Two hormones have been implicated directly in contractility regulation: *prostaglandins* act mainly by enhancing the liberation of Ca^{2+} from intracellular stores, while *oxytocin* lowers the excitation threshold of the muscle cell at which spiking occurs.

The uterine cervix

The cervix is of major importance in retaining the fetus in the uterus, and this function is a reflection of its high connective tissue content, which helps resist stretch. The connective tissue is derived from collagen fibre bundles embedded in a proteoglycan matrix. In order for the fetus to move from the uterus to the outside world, the nature of the pregnancy cervix must change by softening or 'ripening'. The process by which this occurs involves two changes in the intercellular matrix: a reduction in collagen fibres and a marked increase in glycosaminoglycans (GAGs), which decrease the aggregation of those collagen fibres remaining. In the human cervix, *keratan sulphate*, which does not bind to collagen, increases peripartum, at the expense of *dermatan sulphate*, which binds collagen tightly. In consequence, collagen bundles 'loosen'.

Prostaglandins, in addition to their effects on myometrial contractility, also seem to control cervical ripening. A number of clinical trials have demonstrated that both prostaglandin E_2 (PGE_2) and prostaglandin $F_{2\alpha}$ ($PGF_{2\alpha}$) increase the compliance of the cervix when given intravaginally or intracervically. They are able, therefore, to facilitate delivery. Thus, the induction of birth (or evacuation of a late abortus) requires both prostaglandins and oxytocin. Before considering the mechanisms underlying parturition, we will briefly review the physiology of prostaglandin and oxytocin production.

THE HORMONES

Prostaglandins

As we discussed in Chapter 2, prostaglandins are biologically active lipids, probably synthesized in every tissue of the body, including the brain. They are essentially local hormones acting at, or near, their site of synthesis and are inactivated in the lung during one circulation in the bloodstream. At parturition, the endometrium is probably the most important site of prostaglandin synthesis. In Chapter 10, we saw that, in some species, the presence of the conceptus in the luteal phase of the cycle inhibits the synthesis and release of $PGF_{2\alpha}$ by the endometrium, and so prevents luteolysis. This condition endures until parturition when, probably in all species, the disinhibition or promotion of prostaglandin synthesis and release occurs. It is probable that the myometrium, cervix, placenta and fetal membranes also synthesize prostaglandins. Factors that increase prostaglandin synthesis are widely believed to act primarily by altering the stability of membranes binding phospholipase A_2, thus causing liberation of the active enzyme from the lysosomes (Fig. 2.4). Those factors that decrease prostaglandin synthesis probably do the converse, that is, stabilize the lysosomal membrane. It is interesting to note, therefore, that steroid hormones, which may change quite markedly at parturition (see below), exert opposing effects on phospholipase A_2. Some ultrastructural evidence suggests that oestrogens labilize lysosomes, while progesterone stabilizes them. Thus, a rise in the oestrogen : progesterone ratio would result in increased production of arachidonic acid, and hence prostaglandin synthesis (Fig. 2.4).

The oestrogen : progesterone ratio affects not only *synthesis* of prostaglandins but also their *release*. It does this via effects on oxytocin, which stimulates the release of prostaglandins directly from the uterus. In the sheep, oestradiol has been shown to enhance this effect by increasing the number of oxytocin receptors in the endometrium, while progesterone has the reverse effect. This means that a rising oestradiol : progesterone ratio can affect prostaglandin release via an oxytocin-dependent mechanism, independently of any alteration in circulating oxytocin levels. This receptor regulation plays an important role in parturition; indeed, in

women, myometrial oxytocin receptor number doubles at this time. Thus, two routes to increased prostaglandin availability at parturition exist, and both can be induced by an increase in the oestrogen : progesterone ratio. In some species, such a global change in the oestrogen : progesterone ratio has been shown to occur at parturition (see below).

Oxytocin

As we saw in Chapter 2, oxytocin is a nonapeptide, synthesized by magnocellular neurons in the supraoptic and paraventricular nuclei of the hypothalamus (Fig. 2.7). It is transported along their axons to the terminals to be stored for release into the circulation from the posterior lobe of the pituitary (Fig. 2.7b). Oxytocin is released in response to tactile stimulation of the reproductive tract, particularly the uterine cervix. This neuroendocrine reflex has, as its afferent limb: (a) the sensory nerves from the vagina and cervix; (b) the ascending somatosensory pathways in the spinal cord (the anterolateral columns); and (c) an incompletely described projection through the brainstem and medial forebrain bundle that ultimately reaches the hypothalamic magnocellular nuclei. The efferent limb of the reflex is the blood-borne carriage of oxytocin to the uterus, where the hormone exerts its actions on myometrial contractility in interaction with both steroid hormones and prostaglandins. This reflex, often referred to as the *'Ferguson reflex'* (Fig. 12.2), is greatly facilitated in the presence of a high plasma oestrogen : progesterone ratio. The reflex release of oxytocin, which occurs in all species, is very similar to the coitus-induced release of prolactin seen in the rat, which was discussed in Chapter 5. In the next chapter, we will see that stimulation of the nipple during suckling also causes the release of both oxytocin and prolactin via a similar reflex pathway and these events are vital for lactation.

HORMONAL CHANGES AND TIMING OF THE ONSET OF PARTURITION

As prostaglandins are, according to most hypotheses, at the centre of events leading to parturition, it is fundamentally important to understand how prostaglandin synthesis and release are increased at parturition and what influences the timing of these events. We will see that different species achieve this in different ways and that, in all species, fetal and maternal endocrine mechanisms are critical. We will confront the somewhat surprising conclusion that, in many species, the *fetus* itself

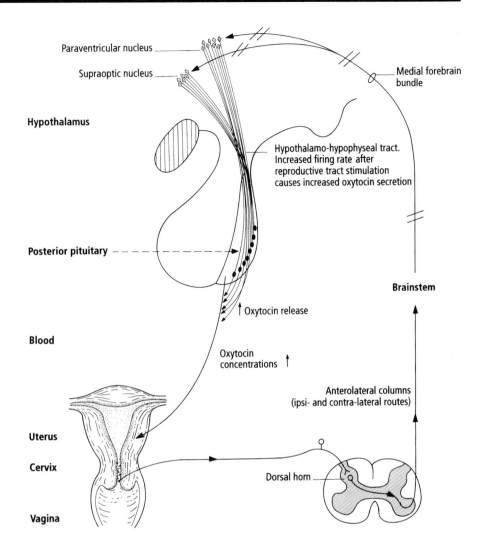

Fig. 12.2 The neuroendocrine reflex (Ferguson reflex) underlying oxytocin synthesis and secretion.

determines the timing. Maturational changes in the fetal hypothalamo-pituitary–adrenal axis provide the key to the way in which this amazing and important task is achieved.

Goats

Goats are representative of those species dependent on progesterone from the corpus luteum throughout pregnancy (see Table 10.1). There is no doubt that parturition in goats is initiated by the sharp fall in circulating concentrations of plasma progesterone that occurs during the last 24 hours before delivery as a consequence of luteal regression. The factor responsible for luteal regression is $PGF_{2\alpha}$, and its concentration in uterine venous blood increases markedly about 48 hours before parturition, preceding the decrease in plasma progesterone levels by about 20 hours. This increased production of $PGF_{2\alpha}$ originates in the placenta and is dependent, in turn, on an increased output of *glucocorticoids* by the fetal adrenal cortex. Indeed, parturition can be induced prematurely by injections of adrenocorticotrophic hormone (ACTH) into the fetus.

These fetal corticosteroids induce *placental aromatizing enzymes* that convert DHA and DHA sulphate (also derived from the fetal adrenal) to oestrogens, which in turn enhance the synthesis and release of $PGF_{2\alpha}$ in the placenta. The $PGF_{2\alpha}$ causes the corpus luteum to regress, and progesterone plasma concentrations plummet, thus removing the 'block' on the myometrium, in which additional $PGF_{2\alpha}$ synthesis and release is enhanced, and myometrial contractions begin. The consequent increased oestrogen : progesterone ratio also facilitates oxytocin release from the posterior pituitary: a phenomenon that is reinforced directly via the neurally-derived input from uterine contractions and cervical dilation as parturition proceeds. Oxytocin, as we saw above, stimulates further release of prostaglandins, and so a positive feedback system is established. Rising fetal corticosteroid levels may also close down production of *caprine placental lactogen* (cPL) during the last 15 days of pregnancy, thereby removing an important luteotrophic stimulus. However, it is not clear exactly how the corticosteroids shut off cPL synthesis nor how important this shut off is to parturition, given the pituitary luteotrophic support in the goat. Fetal cortisol clearly plays a central role in initiating parturition in goats. The mechanism underlying the enhanced output of cortisol by the fetal adrenal will be discussed below in relation to sheep.

Sheep

Although closely related to the goat and with a gestation period of similar length, the sheep is dependent on the placenta and not the corpus luteum for steroid hormone production in the later stages of pregnancy (Table 10.1). However, the fetal pituitary–adrenal axis, as in the goat, has a dominant role in timing the onset of parturition. Thus, fetal hypophysectomy, stalk section or bilateral adrenalectomy all prolong pregnancy indefinitely, while administration in sufficient quantities of ACTH or dexamethasone (a synthetic glucocorticoid) to the fetus induces parturition prematurely. In lambs, there is a marked increase in fetal plasma concentrations of ACTH during the last 14 days of pregnancy, and this rise is associated with a doubling in weight of the fetal adrenal. This increased ACTH output is driven by the hypothalamus, as, within neurons of the parvocellular paraventricular nucleus, there is an increase, from about Day 125 of gestation, in the mRNAs encoding both corticotrophin releasing hormone (CRH) and vasopressin. Accordingly, to the above list of manipulations that delay the onset of parturition can be added neurosurgical lesions of the fetal paraventricular nucleus carried out *in utero*. Together, these data firmly indicate the importance of fetal hypothalamo-pituitary–adrenal interactions in timing the onset of parturition through resultant increases in cortisol secretion.

This steadily rising output of *fetal* ACTH and cortisol prior to normal parturition is followed by an equally marked decrease in progesterone in the *maternal* circulation. However, the relationship between the *rise in corticosteroids* and the *fall in progesterone* is very different from that in the goat. The elevated levels of cortisol in the fetal circulation of the sheep induce activity of the placental enzymes 17α-hydroxylase, steroid $C_{(17)}$-$C_{(20)}$-lyase and probably also aromatases (see Fig. 12.3 & Chapter 2). The

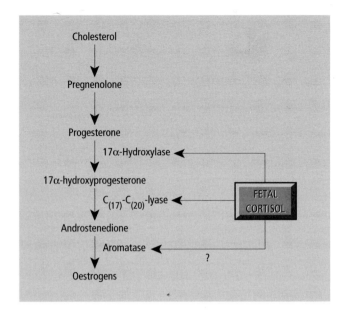

Fig. 12.3 Pattern of oestrogen synthesis in the placenta of the sheep. Cortisol secreted from the fetal adrenal enhances conversion of progesterone to oestrogens by activating the enzymes 17α-hydroxylase, $C_{(17)}$-$C_{(20)}$-lyase and, possibly, aromatases.

effect of this activation is to divert placental progesterone into the synthesis of oestrogens, with a consequent increase in the maternal oestrogen : progesterone ratio. This increase leads to stimulation of the synthesis and release of $PGF_{2\alpha}$ in the decidua and fetal membranes (see 'The hormones' above). Thus, uterine myometrial contractions begin and these induce oxytocin release from the maternal neurohypophysis, which, in turn, reinforces $PGF_{2\alpha}$ release and myometrial contractility. That $PGF_{2\alpha}$ is the final link in this chain of events resulting in myometrial contractions and cervical softening is emphasized by the finding that indomethacin, or other PG inhibitors, block naturally occurring and ACTH-, cortisol- or dexamethasone-induced parturition in the sheep (and goat).

In the goat and the sheep, regardless of the source of hormones in the maintenance of pregnancy, therefore, the fetus times the onset of parturition by increasing the output of adrenal cortisol. The precise mechanisms by which the fetus times this are not understood. Although the relationship between rising fetal cortisol and falling maternal plasma concentrations of progesterone are different in these species, the ultimate result in both is the enhanced synthesis and release of $PGF_{2\alpha}$. A common mechanism of parturition thus exists in these and many other species. What about in the human female?

Human

Despite the background of research summarized above, which points very clearly to the likely mechanisms surrounding parturition in the human female, little is known of the timing mechanism and its relation to alterations in prostaglandin synthesis and release at the onset of parturition. Plasma cortisol concentrations in the fetus and in the amniotic fluid have been seen to rise during the last few weeks of normal pregnancy. For example, corticosteroid sulphate concentrations in the amniotic fluid rise after the 30th week of gestation, and this is paralleled by alterations in the palmitate : stearate ratio that provides an index of lung maturation (see Chapter 11). These data are suggestive of a role for the human fetus, via its adrenal cortex, in timing the onset of parturition. However, fetal anencephaly, with its attendant absence of CRH, is not reliably associated with prolonged gestation. Fetal adrenal hypoplasia may be associated with postmaturity, but not necessarily so: gestation and parturition, within the normal range of variation, have been seen in fetuses with congenital absence of adrenal glands. Infusions of ACTH, or synthetic glucocorticoids, do not apparently induce parturition in the human. Furthermore, fetal cortisol cannot divert placental progesterone into the secretion of oestrogens because the enzyme 17α-hydroxylase does not exist in the human placenta (see Chapter 10). Thus, maternal plasma progesterone concentrations do not fall at parturition. Instead, the fetal adrenal influences placental oestrogen synthesis by providing androgen precursors, notably DHA and DHAS. These androgens are secreted from the specialized fetal zone of the adrenal (which regresses soon after birth: see Chapter 11), and they form 80–90% of the substrate for placental oestrogen synthesis late in pregnancy. However, treatment of women throughout pregnancy with corticosteroids that markedly depress feto-placental androgen and oestrogen production does not delay or otherwise alter the onset of labour. The onset of labour does not appear to be critically dependent, therefore, on the secretory activity of the fetal adrenal, and, reflecting this, there are no clear, consistent and universally observed changes at term in the placental secretion of oestrogens (an indicator of fetal steroidogenic capacity, see Chapter 10). Of course, alterations in the binding of these steroids or local changes in their concentration in uterine tissues (see Chapter 2) may bring about effective local alterations in the oestrogen : progesterone ratio without detectable changes in their circulating plasma levels, but it is not clear whether this happens.

Controversy also surrounds possible changes in prostaglandins in the human before parturition. Many studies have failed to reveal increased $PGF_{2\alpha}$ concentrations in peripheral blood *in advance of labour*. An increase of $PGF_{2\alpha}$ in amniotic fluid has been demonstrated before the onset of labour, and levels do progressively increase as parturition proceeds and cervical dilation increases. The cause of the increase in prostaglandin levels, as will be clear from the above account, is not immediately apparent. However, the observation that the number of myometrial oxytocin receptors doubles around the time of parturition, even in the absence of obvious changes in the oestrogen : progesterone ratio, is consistent with an oxytocin-dependent change in prostaglandin secretion. Surprisingly, experiments in non-human primates, notably the rhesus monkey, have not particularly clarified the mechanisms of parturition occurring in the human: the situation is just as complex and difficult to unravel in this species. Thus, infusion of progesterone, oestradiol and dexamethasone into mother or fetus, or of ACTH into the fetus, has no effect on gestation, but fetal hypophysectomy has been reported to prolong gestation. As in women, cortisol in plasma and $PGF_{2\alpha}$ in amniotic fluid increase prior to parturition, which is itself blocked by prostaglandin inhibitors.

Summary

In a number of mammalian species (not just those exemplified here), the onset of parturition is timed primarily by the fetus via secretions of the adrenal cortex. Although the exact consequences of this increased secretion of fetal cortisol vary, the general result is an increasing ratio of oestrogen : progesterone, which stimulates the synthesis and release of $PGF_{2\alpha}$ in the uterus, the common outcome of the hormonal changes at term. $PGF_{2\alpha}$ is the activator of the mechanical events of parturition, myometrial contractions and cervical ripening. Oxytocin may further enhance the synthesis and release of $PGF_{2\alpha}$ as parturition proceeds. Although, in the human, the beginning and ending of this sequence appear similar (i.e. increased fetal cortisol plasma concentrations and increased levels of $PGF_{2\alpha}$ in maternal amniotic fluid), the intervening changes in the oestrogen : progesterone ratio have not been firmly established. This may mean that such changes occur at a local or cellular level, but evidence in favour of this is sparse. It is generally accepted that maternal mechanisms may contribute to or modulate the timing of parturition controlled largely by the fetus. The nature of these modulating influences, which may, for example, restrict births to certain hours of the day or night, are largely unknown, but might be neural rather than hormonal in origin.

RELAXIN, PREGNANCY AND PARTURITION

The existence of relaxin was first postulated in the 1920s to explain the phenomenon in some species of prenatal separation of the *maternal pubic symphysis* caused by relaxation of the interpubic ligament, hence the name relaxin and its implied role as an aid to parturition.

Relaxin has now been identified as a cytokine, related to insulin (see Table 2.6). There is considerable inter-species variation (>50%) in its amino acid composition and peptide chain lengths.

Source and secretion of relaxin

The major source of relaxin in the pig, rat, mouse, whale and human is apparently the corpus luteum, where it is stored in cytoplasmic granules of the granulosa-derived large lutein cells. Removal of the corpus luteum causes systemic levels of relaxin to fall. It is secreted in relatively small amounts during most of pregnancy, and released in large amounts immediately before parturition, accompanied by degranulation of luteal cells. In pregnant women, relaxin is detectable in the blood as early as weeks 7–10, but maximum plasma concentrations are seen during weeks 38–42. The acute release of relaxin antepartum is probably a consequence of luteolysis. Plasma relaxin levels are apparently not elevated in women during labour induced with either $PGF_{2\alpha}$ or oxytocin. There are substantial species differences in the source of relaxin, for example in guinea-pigs it is produced mainly in the uterus, whereas in rabbits and horses, it is produced in the placenta.

The actions of relaxin

The marked species differences in relaxin structure make unwise the extrapolation of its actions in the rat and pig, which have been most studied, to humans and other animals. Two clear and unequivocal actions have been established in rats and pigs. First, relaxin promotes growth and softening of the uterine cervix, thereby facilitating delivery. Second, relaxin promotes growth and development of the mammary apparatus: in the rat, by stimulating nipple development and, in the sow, by developing the glandular parenchyma (see Chapter 13).

While many other actions have been proposed for relaxin, including separation of the innominate bones in several species, they are less clearly proven.

LABOUR

In the above account, we have discussed those factors that determine the onset of labour, which is the process by which the mother expels the fetus. *Premature labour* in women is said to occur after legal viability (24 weeks) but before 37 weeks of gestation. Labour *at term* occurs after 37 weeks and usually ends before 42 completed weeks. After this time, it is called *post-term*. The process of labour is divided into three stages. The *first stage* begins with the onset of labour (regular painful contractions, and dilatation and shortening of the cervix) and ends when the *uterine cervix is fully dilated*. It may further be divided into a *latent* phase, when the cervix slowly dilates to about 3 cm, and an *active* phase, thereafter when the dilatation of the cervix occurs more rapidly. The *second stage* of labour begins at full dilatation of the cervix and ends with complete *delivery* of the fetus. The *third stage* begins with completion of fetal expulsion and ends with delivery of the placenta.

With the onset of labour, large contractions of the uterine musculature occur. These are regular, occur at shorter and shorter intervals and result in intra-uterine pressures of between 50 and 100 mmHg, compared with about 10 mmHg between contractions. One of the tasks of these contractions is to retract the lower uterine segment and cervix upwards to allow the vagina and the uterus to become one continuous *birth canal*, which allows expulsion of the fetus. The phenomenon of *brachystasis* reflects the property of myometrial cells that helps to achieve this end. Thus, shortening of each muscle cell during contraction is followed, during relaxation, by failure to regain its initial length. With each subsequent contraction, further shortening of the cell occurs, and so, eventually, each myometrial cell becomes shorter and broader, the fundal musculature becomes thicker and uterine volume decreases. The lower uterine segment does not take part in these contractions and remains quite passive during labour. As a result of the muscular phenomenon described above, therefore, the lower segment moves upwards, and is therefore retracted. This event may even be palpated abdominally because the junction between the two segments (the physiological *retraction ring*, marked because of the contrast between thick myometrium above and thin lower uterine segment below) gradually moves upwards. During this time the cervix softens, and, when fully dilated, can no longer be pulled upwards because of its attachment to uterine and uterosacral ligaments and pubocervical fascia.

Moving into the second stage of labour, the fully dilated cervix is drawn up to just below the level of the pelvic inlet. Subsequent uterine contractions, and the resultant decrease in uterine volume, push the fetus downwards and through the pelvis (see Fig. 12.4 for summary). This whole process of labour varies in duration between individuals but usually takes less than 8 hours in multipari and 14 hours in primipari. The first stage occupies much of this time and the second stage should generally last less than 1 hour. A few minutes after delivery of the fetus and clamping the umbilical cord, the placenta becomes detached from the uterine wall as the result of a myometrial contraction. Within a short time, the placenta will be completely expelled by uterine contractions, a process often aided by the midwife or obstetrician by use of pharmacological doses of oxytocic agents or ergometrine and by steadily pulling on the umbilical cord (active management of the third stage of labour).

FETAL MONITORING AND CAESARIAN SECTION

Parturition is a time of vulnerability for the fetus as it undergoes the transition to neonatality. The necessary adjustments that it must make to its own cardiovascular and respiratory systems (see Chapter 11) are preceded by a period of dwindling maternal support as labour progresses. If this period is unduly protracted, the effectiveness of metabolic exchange can decline such as to cause fetal distress and asphyxia. The traditional obstetric approach has been to monitor the fetal heart rate by intermittent auscultation with a Pinard obstetric stethoscope, and to resort to Caesarean section if necessary. Continuous electronic monitoring of fetal heart rate and/or sampling of fetal scalp blood pH is now more widely used in making the decision as to whether the fetus is genuinely hypoxic. However, although introduction of this technique has improved survival rates in difficult cases, its effectiveness in routine cases is debatable. Some studies report a tendency to resort to Caesarean section more readily, thus increasing the use of this delivery procedure disproportionately. However, the likely increase in the use of monitoring procedures in the future, together with advances in understanding the fetal events underlying the traces, will probably benefit obstetric practice considerably.

HIV TRANSMISSION AT DELIVERY

In Chapter 11, we considered the placenta as a feto-maternal barrier, or, more properly, a filter, as it shows selectivity in its transmission properties. Of acute and increasing interest over the past 10 years has been the extent to which HIV can cross the placental filter from an infected mother to infect the fetus. Numerous studies show clearly and unambiguously that transmission of HIV from mother to baby does occur (*vertical transmission*). If neonates, born to mothers who are HIV antibody-positive, are tested for the presence of anti-HIV antibodies, 100% of them test positive. This result is not surprising, as we saw in Chapter 10 that maternal antibodies can cross the human placenta and provide a source of *passive* immune protection against infection. These antibodies decline over a 6-month period, and if, after this time, babies are retested for the presence of antibodies to HIV only a proportion of them remain positive. These antibodies are not of maternal origin, but reflect the baby's own *active* immune response to HIV. Sensitive tests for the direct

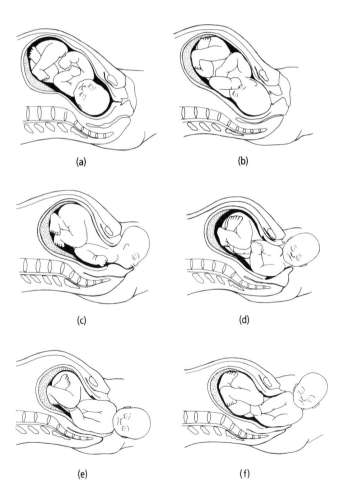

Fig. 12.4 Normal labour showing: (a) engagement and flexion of the head; (b) internal rotation; (c) delivery by extension of the head after dilation of the cervix; (d–f) sequential delivery of the shoulders.

detection of virus confirm that in these babies, but only in these babies, is virus present in addition to antibody. The proportion of babies that carry virus in this way varies in different populations, being highest in East African populations (*c.* 25–35%), intermediate in the USA (*c.* 15–25%), and lowest in Western Europe (*c.* 15–20%). The reasons for this variation are uncertain, but may include the effects of cofactors, such as diet, vitamin A deficiency, other infections and inflammations, intravenous drug abuse, general maternal health and socio-economic status, as well as the virulence of the viral subtype, and the maternal immune status and viral load during pregnancy and parturition. The question is: how has the virus arrived in the baby?

There are, in principle, three routes available: transplacental, parturitional and lactational. It is clear from direct analysis of fetal and placental tissues for virus, that infection can occur *in utero*, and that, in principle, both free virus and cells infected with virus can cross the placenta. However, the suspicion seems to be that this is likely to be a *low frequency route* of transmission. There is little doubt from the results of controlled studies, in which HIV-positive mothers, using exclusively either breast- or bottle-feeding, are compared, that HIV transmission can be halved with bottle-feeding, clearly implicating milk during lactation as a route of infective transmission. However, it is important to note that abandonment of breast-feeding can, in poorer communities, have its own severe health costs, affecting the dietary intake and level of infection in infants, which can erode any clear advantage of reduced HIV transmission.

What about parturition? There is evidence that the birth canal can be a major repository of infectious agents and that maternal bleeding at parturition can also expose neonates to HIV. Among twins delivered to infected mothers, the first born has a higher infection rate than the second. To test the idea that birth canal infections might occur, trials are being undertaken in which disinfectant viricidal swabbing of the birth canal occurs in advance of delivery. Studies comparing Caesarean and natural deliveries have produced conflicting, and as yet unresolved, outcomes. Overall, there is an impression, but no clear evidence, that delivery is a time of high risk for infectious transmission to the baby. That risk is also shared, albeit at a much reduced level, by the midwife and obstetrician. However, unlike the baby, the medical staff can and should protect themselves with gloves, masks, visors and protective gowns. Their reluctance to do so excessively, for fear of intimidating the woman undergoing delivery, is a laudable, if perhaps foolhardy, example of health professionals putting patient care first. Fortunately, there are, as yet, no reported cases of health care workers being infected with HIV during delivery.

For the babies yet to be born to HIV antibody-positive mothers, there may be some good news. A recent multi-centre controlled trial, using matched asymptomatic HIV antibody-positive pregnant women, compared materno-neonatal HIV transmission rates from women who had been treated with either AZT (zidovudine) or placebo. Treatment consisted of oral AZT for up to two trimesters and intravenous AZT infusions during parturition, and their delivered neonates were treated for 6 weeks postnatally (all babies were bottle-fed). The trial was terminated when a reduction in HIV transmission from 25% to 8% was found with use of AZT. Further analysis of the data from this trial should make clear when the AZT was most effective and thus pin-point more precisely periods of maximum risk of viral transmission, and thus the optimal period for use of AZT.

SUMMARY

Once born, inspected, and sexed, the fruit of the past eleven chapters enters into a period of prolonged parental care during which, early on, its nutritional requirements are usually provided by the lactating mother. These topics are the subject of the next chapter.

FURTHER READING

Challis JRG & Lye SJ (1986) Parturition. *Oxford Reviews of Reproductive Biology* **8**, 61–129.

Cupps PT (1991) *Reproduction in Domestic Animals* 4th edition. Academic Press, New York.

Jones CT (Ed.) (1988) *Research in Perinatal Medicine (VII): Fetal and Neonatal Development.* Perinatology Press, Ithaca, New York.

Knobil E & Neill JD (Eds) (1994) *The Physiology of Reproduction* Volume 2, 2nd edition. Raven Press, New York. (Particularly chapter by Challis JRG & Lye SJ, pp. 985–1032.

Künzel W & Jensen A (Eds) (1988) *The Endocrine Control of the Fetus: Physiologic and Pathophysiologic Aspects.* Springer-Verlag, Berlin.

Sherwood OD *et al.* (1993) The physiological effects of relaxin in pregnancy: studies in rats and pigs. *Oxford Reviews in Reproductive Biology* **15**, 43–189.

Lactation and Maternal Behaviour

In this chapter, we will consider the early postnatal events ensuring the survival of the new-born baby. These include the provision of food via the process of lactation and associated nursing, and the relatively extended period of parental care that provides a protective environment in which the young can grow and gradually attain independence.

LACTATION

Among the many changes occurring in the mother during pregnancy are those that involve the breast. In most species, this process is as vital to the success of reproduction as gamete production and fertilization, as failure to lactate will result in early postnatal death of the young. We shall describe the factors that induce, control and regulate breast development, lactation and milk removal by the young, primarily in the human female. We will refer to other species only when data in the human are lacking, or when important differences between species are apparent.

The breast

Anatomy

The mammary gland forms a lobulated mass comprised of glandular (or parenchymatous) tissue, fibrous tissue connecting the *lobes*, and adipose tissue between them. There are 15–20 lobes in the human breast, each consisting of *lobules* of *alveoli*, blood vessels and *lactiferous ducts*. The basic pattern of breast structure shown in Fig. 13.1 is common to all species, even though the number of mammary glands, their size, location and shape vary greatly. The sow, for example, has up to 18 mammary glands (nine pairs), while in the cow and goat their pairs of glands (two and one, respectively) are closely apposed within an abdominal udder. In addition, there is some variation in the pattern of the duct system (Fig. 13.2). Microscopically, the alveolar walls are formed by a single layer of cuboidal to columnar epithelial cells, their shape depending on the fullness or emptiness, respectively, of the alveolar lumen (Fig. 13.3). It is these cells that are responsible for milk synthesis and secretion

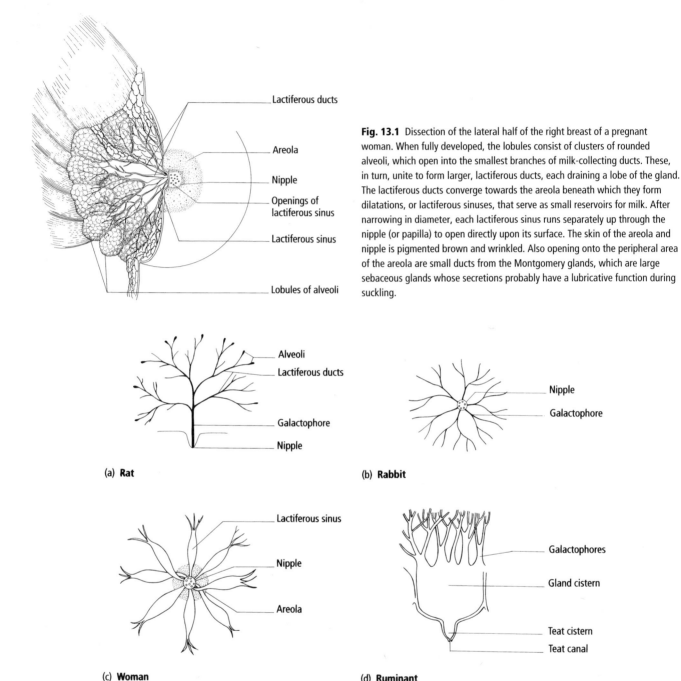

Fig. 13.1 Dissection of the lateral half of the right breast of a pregnant woman. When fully developed, the lobules consist of clusters of rounded alveoli, which open into the smallest branches of milk-collecting ducts. These, in turn, unite to form larger, lactiferous ducts, each draining a lobe of the gland. The lactiferous ducts converge towards the areola beneath which they form dilatations, or lactiferous sinuses, that serve as small reservoirs for milk. After narrowing in diameter, each lactiferous sinus runs separately up through the nipple (or papilla) to open directly upon its surface. The skin of the areola and nipple is pigmented brown and wrinkled. Also opening onto the peripheral area of the areola are small ducts from the Montgomery glands, which are large sebaceous glands whose secretions probably have a lubricative function during suckling.

Fig. 13.2 Different patterns of the ductular system in the mammary glands of four mammals. (a) The rat, in which the lactiferous ducts unite to form a single galactophore, which opens at the nipple. (b) The rabbit, in which a number of lactiferous ducts unite to form several galactophores. (c) The human female, in which one lactiferous duct drains each of 15–20 mammary lobes, dilating as a lactiferous sinus before emerging at the nipple. (d) The ruminant, in which, in the udder, the galactophores open into a large reservoir or gland cistern, which opens into a smaller teat cistern and thence to the surface via a teat canal.

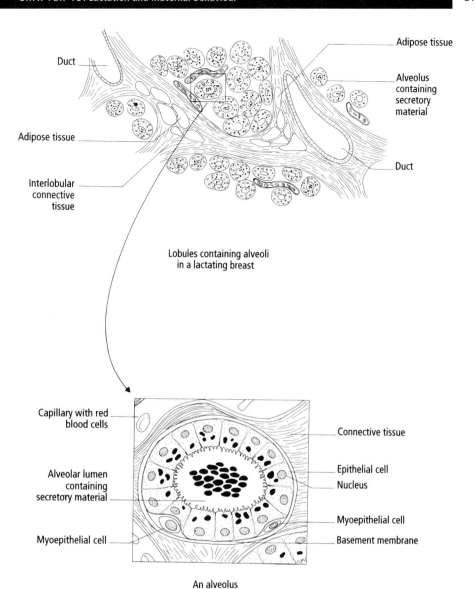

Fig. 13.3 Microscopic structure of (above) lobules in a lactating mammary gland and (below) a high-power view of an alveolus. Note the rich vascular supply from which the single layer of secretory epithelial cells draws precursors used in the synthesis of milk. The myoepithelial cells situated between the basement membrane and epithelial cells form a contractile basket around each alveolus.

during lactation. The *myoepithelial* cells situated between the epithelial cells and the basement membrane have a contractile function, and are important for moving milk from the alveoli into the ducts prior to milk ejection ('Milk removal and the milk ejection reflex' below).

Development

At birth, the mammary gland consists almost entirely of lactiferous ducts with few, if any, alveoli. Apart from a little branching, the breast remains in this state until puberty (see Chapter 1). At this time, and under the action primarily of *oestrogens*, the lactiferous ducts sprout and branch, and their ends form small, solid, spheroidal masses of granular polyhedral cells, which later develop into true alveoli. As menstrual cycles establish themselves, successive exposure of mammary tissue to oestrogen

and progesterone induces additional, if limited, ductal–lobular–alveolar growth, and the breasts increase in size due to the deposition of fat and growth of connective tissue. Adrenal corticosteroids may also contribute to duct development at this time.

Cyclic changes in the breast occur in non-pregnant women and are especially evident premenstrually when there may be an appreciable increase in breast volume (see Chapter 7). In addition, some secretory activity may occur in the alveoli and the resultant small amounts of secretory material can be expressed from the non-pregnant breast during the premenstrual period. The extent of mammary development in non-pregnant women is considerable in comparison with other mammals, including non-human primates, in which appreciable mammary growth is not achieved until the middle or end of pregnancy. In light of this difference, it is not surprising that the hormonal requirements for human breast development during pregnancy also differ from those in other animals. Thus, in other species, notably rodents, a complex of sex steroids, adrenal steroids, growth hormone, prolactin and placental lactogen combine to induce mammary growth, reflecting the relatively poor mammary development occurring before pregnancy. In women, neither placental lactogen nor growth hormone are essential. Rather, during early pregnancy, and under the influence of oestradiol, progesterone and possibly insulin and prolactin, the previously developed ductular–lobular–alveolar system undergoes considerable hypertrophy. Growth factors may play a significant role in regulating mammary growth. Epidermal growth factor and transforming growth factor α, in particular, are able to stimulate the growth of normal mammary cells *in vivo* and *in vitro*, and have been localized to, or are synthesized in, mammary tissues. Their activities seem to be under the regulatory control of mammogenic hormones.

Under these hormonal influences, prominent lobules form in the breast, and the lumina of the alveoli become dilated. Differentiation of the alveolar cells to the form shown in Fig. 13.4 occurs during mid-pregnancy at a time when duct and lobule proliferation has largely ended. The epithelial cells contain substantial amounts of secretory material (Fig. 13.4) from the end of the fourth month of human pregnancy, and the mammary gland is fully developed for lactation, awaiting only the endocrine changes described below for full activation.

The biochemistry of milk

Milk *fat* is synthesized in the smooth endoplasmic reticu-

lum of the alveolar epithelial cells and passes in membrane-bound droplets of increasing size towards the luminal surface of the cell (Fig. 13.4). The droplet then pushes against the cell membrane, causing it to bulge and lose its microvilli. Gradually the cell membrane constricts behind the lipid droplet to form a 'neck' of cytoplasm, which ultimately pinches off to release it, membrane-enclosed, into the alveolar lumen (Fig. 13.4). In contrast, milk *protein* passes through the Golgi apparatus into vacuoles, and is released by exocytosis (Fig. 13.4). Both release processes depend on *prolactin*, receptors for which are present on the alveolar cells.

The composition of milk varies with time *postpartum*. Up to 40 ml/day of a yellowish, sticky secretion called *colostrum* is secreted during the first week postpartum, and, compared with mature milk, contains less water-soluble vitamins (B complex, C), fat and lactose, but greater amounts of proteins, some minerals and fat-soluble vitamins (A, D, E, K), and immunoglobulins (IgGs). During a transitional phase of 2–3 weeks, the concentration of IgGs and total proteins declines, while lactose, fat and the total calorific value of the breast milk increase to yield mature milk, the contents of which are summarized in Table 13.1 (see also discussion on transmission of HIV in Chapter 12).

One or two features of human milk will be emphasized here. The main energy source in this milk is fat, which is almost completely digestible, partly due to the fact that it is present as well-emulsified, small fat globules. Milk fat is also an important carrier for vitamins A and D. *Lactose* (milk sugar) is the predominant carbohydrate in milk. It is less sweet than common sugars and is important for promoting intestinal growth of *Lactobacillus bifidus* flora (lactic acid-producing), as well as providing an essential component (*galactose*) for

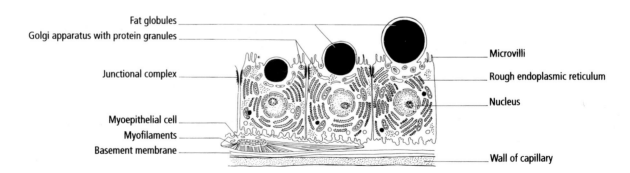

Fig. 13.4 Schematic drawing of the ultrastructure of secretory alveolar epithelial cells. The position of the myoepithelial cell can clearly be seen (compare with Fig. 13.3). Adjoining alveolar cells are connected by junctional complexes near their luminal surfaces, which themselves bear numerous microvilli. The cell cytoplasm is rich in rough endoplasmic reticulum, particu-larly in the basal part of the cell. Many mitochondria are present and the large Golgi apparatus is situated nearer the luminal surface and close to the cell nucleus. The fat globules and protein granules are the cellular secretory products destined for the alveolar lumen. Their manner of extrusion from the cell is also indicated.

Water	Approximately 90 gm%
Lactose	Approximately 7 gm%
Fat	Including essential fatty acids, saturated fatty acids, unsaturated fatty acids
Amino acids	Including essential amino acids
Protein	Including lactalbumin and lactoglobulin
Minerals	Including calcium, iron, magnesium, potassium, sodium, phosphorus, sulphur
Vitamins	Including A, B$_1$, B$_2$, B$_{12}$, C, D, E, K
pH	7.0
Energy value	650 kcal /100 ml

Table 13.1 Some contents of human mature milk

myelin formation in nervous tissue. Lactose is formed within the Golgi apparatus of alveolar cells and is dependent on a combination of α-lactalbumin (the whey protein) and the enzyme galactosyltransferase, which together form lactose synthetase. The sugar passes to the alveolar lumen with the protein granules. Galactosyltransferase activity is stimulated by prolactin.

The initiation and maintenance of milk secretion

It should be clear from the account so far that, in the human, the breasts are sufficiently developed, and the alveoli adequately differentiated, to begin milk secretion by 4 months of pregnancy. But the copious milk secretion characterizing full lactation does not occur until after parturition. Why? The disappearance of oestrogen and progesterone (and perhaps of human placental lactogen (hPL)) from the maternal circulation occurring at or soon after parturition holds the key to this initiation of lactation (lactogenesis). Although prolactin increases in plasma concentration throughout pregnancy and reaches a maximum at term (Fig. 13.5), the breast is simply not responsive to it until steroid levels (and particularly progesterone levels) fall. The steroids (helped perhaps by hPL) appear to inhibit milk secretion by acting directly on mammary tissue, probably on the alveolar cells.

After parturition, prolactin levels also fall abruptly but more slowly. In the absence of suckling, the newly initiated milk secretion will last, if scantily, for 3 or 4 weeks, during which period blood prolactin concentrations remain well above normal. However, if prolactin levels are to remain elevated and full lactation is to continue with copious milk secretion (galactopoiesis), suckling and attendant nipple stimulation are essential. Suckling achieves this release of prolactin from the anterior lobe of the pituitary via a neuroendocrine reflex, and denervation of the nipple prevents prolactin release in response to nipple stimulation (see also discussion of suckling-induced delay of implantation, Chapter 9). The afferent limb of this reflex consists of neural pathways convey-

ing sensory information from the nipples up the spinal cord in the anterolateral columns and via the brainstem to the hypothalamus (Fig. 13.6). In the hypothalamus, two efferent components are at work (see Chapter 5). First, there is a reduction in dopamine (prolactin inhibitory factor (PIF)) secretion into the portal vessels during nipple stimulation. Second, experiments in suckling rats have revealed a marked increase in the secretion of the potent prolactin releasing factor vasoactive intestinal polypeptide (VIP) into the portal vessels. Increased amounts of VIP mRNA are seen in the

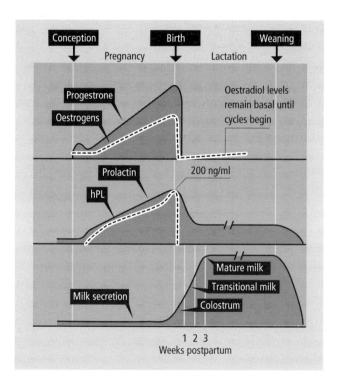

Fig. 13.5 The sequence of hormone changes in the maternal circulation which underlie the onset of lactation in women. Withdrawal of oestrogen and progesterone is critical and removes a block to prolactin-induced milk secretion in the gland. Withdrawal of hPL may have a similar function but this is less certain.

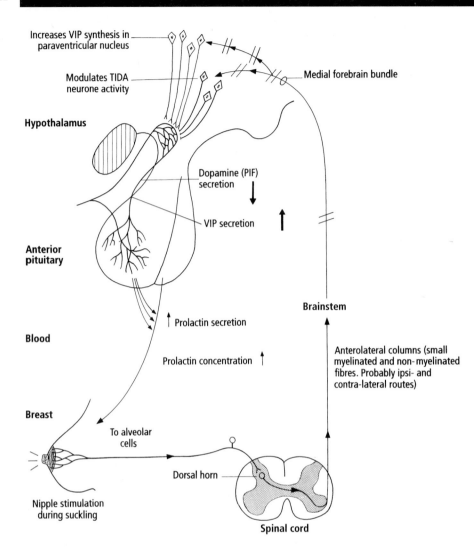

Fig. 13.6 Somatosensory pathways in the suckling-induced reflex release of prolactin. The exact route taken by sensory information between brainstem and hypothalamus is not certain. Although tuberoinfundibular dopamine (TIDA) neuron activity is modulated as a result of the arrival of this somatosensory-derived input, the increased secretory activity of VIP-containing neurons in the paraventricular nucleus is of crucial importance in driving prolactin secretion during suckling.

neurons of the medial parvocellular paraventricular nucleus, and axons from these terminate in the palisade zone of the median eminence in association with portal capillaries. The reduction in tuberoinfundibular dopamine neuron activity and hence dopamine release during suckling also sensitizes the lactotrophs to VIP, indicating that the two mechanisms may interact positively. The amount of prolactin released is determined by the strength and duration of nipple stimulation during suckling. Suckling at both breasts simultaneously, when feeding twins for example, induces a greater release of prolactin than occurs during stimulation of one breast.

The circulating plasma concentration of prolactin during lactation appears, more than any other factor, to determine the amount of milk secreted in women. Thus, declining milk secretion in women can be boosted, with consequent breast engorgement, by treatment with thyrotrophin releasing hormone to release prolactin (see

Chapter 5). Clearly, the suckling stimulus or nipple stimulation fulfils a most important function in galactopoiesis. In other species, additional hormones (growth hormone, insulin and adrenal steroids among them) may be essential for the successful maintenance of lactation.

The fact that nipple stimulation during a feed (suckling) induces prolactin release, which subsequently induces further milk secretion, suggests that *the baby actually orders its next meal during its current one.*

Summary

We have seen that development of the mammary gland's ductular–alveolar system and milk-secreting capacity during pregnancy is under the influence of adrenal, ovarian and placental steroids. The initiation of milk secretion depends on the presence of high levels of pro-

Hormone or activity	Effect
Early to mid-pregnancy	
Oestrogen, progesterone and corticosteroids	Ductular–lobular–alveolar growth. Considerable branching of the duct systems until mid-pregnancy. Followed by considerable differentiation of epithelial stem cells into a true alveolar secretory epithelium
Oestrogen and progesterone (hPL?)	Little or no milk secretion occurs due to the inhibitory effects of these hormones on prolactin stimulation of alveolar cells. Continues until ...
Late pregnancy and term	
Oestrogen and progesterone	Pronounced alveolar epithelial cell differentiation
Steroid and prolactin levels high Steroids begin to fall	Colostrum secretion
Parturition	
Oestrogen and progesterone fall precipitously. Prolactin levels decline but basal concentrations remain high	Stimulation of active secretion of colostrum and, over 20 days or so, secretion of mature milk. Full lactation initiated
Suckling	Induces episodic prolactin release at each feed. Maintains milk secretion by promoting synthesis of lipids, milk (particularly α-lactalbumin) and lactose

Table 13.2 Summary of events leading to full lactation in the human

lactin and withdrawal of oestrogen and progesterone. The maintenance of milk secretion then depends, in the human female, solely on the continued secretion of prolactin, maintained by nipple stimulation during suckling. These events are summarized in Table 13.2. In the next section, we will examine how the suckling infant removes milk from the breast, and the *milk ejection reflex*, which subserves this task.

Milk removal and the milk ejection reflex (MER)

Milk removal involves transport of milk from the alveolar lumina to the nipple (or teat) where it becomes available to, and is removed by, the suckling infant. The MER, which underlies this function, has much in common with the reflexly induced release of prolactin described above.

Neuroendocrine mechanisms underlying the MER

Stimulation of the nipple during suckling probably represents the most potent stimulus to milk ejection. The sensory information so generated travels via the spinal cord and brainstem to activate *oxytocin neurons* in the paraventricular and supraoptic nuclei in the hypothalamus (Fig. 13.7). This input boosts not only the synthesis

of oxytocin, but also its release from the posterior pituitary into the bloodstream. On reaching the mammary gland, oxytocin causes contraction of the myoepithelial cells, which surround the alveoli, to induce the expulsion of milk into the ducts and a build up of intramammary pressure (*milk let-down* or 'draught'), which may cause milk to spurt from the nipple or teat.

Although touch and pressure at the nipple are very potent stimuli to oxytocin release and milk let-down, the MER can be *conditioned* to occur in response to other stimuli, such as a baby's hungry cry or, in cows, the rattling of a milk bucket. Such a *conditioned reflex* release does *not* occur in the case of prolactin, where nipple stimulation seems to be the only effective inducer. The fact that the cow's udder contains, in the gland cistern, all the milk obtainable during milking emphasizes that let-down is due to the increase in mammary pressure resulting from the expulsion of milk from alveoli to fine ducts, rather than any sudden increase in milk secretion as was once thought. The same situation applies to the human female.

In Chapter 12, we described how stimulation of the female's reproductive tract, particularly the vagina and cervix, also induces the release of oxytocin. The oxytocin release so induced explains the phenomenon of milk ejection during coitus in lactating women and the rather

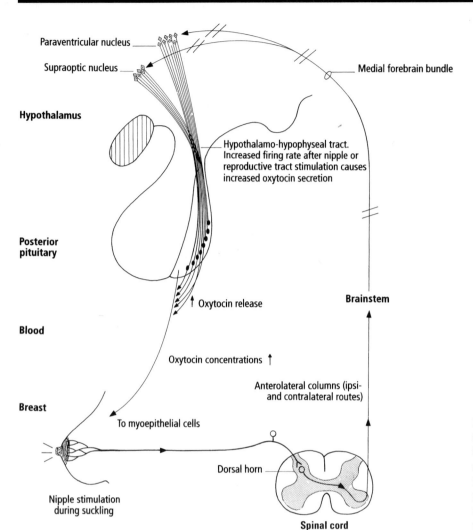

Fig. 13.7 Somatosensory pathways in the suckling-induced reflex release of oxytocin. The brainstem–hypothalamic route taken by the sensory information is uncertain, but it probably involves an important relay in the midbrain peripeduncular nucleus before travelling in the medial forebrain bundle to reach the magnocellular nuclei.

ancient, enduring and erstwhile puzzling practice of blowing air into a cow's vagina to induce milk draught! The MER is particularly susceptible to inhibition by physical and psychological stresses; discomfort immediately after parturition or worry and uncertainty about breast-feeding are potentially important inhibitors of the successful initiation and early maintenance of lactation. The way in which 'stress' inhibits the MER is not clear, but may involve inhibition of oxytocin release, and/or the release of catecholamines, such as adrenaline, and activation of the sympathetic nervous system. Constriction of mammary blood vessels induced by adrenergic stimulation might limit access of oxytocin to the myoepithelial cells.

The afferent route taken by sensory information arising at the nipple (or reproductive tract) is quite well established from research on rodents. It involves the peripheral sensory nerves, which enter the spinal cord via the dorsal roots, and a number of synaptic relays in the dorsal horn before transmission up the anterolateral columns. These pathways contain fibres destined for various sites in the thalamus, spinal cord and, in particular, the brainstem reticular formation. From this point, the route to the hypothalamic, magnocellular paraventricular and supraoptic nuclei (Fig. 5.2) is not entirely clear, but the *midbrain peripeduncular nucleus* appears to be an important relay. Fibres then run via the *medial forebrain bundle*, which courses through the lateral hypothalamus, to reach the magnocellular nuclei.

Suckling

There has, for a number of years, been considerable debate as to the mechanics of suckling. Many workers hold that infants obtain milk from the breast by actively sucking, and that an airtight seal between lips and breast is essential for the negative pressure-dependent transfer of milk. However, X-ray cinematographic evidence suggests that this may not be the case and that milk is obtained by expressing it from the nipple or teat, sucking merely aiding the process. Thus, in the human, the nipple and areola are drawn out to form a teat, which is compressed between the infant's tongue and hard palate. The milk is then stripped out of this 'teat' by the tongue compressing the nipple from base to apex against the hard palate. Pressure on the base of the teat is then released, allowing its rapid refill with milk due to the oxytocin-induced increase in intramammary pressure that subserves let-down. There are undoubtedly species differences in this process, and young also readily adapt to alternative means of obtaining milk. Bottle-feeding using stiff teats, for example, requires more sucking than stripping, and calves or human infants learn this skill rapidly.

Summary

Removal of milk from the breast is dependent on the suckling-induced MER. By this means, stimulation of the nipple and other cues associated with nursing induce the release of oxytocin from the neurohypophysis. This hormone stimulates contraction of the myoepithelial cells, which surround each alveolus, forcing milk out of the alveoli into the smaller lactiferous ducts. The resulting increase in intramammary pressure results in milk let-down, causing milk to spurt from the nipple and to be removed from the breast by the suckling infant. The neural pathways mediating the MER are not understood completely at present, but are likely to be complicated, particularly when mediating psychological factors associated with the induction or inhibition of the MER.

Cessation of lactation

Involution of the breast

After the cessation of lactation, involution of the mammary gland takes about 3 months. When nursing is discontinued, milk accumulates in the alveoli and small lactiferous ducts causing distention and mechanical atrophy of the epithelial structures, rupture of the alveolar walls and the attendant formation of large hollow spaces in the mammary tissue. The alveolar distention also results in alveolar hypoxia and reduction in nutrient supply because mammary capillaries are compressed. Milk secretion is, therefore, greatly suppressed not so much by the fall in plasma prolactin (as suckling frequency diminishes), but as a consequence of the effects of local mechanical factors. As desquamated alveolar cells and glandular debris become phagocytosed, the lobular–acinar structures become smaller and fewer, and the ductular system in the breast again

begins to predominate. The alveolar lumina decrease in size and may eventually disappear, and their lining changes from the secretory single-layered to a non-secretory two-layered type of epithelium, previously seen prior to pregnancy. This whole involutional process is more intense if nursing is stopped suddenly rather than continuing at reduced frequency during gradual weaning.

Although these changes in the mammary gland are pronounced, they are markedly different to those occurring in postmenopausal women. In the latter, there is clear structural atrophy of the breast rather than a transition to a period of inactivity. The breasts invariably remain larger after lactation than they were prior to pregnancy because of the increased deposition of fat and connective tissue that occurs between these two time points.

Suppression of lactation

It may be necessary to suppress lactation in women for a variety of reasons. Nursing may be contraindicated for clinical reasons, for example the mother may be HIV antibody-positive (see Chapter 12) or may simply prefer not to breast-feed her baby. Stillbirth or abortion after 4 months of pregnancy will be followed by unwanted and unnecessary lactation. A number of traditional methods for suppressing lactation may still be employed today, including breast-binding, application of ice packs, or treatment with sex steroids to antagonize the effects of prolactin. However, the most widely used lactation suppressant today is bromocriptine, the dopamine D_2 receptor agonist, which markedly depresses prolactin, and hence milk secretion (see Chapter 5).

Fertility during lactation

Lactation can continue for months. Menstruation and ovulation return more slowly in lactating than non-lactating women postpartum, normal reproductive function usually re-establishing itself by 3–6 months. However, menstruation itself is a poor indicator of fertility during this period, as many women bearing successive children within a year can testify! Conception often occurs in lactating women without an intervening menstruation. Neither ovulation nor menstruation normally occur before 6 weeks postpartum, but approximately half of all contraceptively unprotected, nursing mothers become pregnant during 9 months of lactation. The reasons for lactational amenorrhoea and anovulation are at present unclear. The most popular hypotheses favour a critical role for prolactin in suppressing the initiation of cyclic release of gonadotrophins. These are discussed more fully in the context of hyperprolactinaemia (which also characterizes lactation) in Chapter 5 and of facultative delayed implantation (see Chapter 9).

MATERNAL BEHAVIOUR

In Chapter 1, we gained some insight into how important the influence of behavioural interactions between a growing infant and its parents could be in forming a sexual and gender identity. This example illustrates the vital role that parents play in ensuring not only the normal development but also the very survival of their offspring. Early on, the mother is of special importance and she must display a whole range of interrelated patterns of maternal behaviour in order that her offspring is given protection, warmth, food and affection. The infant, however, is not merely the passive receiver of all this attention, but an active participant in a two-way interaction, eliciting by its own actions appropriate responses from its mother. Our understanding of how efficient parental care operates, particularly in the human, is far from complete. Here, we will examine some of the important features in a comparative setting in order to reveal some of the general principles involved.

General considerations

Maternal behaviour may be considered to occur in three sequential stages: (1) behaviour preparatory to arrival of the young displayed during gestation, for example nest building; (2) behaviour concerned with the care and protection of the young early after parturition and associated with lactation; (3) behaviour associated with the progressive independence of the young and associated with weaning.

During gestation many mammals enter a phase of nest-building activity that is highly characteristic for each species. Very often at this time, the female may show a marked increase in aggression and defend the area in which the nest has been made. Females of few species, and human females may be among them, will accept the sexual advances of males as gestation proceeds.

After birth, the females of all mammalian species are critical to the survival of their young, because they provide the essential food supply during lactation. The other roles of the mother during the early postnatal period vary from species to species depending on the stage of development of the young at birth. Those

species in which the young are very immature, notably the marsupials, show little change in their behaviour at parturition. The young crawls into the mother's pouch, attaches to a teat, and the mother, apart from cleaning her pouch more often, shows little additional maternal behaviour, apparently not recognizing her own as distinct from other young at this time.

Young who are born naked and blind (e.g. mice, rats, rabbits, ferrets and bears), so-called *altricial young*, have food and warmth provided by their suckling mothers in nests. Maternal retrieval forms an important element of behaviour in these species, often elicited and directed by ultrasonic calls from misplaced young.

Semi-altricial young are those born with hair and sight (e.g. carnivores, non-human primates), but have poorly developed motor skills and may need to be carried by their parents. The parents may have nests or dens in which the female stays with the young most of the time (e.g. the canids), the males returning periodically to the nest to regurgitate food, initially for the non-excursive mothers, but later for the young as well. Primates do not normally leave their young in nests, but carry them around, the infants having well-developed clinging reflexes. Males often help in this task.

Ungulates, hystricomorph rodents and aquatic mammals give birth to *precocious* young, which can move about well, and, to some extent, fend for themselves. The mother–infant bond in these species seems to reflect more a need for contact rather than nourishment, and mothers appear to recognize their own young quickly, and thereafter will feed only them.

As infants grow, there is a gradual change in both their behaviour and that of their mothers, which ultimately results in the attainment of independence. Infants, during this time, tend to move away from the mother more often and to greater distances. Mothers, on the other hand, retrieve them less and encourage this exploration by rejecting them more often. Suckling occurs less frequently, lactation ceases and the infant is weaned. In solitary species, the young move away from their mother quite rapidly, the females stop maintaining the nest and the temporary family aggregation disintegrates. In more social, group-living species, the young become independent of, and are rejected by, their mothers, but are progressively integrated within the group.

Maternal behaviour comprises, therefore, a complex and variable pattern of interaction between mother and young, well adapted to the social and environmental context in which it occurs. Despite, or perhaps because of, this variety, little is known in detail of the mechanisms underlying the onset and maintenance of the mother–infant bond in many species.

Non-primates

The onset and maintenance of maternal behaviour

A prevailing belief, in studies of maternal behaviour, has been that the distinctive profile of pregnancy hormones, progesterone, oestradiol and prolactin/PL, must in some way be involved in nest building prior to birth, and in the prompt appearance of maternal behaviours directed so selectively towards the mother's own young after birth. However, it is clear that hormones are not an *essential requirement* for the display of maternal behaviour. Thus, ovariectomized, hypophysectomized female rats, and even castrated or intact male rats, will all develop elements of the 'maternal' behaviour pattern when exposed to pups for a period of about 4–7 days. This 'pup-stimulation', in which attributes of the pups evoke behavioural changes, is often called 'sensitization' and can occur independently of any hormonal factor. It is also generally accepted that continuing maternal behaviour in the normal, postpartum lactating female is not dependent on hormones, as postpartum ovariectomy and hypophysectomy are not followed by any *decline* in behaviour (other than via endocrine effects on lactation, which must be taken into account). Thus, hormones do not seem to be *required* for either *initiation* or *maintenance* of maternal behaviour. Neither do they seem to influence the naturally occurring *withdrawal* of maternal behaviour, as both sensitized and lactating 'mothers' show a similar decline in their maternal care over a period of 10–20 days postpartum, avoiding the nursing of pups and increasing their rejection.

It would be quite wrong, however, to suppose that hormones have *no* influence on elements of maternal behaviour. Non-postpartum females (or males) require several days of exposure to pups before they display such behaviour, while *postpartum* females display all elements of maternal behaviour immediately after delivery. Perhaps the hormones of gestation, and particularly the changes occurring prior to parturition, are important determinants of the *prompt onset* of maternal behaviour. Sequences of injections of oestradiol, progesterone and prolactin, to mimic the changes seen during pregnancy, do induce a more rapid onset of maternal behaviour in virgin females presented with pups. Oestradiol appears to be the most important hormone in this regard. However, although eliciting a more rapid response, this treatment has never adequately mimicked the *immediacy* of maternal care following normal delivery. Clearly, there is something special about parturition itself that renders the mother uniquely sensitive to the new-born. Experiments on sheep have shown that this is indeed the case.

Non-parturient ewes, or parturient ewes 2 hours or more after having given birth, will not accept or nurse an orphaned lamb even if they are oestrogen primed. They can be induced to do so with 50% or so success by being made temporarily anosmic by use of a nasal spray, or alternatively by draping the pelt of the ewe's own dead lamb over an orphaned lamb that requires fostering. Olfactory cues from the lamb apparently, then, prevent it from being nursed by any mother, other than its own. The parturient ewe will accept an alien lamb and nurse it along with its own, provided it is presented within 2 hours or so of the ewe having herself given birth. The reason for this seems to be the ewe's altered olfactory responsiveness to the lamb during the immediate postpartum period. There are two possible mechanisms. Perhaps the ewe can selectively recognize the odour of her own lamb and respond only to it. Alternatively, the ewe may be functionally anosmic in the immediate *postpartum* period and, therefore, does not respond negatively to an alien lamb by rejecting it. The latter explanation seems more likely, but in either case, how is this olfactory mechanism brought into play? In an ingenious experiment, stimulation of the cervix and distension of the vagina (using a vibrator and the bladder of a rugby football, respectively) in oestrogen-primed non-parturient ewes caused the immediate (within minutes) display of maternal behaviour towards alien lambs (Fig. 13.8). The same immediate response has now been seen to occur in rats: cervical stimulation of virgin, oestrogen-treated females caused maternal behaviour within minutes of exposure to pups.

Neural mechanisms underlying maternal behaviour

The neural basis of maternal behaviour has been studied mainly in the rat and resembles that underlying sexual behaviour. Thus, the medial preoptic area appears to be of major importance, as bilateral lesions of this area severely impair pup retrieval, nest building and nursing responses in *postpartum* female rats. The medial preoptic area may also be important in mediating the effects of oestradiol on maternal behaviour, as implantation of oestradiol in this area facilitated maternal behaviour in hypophysectomized/ovariectomized female rats.

The *medial amygdala* is an especially important structure in mediating the use of olfactory information in the regulation of maternal behaviour. It receives major projections from the olfactory system, and particularly from the *accessory olfactory bulb*, which is the principal recipient of the special kind of olfactory information processed by the *vomeronasal organ* in non-primate species. The medial amygdala, in its turn, projects richly to the medial preoptic area, either directly or via the *bed nucleus* of the *stria terminalis*. Lesions of the medial amygdala

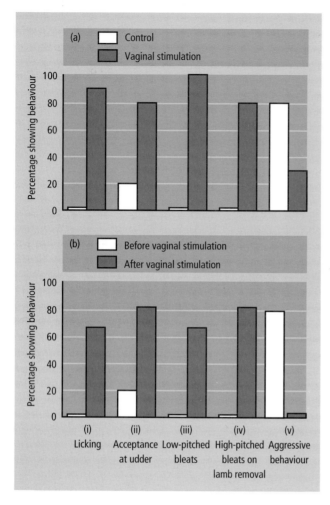

Fig. 13.8 (a) The effects of vaginal stimulation on the maternal behaviour of non-pregnant ewes. After stimulation, ewes: (i) lick the alien lambs; (ii) allow them to suckle at the udder; (iii) emit low-pitched bleats characteristic of a maternal ewe; or (iv) emit high-pitched bleats, indicating distress, if the lamb is removed; and (v) exhibit a marked decrease in aggression towards the lamb. (b) Here, the controls in (a), who showed little or no maternal behaviour, were subjected to vaginal stimulation at the end of the observation period. As can be seen, their change in behaviour towards the alien lamb immediately afterwards is dramatic.

facilitate maternal behaviour by reducing its latency of onset in female rats exposed to pups. Thus, removing either the receipt of olfactory information by peripheral anosmia or its central transfer by amygdalectomy, reduces the time required for pup stimulation to sensitize maternal responses.

The dramatic dependence of maternal behaviour on olfactory processing in sheep has been studied in great detail by measuring moment-to-moment changes in transmitter release within the olfactory bulb of post-parturient ewes in the process of becoming maternal.

Marked changes in oxytocin release are evident around parturition, as might have been expected given the effects of cervical stimulation on the onset of maternal behaviour. Noradrenaline release is also markedly altered within a similar time-frame. Both these changes in neurotransmitter release in systems afferent to the olfactory bulb seem to underlie the olfactory recognition by a mother of its offspring. Such processes ensure the rapid onset of maternal responses that are vital if the lamb is to gain access to its mother's milk and survive the difficult first few weeks of life. Virtually nothing is known about the neuroendocrine mechanisms of maternal behaviour in primate species.

Summary

The sequence of events that influence the display of maternal behaviour would appear to be: (a) exposure to hormones, particularly oestradiol, during late pregnancy; (b) parturition or cervical stimulation itself; and (c) continuing exposure to the new-born and growing infant. Maternal behaviour will be displayed in the absence of either (a) or (b), but with reduced success. The realization that cervical stimulation has such dramatic effects has already had an impact on sheep farming. It has also provided a stimulus to studies of maternal behaviour in women as well, as the data might have implications for the success of mother–infant bonding following non-vaginal deliveries. Indeed, mother–infant bonding after Caesarean delivery may be different to that seen after normal delivery, but whether or not this is related to differences in cervico-vaginal stimulation has not been studied directly. In humans and other primates, little is known of hormonal or other determinants of maternal behaviour, but some work, particularly on rhesus monkeys, has revealed a number of factors that contribute to the development of the mother–infant bond and its subsequent severance. These findings may have particular relevance to our understanding of human maternal behaviour. Although species differences should never be minimized, comparative studies of maternal behaviour may serve to focus our attention on critical, and often analogous, elements of behaviour, and their phases of development in humans.

Primates

Maternal behaviour in non-human primates and its relationship to human maternal behaviour

Baby monkeys spend the majority of their first few months or years (in apes) of life clinging tenaciously to their mothers, and occasionally fathers, aunts and juveniles, in characteristic ventro-ventral, back-riding or arm-cradled positions. During this period, the mother gives relatively little active physical support to her young, particularly if they cling ventro-ventrally, and so a newly-born monkey's motor capabilities are essential to successful early maternal contact. Similarly, a well-developed 'rooting reflex', also seen in human babies, by which the head turns towards a tactile stimulus around the mouth, particularly the cheeks, ensures that the infant gains the nipple in order to suckle. These clinging and rooting reflexes are present immediately after birth, and are, in turn, elicited by cues from the mother, largely because she is the individual most likely to be in proximity to the infant.

'Contact comfort' also seems to be an important determinant of an infant's early goal-directed movements, as experiments using surrogate mothers have shown. Thus, a 'cuddly' towelling-covered wire model is much preferred by a baby monkey to a bare wire surrogate, even if the latter can provide milk. Similarly, a warm, moving, milk-providing surrogate will be preferred to one without these attributes. Although it is difficult to say why babies prefer, and derive pleasure from, such cues, it presumably involves feelings of security and comfort. There are many analogous instances when the behaviour of human infants towards a warm, rocking mother or soft blanket is considered.

Clinging, contact comfort and rooting behaviours emphasize the baby's contribution to the interactions at the earliest moments after birth, and are of importance in establishing the bond with the mother and securing the immediate needs of warmth, contact and food with only minimal help from her. The mother must also keep her infant clean and protect it. Such behaviour is clearly elicited quite specifically by the infant, which emphasizes that the close contact at suckling forms the focal point around which subsequent patterns of maternal behaviour develop. Although the baby initiates contact by clinging, the mother clearly derives considerable rewards too, as evidenced by the fact that a mother will carry her dead baby around for some days and show distress when it is removed. It is extremely difficult to assess what the nature of this reward is.

In social primates, females have plenty of opportunity to learn and 'rehearse' skills they will subsequently need as mothers. They will watch other monkeys, particularly their mothers, holding infants and may even 'practice' by holding siblings themselves. Experiments demonstrating that females reared in a socially deprived environment make poor, aggressive and rejecting mothers are well known. There may also be

important parallels in disturbances of parental care in humans.

Initially, the infant does not respond to its mother as an individual, and its filial responses can be elicited by a wide range of stimuli (fur, nipples, etc.), usually but not necessarily associated with its own mother's body. Eventually the range of stimuli eliciting responses in the baby becomes narrowed, through the developing process of maternal recognition, to those from its mother alone. Similar processes occur in human babies. Olfactory cues seem to be of particular importance in the mechanisms by which babies recognize their mothers and vice versa.

Communication between mother and infant takes several forms, and these reinforce both mutual recognition and the mother–infant bond. Thus, we have already seen that the mother gains some positive stimulation from carrying her infant, and qualities of the latter's coat, its size, shape and smell may contribute to its 'attractiveness' in this context. Baby monkeys also make characteristic vocalizations, for example 'whoos' when separated from their mothers and 'geckers' when frightened or denied access to the nipple. The mother usually responds rapidly to such vocalizations with physical contact, giving an immediate soothing effect. There are obvious parallels with human babies who cry when in pain, frightened, hungry or in a temper, and who are soothed by close contact with their mothers. Furthermore, mothers can distinguish the different types of crying and learn to respond appropriately to their baby's needs. Undoubtedly, there are many more subtle and, as yet, poorly defined means of communication between mother and infant: facial expressions (e.g. 'grins' in monkeys, and smiles in human babies), which convey different things at different times, also probably contribute in an important way to the mother–infant bond.

As an infant grows and begins to move away from its mother, it is essential that it understands and complies with its mother's signals concerning potential hazards, for example proximity of a predator. Thus, communication between mother and infant changes in time, as does the bond between them, in parallel with, and often as a consequence of, the infant's cognitive and motor development. This, of course, one knows intuitively to be the case, but it cannot be overemphasized, as the very gradual process of gaining independence from the mother and exploring the environment, including play with other infants, is vital, in its turn, for cognitive development. Extreme fear of strangers and novel surroundings could easily result in an infant never leaving its mother, and, thus, failing to gain experience of the wider world in which it must live. Indeed, monkeys reared in isolation show great fear of novel stimuli when released, probably because their cognitive development has been restricted and impaired. It is in this context, then, that the delicate balance between a mother's rejection of the infant and the latter's curiosity and exploratory tendencies are of great importance. The mother's proximity and availability allow the infant to resolve the conflict between explore–retreat tendencies and so increase its familiarity with strange objects while assessing their safety or hostility (Fig. 13.9). If, during this time, a more prolonged separation of the infant from its mother is imposed, by taking her away, devastating effects can follow. The baby shows considerable distress, withdraws into a hunched, depressed posture, and decreases its motor activity. Reuniting the pair is followed by a period of intense contact, but with eventual and gradual rejection of the infant by the mother to re-establish pre-separation patterns and levels of interaction. The longer the separation the more severe the effects, and the mother's behaviour is critical in restoring the infant's security, just as it is in maintaining it during normal mother–infant bonding. The fact that, under some circumstances, a period of separation can have long-term effects on behaviour, particularly in terms of fear responses to strange objects in an unfamiliar environment, has led to the belief that such events can occur in human babies. It is important to emphasize, however, that because mother–infant separation *can* have long-term behavioural effects, it *need* not necessarily have them and many factors may influence the final outcome.

Human mother–infant interaction

It is beyond the scope of this book to do credit to the wealth of data on mother–infant interaction in humans. However, one or two aspects will be presented as they demonstrate striking parallels with and amplify some of the data described above. Despite the almost folklorish assertion that new-born babies cannot see, it is now quite clear that they can, and furthermore, that they can direct their attentive responses selectively to specific visual stimuli. In particular, stimuli associated with the human face seem especially important and there is good evidence for facial mimicry in babies just a few hours old. The fact that mothers tend to look intently at their baby's face from the earliest moments after birth therefore indicates a reciprocity in mother–infant interaction of considerable importance in establishing the bond between them, leading subsequently to their mutual recognition. Observations of early maternal behaviour after home deliveries in California show that, immediately postpartum, mothers pick up their infant, stroke its face and start breast-feeding while gazing intently

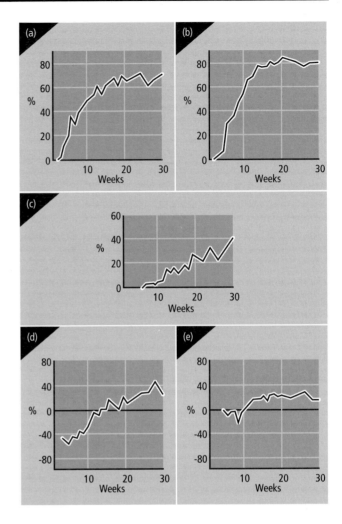

Fig. 13.9 The course of mother–infant interaction in small captive groups of rhesus monkeys. (a) Total time infant spends off mother as a percentage of total time watched. (b) Time spent out of arm's reach (> 60 cm) of mother. (c) Relative frequency of rejections (ratio of numbers of occasions on which infant attempted to gain ventro-ventral contact and was rejected, to number of occasions on which it made contact on mother's initiative, on its own initiative or attempted unsuccessfully to gain contact). (d) Infant's role in ventro-ventral contacts (number of contacts made on infant's initiative, as a percentage of total number made, minus number of contacts broken by infant, as a percentage of total number broken). (e) Infant's role in the maintenance of proximity. Weeks indicates the first 30 weeks of life.

Note that, initially, the infant stays on the mother all the time, clasping her ventro-ventrally, but gradually it spends more time off her, both at hand and also out of arm's reach. Early on, the mother is primarily responsible for the close contact, restricting the infant's sorties by hanging on to a tail or foot. During this time, therefore, the infant is responsible for breaking, and the mother for making, contact. Later, however, the infant becomes primarily responsible for making contact, as the mother rejects its approaches more often and initiates contact less often.

into its face. In hospital deliveries, the pattern differs only slightly, the mothers particularly exploring their baby's extremities, but still with considerable emphasis on eye-to-eye contact. It is the contention of many workers that this early period of intense eye-to-eye contact and physical exploration, which often forms quite a consistent sequence, may be very important in mediating the early formation of the mother–infant bond. But the interaction is not one way, the baby emits signals to the mother that evoke her maternal responses and willingness to nurse.

Reference was made above to the effects of early separation on subsequent mother–infant interaction in rhesus monkeys. Some evidence indicates that enforced early separation of women from their new-born babies for periods up to 3 weeks, for example as might occur after premature delivery, can be associated with differences in *attachment behaviour* (bonding) when compared with mothers similarly separated from their babies, but allowed additional contact during the first few days after birth. Modern paediatric practice takes account of such findings, and, where possible, ensures as much contact as is practicable between mothers and their young. Again, it must be emphasized that although postnatal separation of mother and baby can have delayed and long-lasting effects, and these have been reported to include child abuse, it need not necessarily do so. Many other factors may intervene to alter the predicted outcome.

In addition to the stimuli associated with the mother's face being important for eventual maternal recognition by the infant, there are data to suggest that olfaction can be used by neonates to differentiate between their own and another mother. In experiments in which an infant was presented with breast pads (which had absorbed milk) from its own mother or from an 'alien' lactating female, significantly more time was spent, by 6 days of age, turning towards its own mother's pad. Babies, particularly when several weeks old, attune to their mother's facial expressions when they talk and coo. In one study, a 4-week-old infant, with a blind mother who had never been sighted and displayed a mask-like face during speech, tended to avert his eyes and face from the mother when she leaned over to talk to him. When interacting with other, sighted, individuals, however, this was not the case. The normal interaction had, therefore, been distorted, but not completely so, as other modes of communication (verbal–auditory, in particular) were used successfully to overcome this interaction deficit.

The essential contribution of the baby to the bond between it and its mother is highlighted when babies display behaviour that disrupts the relationship. A baby who cries and shows avoidance responses when picked up may very easily induce feelings of frustration, confusion and anxiety in the parents. They may, in fact, feel rejected by the infant, quite the opposite to what is usually encountered when examining the occurrence of rejection in a mother–infant dyad. The behaviour displayed subsequently by the 'rejected' parents will affect their infant's developing behaviour and so the path is potentially set for an unsatisfactory and enduring pattern of interaction between them.

Summary

There is still much to learn about the mechanisms regulating the onset, course and maintenance of human parental care. The brief account above should not be taken as dogma. It is not intended to represent the 'way' a mother or father should behave towards their baby, nor that adverse consequences will result if they do not behave in this way. However, by studying the behaviour of both human and non-human primate mother–infant pairs, we may begin to understand what factors contribute to the success and richness of the mother–infant bond. Equally, we may discover what contributes to its breakdown, and how failure to establish an adequate bond at an appropriate time leads to disturbed behaviour in the parents, or the infant, or both, later on. Given the pervasive effects of good or bad parental care on subsequent social and, indeed, parental behaviour, the importance of such research is obvious.

FURTHER READING

Bowlby J (1969) *Attachment and Loss: Attachment* Volume 1. Hogarth, London.

Bowlby J (1973) *Attachment and Loss: Separation* Volume 2. Hogarth.

Ciba Foundation Symposium 33 (1975) *Parent–Infant Interaction.* Elsevier, Amsterdam (New Series).

Everitt BJ & Keverne EB (1986) Reproduction. In: *Neuroendocrinology* (Eds Lightman SL & Everitt BJ), pp. 472–537. Blackwell Scientific Publications, Oxford.

Harlow HF & Suomi SJ (1970) Nature of love-simplified. *American Journal of Psychology* **25**, 161–168.

Harlow HF & Zimmerman RR (1959) Affectional responses in the infant monkey. *Science* **130**, 421–432.

Hinde RA (1974) *Biological Bases of Human Social Behaviour.* McGraw-Hill, New York.

Keverne EB, Levy F, Poindson P & Lindsay DR (1983) Vaginal stimulation: an important determinant of maternal bonding in sheep. *Science* **219**, 81–83.

Knobil E & Neill JD (Eds) (1994) *The Physiology of Reproduction* Volume 2, 2nd edition. Raven Press, New York. (Particularly chapters by: Imagawa W *et al.*, pp. 1033–1065; Tucker HA, pp. 1065–1099; Wakerley JB, pp. 1131–1179; McNeilly AS, pp. 1179–1213.

Larson BL & Smith VR (Eds) (1974) *Lactation. A Comprehensive Treatise.* Volumes I & IV. Academic Press, New York.

McNeilly AS (1986) Reproductive disorders. In: *Neuroendocrinology* (Eds Lightman SL & Everitt BJ), pp. 563–588. Blackwell Scientific Publications, Oxford.

Numan M (1994) Maternal Behavior. In: *The Physiology of Reproduction* (Eds Knobil E & Neill JD), Volume 2, pp. 221–302. Raven Press, New York.

Rutter M (1977) *Maternal Deprivation Reassessed*. Penguin.

Trivers RL (1972) Parental investment and sexual selection. In: *Selection and the Descent of Man* (Ed. Campbell B), pp. 136–179. Aldine Publ., New York.

Walser ES (1977) Maternal behaviour in mammals. *Symposia of the Zoology Society of London* **41,** 313–331.

Wilson EO (1975) *Sociobiology. The New Synthesis.* Belknap, Cambridge, MA.

Fertility

CONTENTS

In the preceding chapters, we have attempted to describe the important mechanisms that underlie the establishment of sex, the attainment of sexual maturity, the production and successful interaction of male and female gametes, the initiation and maintenance of pregnancy and the production and care of the new-born. These processes absorb much of our physical, physiological and behavioural energies. Reproduction and its associated activities permeate all aspects of our lives. This ramification of sex, in turn, makes it highly susceptible to social and environmental influences, and these influences can be of adaptive value, as we have seen at several points in the book. However, while the adaptive value of sensitivity to environmental and social signals seems undeniable for the reproductive efficiency of the species as a whole, it is not necessarily so for the individual. In contemporary human society, the increasing emphasis on individual rights and welfare has prompted a more sympathetic attitude towards the fertility of the individual, both its limitation and its encouragement. Moreover, sexual interaction among humans has ceased to be related directly and exclusively to fertility, but rather has itself assumed a wider social role. Sexuality and its expression has, in its own right and independent of any reproductive considerations, become an important element within human society. In this final chapter, therefore, we stand back from a detailed consideration of the mechanisms of reproduction and look instead at human fertility patterns, and at the factors, both natural and artificial, that may influence them. We draw heavily on examples of the reproductive mechanisms described earlier to illustrate our points, and so the chapter cross-refers extensively.

To begin with, some definitions are necessary. The *fertility* of an individual, a couple or a population refers to the number of children born and the *fertility rate* of a population is the number of births in a defined period of time divided by the number of women of reproductive age. Thus, fertility is a measure of *actual outcome* of the reproductive process. In contrast, *fecundability* is defined as the probability of conceiving in a menstrual cycle in a woman who has regular periods and engages in regular unprotected intercourse, whether or not conception goes to term. Finally, *fecundity* is a measure of the capacity to conceive *and* produce a live birth under the same circumstances. Both fecundability and fecundity are therefore measures of the *reproductive potential* of women. As the conditions for defining these terms are rarely met, they are somewhat notional; even when they are met, the difficulty of determining whether or not a woman has conceived in any given cycle until at the earliest 10–14 days post-conception, means that fecundability will be underestimated due to very early pregnancy loss. In practice, when expressing values for the fecundability and fecundity of women or populations, it is assumed that women wishing to conceive are ovulating and copulating around mid-cycle; then effective values are produced by dividing either clinically recognized pregnancies (human chorionic gonadotrophin (hCG) test, Chapter 10 and/or ultrasound scan: Fig. 14.1) or live births by the number of years of attempting fertility. In reality, the fertility of a population or individual rarely reaches its theoretical maximum or fecundity.

Clearly, therefore, constraints are placed on reproductive efficiency. It is the nature of these constraints that concerns us in this chapter.

The effects of age and senescence on fertility and fecundity will initially be considered. We will see that, for women, puberty and menopause mark the limits of the period of fertility. However, within this period, most women are not continuously reproducing. Traditionally three factors have limited their fertility: namely, social or religious practices and traditions, pathology (infertility or subfertility), and the use of contraception and abortion. Each of these will be considered in turn.

AGE, SENESCENCE AND REPRODUCTIVE CAPACITY

Ageing should be distinguished from senescence. Ageing begins at conception and is a description of what happens with the passage of time. *Senescence* reflects deterioration related to dysfunction and disease. Typically, reproductive senescence begins during middle age, but its onset and time course varies widely among individuals of the same age. In women, a major change occurs at the *menopause*, defined as the last menstrual cycle, and this results from the exhaustion of functional ovarian follicles and is thus primarily part of the ageing process. However, other senescent changes in the reproductive potential of women may precede this landmark. Most men, in contrast, do not experience gametic exhaustion or a sudden fall off in fertility as part of the ageing process, there being no evidence for a male equivalent of the menopause, despite popular claims to the contrary. They do, however, experience senescent changes, which, as for women, show a great variation among individuals.

Women

The fertility of a woman varies with her age. Menstrual cycles begin during puberty, and *menarche* marks the earliest expression of the potential fertility of the female with its implication that a full ovarian and uterine cycle has been achieved. Failure of menarche indicates a diagnosis of *primary amenorrhoea*. This may result from a failure of normal maturation of the underlying neuro-endocrine mechanisms (see Chapter 6), from primary defects in the gonad, such as dysgenesis or agenesis due to chromosomal abnormality (e.g. Turner's syndrome and true hermaphroditism, Chapter 1), or from primary defects in the genital tract, such as lack of patency between uterus and vagina (*cryptomenorrhoea* or *hidden*

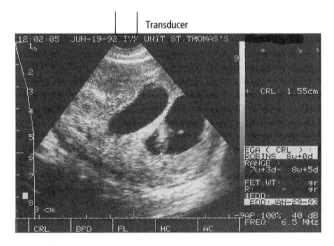

Transducer

Fig. 14.1 Transvaginal ultrasound scan of a twin gestation in an 8-week pregnant woman. Both sacs can be seen on this longitudinal section, although only one embryo (in the lower right hand sac) is visible in this plane. The gestational age of the embryos can be determined very easily from the measurement of the crown : rump length (between the white caliper crosses).

menses), or, indeed, the absence of internal genitalia as seen in the testicular feminization syndrome (see Chapters 1 & 2). However, even in normal females, early menstrual cycles are rarely regular. Some cycles are anovulatory and may lack, or have abbreviated, luteal phases (Fig. 14.2). In general, the follicular phase tends to become shorter with age and the luteal phase tends to lengthen, for reasons that are unclear.

Fertility is highest in women in their twenties and declines thereafter (Fig. 14.3). This decline could, in principle, be due to a reduction in fecundity or to changes in sexual behaviour resulting in older women having a reduced chance of, or inclination for, unprotected intercourse. There is little evidence for the latter; indeed, some evidence points to increased sexual activity by women with age. Measures of the risk of childlessness in women of different ages at marriage, who then try actively to have offspring, are also shown in Fig. 14.3, and suggest a real decrease in fecundity. It does seem clear that the ability to sustain a pregnancy through to successful parturition declines slowly to age 35 years and more rapidly thereafter when an increasing frequency of failed ovulation, perinatal or neonatal mortality, low birth weights, maternal hypertension, and congenital malformations especially, but not exclusively, due to fetal chromosomal imbalance are encountered.

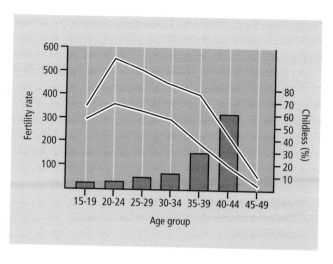

Fig. 14.3 Rates of fertility and childlessness by age of woman. The fertility rate data were collected from populations of married women in which, it is alleged, no efforts were made to limit reproduction. The range reflects the different circumstances of the populations (drawn from the developed and developing world), but the overall pattern is the same regardless. These data approach a measure of fecundity by age in humans. The histograms show the proportions of women remaining childless after first marriage at the ages indicated despite continuing attempts to deliver a child. Note the sharp rise above 35 years, implying a fall in fecundity from this time onwards.

Accordingly, it is not surprising that irregular menstrual cycles may begin to reappear in some women in their early forties and mark the onset of the *climacteric*, a period of reproductive change that may last for up to 10 years before the last menstrual cycle (the *menopause*). This *secondary amenorrhoea* occurs at a mean age of 52 years in the USA. Symptoms associated with the climacteric can include mood changes, irritability, loss of libido and hot flushes. The climacteric reflects declining numbers of ovarian follicles and their reduced responsiveness to gonadotrophins. The final cessation of reproductive life is, thus, a function of ovarian failure. Premature loss of oocytes, and thus *premature menopause*, occurs in about 2% of women, in some as early as their late teens and early twenties. The causes of premature menopause are unclear, although in some women a familial element exists. With the loss of follicular activity, secretion of follicular inhibin and oestrogen declines, although androgen output tends to rise (Table 14.1). The fall in negative feedback stimulus leads to elevation of gonadotrophin levels, the increase in follicle stimulating hormone (FSH) preceding that of luteinizing hormone (LH) (Fig. 5.8).

The physical, functional and emotional changes of the climacteric and menopause are driven directly or indirectly by these ovarian changes. Oestrogen withdrawal is responsible for: vasomotor changes, such as 'hot

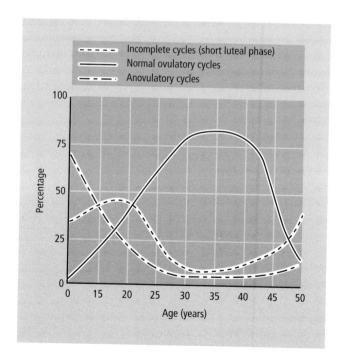

Fig. 14.2 Relative incidence of three types of menstrual cycle with age of woman.

Table 14.1 Steroidogenesis in the postmenopausal woman

Hormone	Change from resting plasma concentrations (%)	
	Dexamethasone suppression (% decrease)	hCG stimulation (% increase)
Oestradiol	> 50	0
Oestrone	> 75	10
Progesterone	> 75	10
17α-OH Progesterone	53	42 *
Dehydroepiandrosterone	65	60 *
Androstenedione	70	10
Testosterone	40	60 *
Dihydrotestosterone	> 30	–

Dexamethasone inhibits adrenal cortical steroidogenesis by depressing ACTH output, so the degree of reduction in plasma steroid levels shown suggests the degree of adrenal contribution which is high. hCG provides a gonadotrophic stimulus to the ovaries, so the degree of increase in plasma steroid levels shown suggests the degree to which the ovaries are able normally to secrete the hormones. In the second column, there are only three significant changes from baseline (asterisked).

Regardless of the time of the menopause, some women can none the less, if provided with an oocyte from a donor woman, carry a pregnancy to term successfully. The early part of this pregnancy requires administration of exogenous hormones: first, to build up the regressed reproductive tract (see Chapter 7) and, second, to mimic luteal support until the fetoplacental unit takes over endocrine control (see Chapter 10). The capacity of women in their forties or older to give birth after *oocyte donation therapy* emphasizes that the primary reproductive ageing process is oocyte and follicle loss. However, it is probable that senescence in other tissues will reduce the likelihood of a successful outcome in such patients.

Men

Spermatozoa can be produced well beyond 40 years of age. Loss of libido, impotence and failure to achieve orgasm do occur with higher frequency from 30–40 years onwards (Fig. 14.4), but mostly these appear to result from other senescent or iatrogenic factors, for example diabetes, pharmacological control of high blood pressure and neurodegeneration (see Chapter 8). There is little evidence for a significant fall in testosterone or the expected rise in gonadotrophins that would follow it (Fig. 14.4) and even men in their eighties have testosterone levels well above the range found to be supramaximal for behavioural effects (Fig. 7.16). Some (disputed) evidence for a fall in the proportion of free : bound testosterone has been proposed to lead to a decline in *available* androgens, and used as a justification for androgen therapy. However, this area is controversial.

flushes' and 'night sweats'; changes in ratios of blood lipids, associated with an increased risk of coronary thrombosis; reduction in size of the uterus and breasts; a reduction in vaginal lubrication and a rise in the pH of vaginal fluids, in consequence of which discomfort during intercourse (*dyspareunia*) and *atrophic vaginitis* may occur. The anti-parathormone activity of oestrogen is also lost at this time, resulting in increased bone catabolism, osteoporosis and more brittle bones. All these symptoms of the menopause may be prevented or arrested by oestrogen. Although there is not uniform agreement about the safety and advisability of using oestrogen therapy in postmenopausal women, the adverse risks of coronary heart disease and osteoporosis probably outweigh substantially the perceived risk of breast or uterine cancer.

A number of behavioural changes may occur during the climacteric, for example depression, tension, anxiety and mental confusion. However, these changes may not be related directly to steroid withdrawal but rather may result secondarily from difficulties in psychological adjustment to a changing role and status, and in part from insomnia due to night sweats. Loss of libido is common and may be related to the changes in steroid hormone secretion. In Chapter 7, we reported that treatment with androgens is remarkably successful in restoring libido in some postmenopausal women, although this treatment is not prescribed routinely.

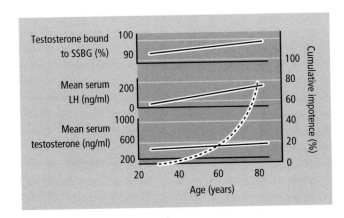

Fig. 14.4 Ageing in males: note that there is no obvious fall in testosterone, nor a marked rise in LH with age (compare those for women in Fig. 5.8). The percentage of testosterone bound to sex steroid binding globulin does show a slight rise with age in some studies. In contrast to these very small changes, the cumulative incidence of impotence in men rises markedly from 40 years of age.

SOCIAL CONSTRAINTS ON FERTILITY

There is a rich literature on the sociology, anthropology and history of reproduction, which is too large for this book to cover comprehensively. It is an important body of knowledge, as all studies show that the simple availability of scientific and medical technology, and educational information about it are not in themselves sufficient to change sexual and reproductive behaviours. A further element, namely the *motivation* required for a change in behaviour to occur, is critically dependent upon an individual's social network, values and beliefs, and the peer pressure that reinforces these. Among the important social variables that may influence fertility in either direction are:

1 the accepted social roles of men and women, the perceived and legal balance of power between them, and, in particular, the extent to which women are educated;

2 the age of women at marriage and at birth of the first child, and the maternal mortality rate;

3 the accepted size of the family and the preferred sex ratio, which may also be influenced by inheritance patterns;

4 the desirability of spacing children and the anticipated child mortality pattern;

5 the perceived economic advantages of a given family size;

6 the permitted or expected frequency of intercourse in relation to the point in the menstrual cycle, the time of year or religious calendar, age of partners, the delivery or suckling of children;

7 the extent to which maternal lactation (with its consequent hyperprolactinaemia and depression of fertility: see Chapters 5 & 13) is replaced by use of milk substitutes or surrogate mothers;

8 the acceptability of sexual interactions outside (or to the exclusion of) the usual framework in which successful pregnancy might result, emphasizing the sexual rather than the reproductive function of coition (e.g. prostitution, homosexuality);

9 the role of celibacy;

10 the acceptability of reproduction outside of tightly controlled social structures (single or unmarried mothers);

11 the incidence of divorce and the attendant delay before remarriage;

12 the strength with which religious beliefs are held by, or imposed on, the individual.

Each of these factors will affect, to varying degrees, the overall fertility of the individual woman, couple or society, and the various factors will interact with each other. For example, high economic expectations coupled with low infant mortality and an equal role for partners of each gender tends to reduce fertility both by delaying birth of the first child and reducing overall family size. Perhaps the single largest change that has occurred in most societies over the past century is the fall in expected infant mortality arising from improved diet and hygiene, and the introduction of antibiotics and prophylactic immunization. However, this fall has led to a corresponding rise in the numbers of females surviving beyond puberty and, thus, to an increased fecundity of the population as a whole. The time lag between the decline in infant mortality and the reproductive adjustment to it, via the compensatory social factors listed above, has lead to a world population explosion. Indeed, it is not clear that social adjustment alone is adequate to correct this imbalance. The increased use of artificial constraints on fertility in various societies has become an essential part of the social response to limit population growth and to regulate individual fertility according to desired or imposed social patterns.

ARTIFICIAL CONTROL OF FERTILITY

The three socially accepted artificial controls over fertility available to varying extents in different societies are *contraception*, induced *abortion* and *sterilization*. Not all of these controls are 100% effective and, thus, they should be seen as regulating fertility by delaying births or increasing the interval between births. Their relative cost-effectiveness for both the population as a whole and the individual in particular is, of course, of the greatest importance.

Sterilization

Sterilization should be 100% effective and should be entered into as an irreversible procedure. It is, therefore, the method of fertility limitation selected by individuals or couples who have achieved their desired family size or who, for eugenic or health reasons, must or should avoid reproduction.

In men, sterilization involves *vasoligation* or *vasectomy* (ligation, or removal of part, of the vas deferens: see Chapters 7 & 8). The operation can be done under local anaesthetic as an outpatient; it is quick, and leads to an aspermic ejaculate within 2–3 months. Performed competently, vasectomy provides an efficient method for limiting fertility. However, the obstruction caused by the ligation of the vas deferens can lead to a local granuloma, due to the build up of spermatozoa and the small volume of fluid that continues to pass out of the epididymides. Certain consequences follow: first, there may be chronic or intermittent tenderness in the scrotal region; second, leakage of spermatozoal debris into the systemic circulation from the site of inflammation induces, in many men, an immune response to their own spermatozoa; third, in several species of experimental animal, a progressive decline in spermatogenic output occurs, and it is not clear whether this is accompanied by *orchitis* (inflammation of the testis), but men seem to be relatively less affected in this way. Some success with the surgical reversal of vasectomy has been achieved, but recent developments in the technology of assisted conception now make the infertility resulting from vasectomy potentially fully reversible (see 'Oligospermia' below). It has also been suggested that men could produce samples of semen pre-vasectomy for cryostorage and potential use later. Psychological factors, particularly arising from the erroneous equation of *fertility* with *potency*, may result in anxiety that leads to impotence in some men. This fear also prevents many men from accepting the procedure. Sexual arousal and ejaculation are, of course, independent of sperm release.

In women, sterilization involves ligation, electrocoagulation or removal of all (*salpingectomy*) or a section (*fimbriectomy*) of the oviducts. A general anaesthetic is usually required, although regional anaesthetic can be used. The surgical approach is generally made with the use of a *laparoscope* through the abdominal wall (*laparotomy*). Although the procedures require surgical skill, the patient can be brought in as a day patient. The operation has a very low failure rate (<0.3%) and, although it also cannot be easily reversed surgically, the new techniques of assisted reproduction (see later) allow the infertility to be circumvented. Sterilization may also be accomplished by hysterectomy, especially in premenopausal women troubled by irregular, painful or heavy menses or pathology of the uterus, such as fibroids or premalignant disease. Unwanted effects of sterilization can include: incidental surgical trauma at the time of the procedure; persistent pain at the site of the ligation; the very slight risk of mortality attendant on use of anaesthetics; and psychological disturbance associated with loss of fecundity and the perception that femininity or womanhood may also have been damaged.

Contraception

Contraception differs formally from sterilization only in its potential or actual ease of reversibility and thereby in the control that the individual exerts over its use. It is,

		World-wide	UK: motivated woman above 25 with regular partners
Sterilization	Male	0–0.2	0.02
Sterilization	Female	0– 0.5	0.13
Implanted progestogens		0–1	–
Combined pills	high oestrogen	0.1–3	0.16
	low oestrogen	0.2–3	0.27
Progesterone only pill		0.3–4	1.2
IUCD		0.3–2	–
IUCD containing progesterone		< 0.5	–
Diaphragm/cap		2–15	1.9 (to 6)
Male condom		2–15	3.6 (0.7 to 32!)
Female condom		–	4.0 (?)
Spermicides alone		4–25	11.9
Coitus interruptus		8–17	6.7
Rhythm		6–25	15.5
No method	25-35-year-old woman	80–90	–
	>35-year-old woman	40–50	–

* Expressed in terms of the number of pregnancies encountered over 100 women years (= 10 women for 10 fertile years, or 100 women for 1 fertile year, etc.) To calculate how many years of regular use by a couple of the method would be required for a chance of one pregnancy, divide 100 by the user failure rate e.g. for women of 25 - 35 using no method, there is a very high chance of pregnancy within 1 year.

Table 14.2 User failure rates for some contraceptives*

Table 14.3 Main method of contraception used by women (16–49 years) in the UK (1991)

Sterilization	25%
Pill	23%
Condom	16%
Abstinence	16%
Pregnant/wanting pregnancy	9%
IUD	5%
Sterile	4%
Diaphragm/cap	2%

Table 14.4 Age breakdown in use of contraceptives by UK women (1991)

	Age				
	16–17	18–19	20–24	30–34	40–44
Sterilization	0	0	1	21	50
Pill	16	46	48	25	4
Condom	10	15	14	17	13

thus, the artificial control of choice for those who wish to delay the expression of fertility or to exercise it with discrimination. The methods currently available are listed in Table 14.2, together with estimates of their efficiency in terms of their impact on pregnancy rates. The reported rates vary widely depending upon the education, motivation and experience of the users. The table also indicates that the type of contraceptive used varies geographically, a variation arising from the differential access to technology, good hygiene, education and medical follow up in different parts of the world. A breakdown of the use of contraceptives in the UK is given in Table 14.3, while data in Table 14.4 reveal the marked age dependency of this usage. This variation reflects the differing objectives of fertility limitation at different ages. Even the most cursory glance through these data makes evident the complexity of contraceptive use, in turn, reflecting the variety of needs, resources and beliefs available to those exercising usage.

Natural methods

Contraceptive approaches that rely on coital technique, rather than the use of technical or pharmaceutical aids, straddle the boundary between social and artificial approaches to fertility control. In consequence, such approaches seem to pose particular problems for theologians and lawyers when prescribing or proscribing sexual behaviours. For example, of the techniques listed below, only the rhythm method is sanctioned by the Roman Catholic church on the grounds that the other

approaches are 'unnatural'. All the methods described below occur naturally in the biological sense, if not in the theological sense.

The rhythm method: For many centuries common belief wrongly equated menstruation with the period of maximum fertility. In fact, of course, ovulation occurs at approximately mid-cycle. However, given the potential life of spermatozoa in the cervix of several days and a 24-hour life of the oocyte (see Chapter 8), coition should be avoided through much of the first part of the cycle (Fig. 14.5). Moreover, as shown in Fig. 14.5, the timing is based on a count back from the small rise in basal body temperature of 0.2–0.6°C, which occurs *post-ovulation*. Thus, the approach is retrospective and will only be safe if the woman's cycle is sufficiently constant, which it often is not, especially given its susceptibility to emotional or stressful disturbances (see Chapter 5). These factors, and the high motivation required by the sexual partners, makes for a high failure rate, being some 20-fold less effective than oral contraceptives *at their worst* (Table 14.2). The devising of a simple test for incipient ovulation is unlikely to improve the success of the method more than marginally, given the nature of its limitations outlined above. Additionally, there are (disputed) claims of a

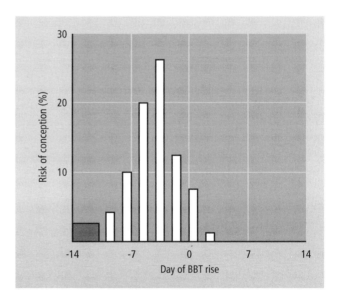

Fig. 14.5 The percentage risk of conception at different times in the unprotected cycle of women using the rhythm method, plotted as days before or after the day of rise in basal body temperature (BBT) (Day 0). It is clear that coital interactions during much of the first part of the cycle carry a substantial risk of pregnancy. Only the period after the BBT rise until the end of the period of menstrual flow is truly safe. This extended period of risk is due to the potential longevity of spermatozoa in the female tract and variability in the length of the follicular phase in some women.

significantly increased frequency of genetic abnormalities in embryos derived by fertilization of aged eggs occurring in couples using this method of fertility control (see Chapter 8; also 'Spontaneous pregnancy loss' below).

Coitus interruptus: The withdrawal of the penis from the vagina during copulation but prior to ejaculation has been, and remains, one of the most frequently used forms of contraception in Europe, and the technique is credited (together with abortion) as being responsible for much of the decline in birth rate at the time of the industrial revolution. Failures will occur, due both to lack of adequate control by men at the moment of ejaculation and by insemination with spermatozoa that leak from the urethra before ejaculation or persist from a prior ejaculation.

Masturbation and other forms of sexual interaction: Mutual masturbation is one commonly employed means of reducing the incidence of pregnancy, and is highly effective as such. It is also free of risks of transmission of HIV as long as fresh semen or vaginal fluids are not subsequently rubbed manually into vagina, anus or skin lesions. Oral and anal sex are commonly used in many societies to gain sexual pleasure without risking fertility. While the evidence linking oral sex to HIV transmission is weak and disputed, other genital infections, such as herpes, hepatitis and gonorrhoea are readily spread by this sexual practice. Oral sex (unprotected by use of a condom) is, however, clearly low risk for HIV transmission when compared with unprotected vaginal or anal sex. In the latter case, the epithelia lining the anus and rectum are more easily damaged than the vaginal epithelium, making bleeding more frequent and infection transmission to both the penetrated and penetrating partners more common. High strength condoms (see below) substantially reduce the risk of transmission of infection anally.

Caps, diaphragms and spermicidal foams, jellies, creams and sponges

The combination of both a physical seal at the cervix between the vagina and the uterus, and a spermicide at the site of seminal deposition has been used for over a century. The method became popular initially as, for the first time, it gave women some control over the use of contraception. However, the method has decreased in popularity with the development of more modern methods of contraception for females.

The diaphragm, via its sprung margin, occludes the top of the vagina including the cervix. The smaller 'Dutch' or cervical cap fits directly into and around the cervix, where it is held by suction. These devices can be fitted well in advance of intercourse. As neither of them gives a perfect

barrier, but will reduce the chances of spermatozoal passage up the genital tract, they should only be used together with spermicides, which must be delivered into the vagina and/or placed in the device just prior to its insertion. Spermicides are available as foams or pessaries. Spermicides and spermicidal impregnated sponges should *not* be relied on *alone* for contraception.

These contraceptive approaches give more control to the female partner and their use is relatively independent of intercourse compared with the condom (see below). They also offer protection from pelvic inflammatory disease. Their disadvantages are the requirements for careful fitting and training in its use, high motivation, good hygiene and a willingness of women to touch their own internal genitalia. When used by motivated women, they can be very effective, but can also be of relatively low efficiency (Table 14.2).

Condoms

Condoms are probably the commonest form of mechanical contraceptive in use, particularly since their promotion with the appearance of HIV (see Chapter 7). The improved strength, lubrication and design of modern condoms has considerably enhanced their efficiency and durability. Condoms are cheap, readily available, relatively easy to use and also give protection against venereal diseases. Good use is crucial to their success (see Table 14.5).

More recently, the *female condom* has been marketed (under the brand name *Femidom*). It resembles an extra large lubricated condom with a rimmed structure at the closed end, akin to the rim of a diaphragm, which fits into the vagina, thus protecting the cervical os in the vaginal vault. It combines the protective advantages of the penile condom with many of the advantages of the diaphragm. Unlike the diaphragm, it does not require fitting or complex training in its use. Its user friendliness and effectiveness remain the subject of research.

Steroidal contraceptives for women

Since the initial development of 'the female pill' in the 1960s, a range of steroid-based contraceptives has been developed (Table 14.6). Synthetic steroids, as described in Chapter 2 (Table 2.7), are used in these preparations as their half-lives in the body are longer and their effects therefore sustained. The general underlying principle is the suppression of ovulation by the negative and anti-positive feedback effects of progesterone (with or without oestrogen) on the pituitary and hypothalamus (see Chapter 5). Additionally, progesterone can exert direct anti-fertility effects on the female genital tract. When considering these methods, it is important to remember that, 'naturally', female mammals living with

Table 14.5 Condom use

Do	Don't
Use kite-marked only	Don't use after 5 years old or exposed to excessive heat or UV light
Use water-based lubricants (KY jelly, glycerol, spermicides such as nonoxynol-9*)	Don't use oil-based lubricants (vaseline, baby oil, suntan lotions)
Use only once	Don't allow penis to become flaccid inside partner
Take care not to tear or snag condom	
Put on penis before **any** genital/anal contact	Don't have genital/anal contact after condom removal, unless penis washed
Use spermicides/bacteriocides/viricides (foams, pessaries, sponges)	
Use condoms if risk of HIV, hepatitis, herpes, gonorrhoea, chlamydia, and cervical dysplasia	Don't use ordinary strength condoms if having anal intercourse

* Note: frequent use of nonoxynol can result in a mild inflammatory response of vaginal epithelium which can increase risk of viral transmission

Table 14.6 Steroid-based contraceptives for women

	Oral COC*	POP†	Injectables	Implant Norplant	Vaginal ring or IUCD
Administration					
Frequency	Daily	Daily	2-to 3-monthly	5-yearly	3 months
Relative progestagen dose	Low	Ultra-low	High	Ultra-low	Ultra-low
Blood levels	Rapidly fluctuating	Rapidly fluctuating	Initial peak then decline	Constant	Constant
How does it work?					
Ovary: ovulation supressed§	+++	+	++	++	+
Cervical mucus: sperm penetrability down	Yes	Yes	Yes	Yes	Yes
Endometrium: receptivity to blastocyst down	Yes	Yes	Yes	Yes	Yes
User failure rates	0.2–3	0.3–4	<2	0–1	3
Menstrual pattern	Regular	Often irregular	Irregular	Irregular	Irregular
Amenorrhoea during use	Rare	Occasional	Common	Common	Common
Reversibility					
Immediate termination possible?	Yes	Yes	No	Yes	Yes
By woman herself at any time?	Yes	Yes	No	No	Yes/No
Time to first likely conception from first omitted dose/removal	3 months	c. 1 month	3–6 months	c. 1 month	c. 1 month

* COC, combined (oestrogen and progestagen) oral contraceptive.
† POP, progesterone only pill.
§ By two mechanisms - no preovulatory follicles formed and/or no LH surges occur.
(Data adapted from J. Guillebaud, Contraception: your questions answered, Churchill Livingstone).

males will not experience a series of oestrous cycles but will be pregnant or nursing for much of their fertile lives, and so will experience extended exposure to higher levels of oestrogens and progestagens. Humans, in contrast, are *atypical in being reproductively cyclic*, especially in modern developed societies. The steroidal contraceptives, in essence, mimic the continuous exposure to steroids experienced during pregnancy, during which of course the hypothalamic–pituitary axis is suppressed (see Chapters 5 & 10), and so, paradoxically, may take the endocrine status of women closer to that of most other female mammals.

The variables to consider in any steroidal contraceptive are: the *nature* of the steroidal components (oestrogens + progestagens, or progestogens only?); the *potency* and *dose* of the synthetic steroid(s) used; and the *duration of exposure* to steroids of different types or doses. A maximal contraceptive effect and complete continuous amenorrhoea is provided by *high* doses of *both* types of steroid *continuously* present. However, against this optimal contraceptive regimen must be balanced: the side-effects of the steroids (as also observed during pregnancy or even the luteal cycle: Chapters 7 & 10); a woman's perception and concern about what is happening to her reproductive system when it is closed down completely; and the requirement that reproductive capacity be recovered reasonably rapidly when contraceptive practice ceases. How can each of the potential variables be adjusted to maintain effective contraception while minimizing or eliminating the influence of these non-contraceptive concerns?

Progesterone alone can suppress ovulation, but it does so with variable efficiency unless present constantly. Oestrogen addition exerts an additional negative feedback effect of its own and also promotes the development of progesterone receptors (see Chapter 7), which renders the progestins in the contraceptive more effective. Thus, the *combined oral contraceptive* (COC) uses, as its name implies, both types of steroid. Usually, this contraceptive is taken for 21 days, during which time follicular growth and the output of endogenous gonadotrophins and ovarian oestrogen are completely suppressed (Fig. 14.6). The exogenous hormones develop and maintain the endometrial lining. This period of suppression is then followed by a 7-day break, with either no pills or placebo pills. During this period, the endometrium breaks down, which leads to a *withdrawal bleeding* that *simulates* menstruation. This withdrawal bleeding is important to many women psychologically, as it 'confirms' their cyclicity and lack of pregnancy (although, of course, it does not reflect a true ovarian cycle). Without it, some women will discontinue contraceptive use, and so it is of importance for *compliance*, an essential element of a contraceptive's effectiveness. However, the absence of exogenous steroids for 7 days results in a reawakening of the woman's own hypothalamic–pituitary axis, a rising output of gonadotrophins, and follicular development (Fig. 14.6). Indeed, towards the end of the 7-day pill-free period, some women can come dangerously close to ovulating. A failure to take the first active pill of the new series, or interference with its effectiveness (e.g. vomiting it up), can lead to contraceptive failure. This 21 days on, 7 days off regimen is

called *monocyclic*. For more assured contraceptive efficiency, and for women who experience severe withdrawal symptoms (quasi-menstrual tension) or for whom a regular bleed is socially or medically difficult, a recommended alternative is the use of the pill on a *bi- or tricyclic* regimen, in which the durations of active pill : placebo are 42 : 7 or 63 : 7 days, respectively. Such regimens, of course, lose the 'regular cycle' appearance, important for some pill users. (*Note:* these regimens should not be confused with bi- and tri*phasic* pills; see below).

It will be clear from this discussion that the regimen to be used is best tailored to the psychological, physical and social condition and needs of each woman. This personal tailoring also extends to the potency and dose of the constituent steroids used. Individual women react differently to the same preparation, some being more sensitive than others and so requiring lower doses or less potent preparations. The level and balance of oestrogens and progestagens varies in different preparations, thereby offering a range of contraceptives to suit women with different clinical histories and physiologies. As a rule of thumb, it is best to go for the lowest dose of steroids possible, and this may be estimated as being just above the dose that gives *breakthrough bleeding*, an endometrial blood loss occurring *while the woman is taking an active preparation*. Breakthrough bleeding reflects inadequate support for the endometrium by the exogenous contraceptive steroids, and provides an indication of the woman's sensitivity to their action.

Other variants of the oral contraceptive include changing the ratio of oestrogen to progesterone in different pills during the 21-day cycle. The intention here is to mimic more closely the natural cycle. The now discontinued *sequential oral contraceptives* used oestrogens only for 7–14 days and a combined oestrogen–progestagen mix for 14 or 7 days, followed by 7 pill-free days, but required higher doses of oestrogen and were less contraceptively efficient. In contrast, *biphasic and triphasic oral contraceptives* combine the mimicry of a normal cycle attempted in sequential preparations with the contraceptive efficiency of monocyclic COCs. Thus, in biphasic preparations, both oestrogen and progestagen are given for the first half of the cycle, but the progestagen dose is stepped up at mid-cycle. The contraceptive efficiency is as good as in the conventional combined tablets but the doses of steroid used are lower. In triphasic preparations, 5 or 6 days have low oestrogen and low progestagen, a futher 6 or 5 days have both oestrogen and progestagen slightly elevated, and the remaining 10 days have low oestrogen and doubled progestagen. These preparations place increasing emphasis on the capacity of progestagens to block the positive feedback effect of

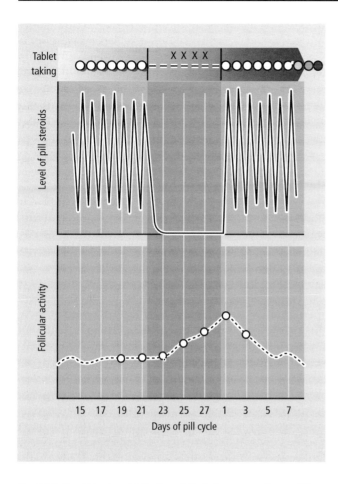

Fig. 14.6 Steroid levels and follicular activity during monocyclic use of the combined oral contraceptive. The days of pill taking are indicated in the centre by small pill symbols, while the pill free days are dashed. 'X' indictates the days on which withdrawal bleeding occurs. The panel above shows the profile of pill steroids in the blood: note how they reflect the episodic (daily) taking of the pill and how rapidly they fall when pill taking ceases. The lower panel shows follicular activity, which can be measured by ultrasound scan (see Fig. 14.9) or be reflected in the output of endogenous steroids. Activity is low while the pill is taken, but note how rapidly follicles develop when pill taking ceases, and how negative feedback is reduced and gonadotrophin output rises. Resumption of pill taking shows an immediate suppression effect. It is particularly important to note that a failure to resume pill taking on time leaves a woman potentially very vulnerable to ovulation and pregnancy.

oestrogen and less on direct negative feedback effects (see Chapter 5). In consequence, oestrogen levels are lower and a more normal 'cycle' is achieved. The bi- and tricyclics are particularly popular in North America.

Progestagenic contraceptives may be taken daily by an oral route or given by depot injection subcutaneously for continuous release over a period of 8 weeks to 5 years (Table 14.6). *Progesterone only pills* (POPs) are taken continuously, contain low doses of progestin, and work primarily by effects on cervical mucus and perhaps the

endometrium (see Chapter 7). However, in a significant proportion of users (*c.* 20%), ovulation is suppressed, and, in around 40%, follicular–luteal activity is abnormal. In all users, irregular bleeding may occur. Because the dose of progesterone used is small and the pill is taken daily, side-effects are relatively few compared with COCs, although weight gain can be problematic due to progesterone's anabolic effects (see Chapter 7). However, the effect of the progesterone on cervical mucus lasts for only 22–26 hours. Therefore, if the woman is significantly delayed in taking her daily pill, fertility returns. This requirement for a highly organized life style and level of commitment is reflected in the much higher failure rates among younger women.

Injectable depot progestagens are extremely effective contraceptives and used by millions of women world-wide, although less so in more developed countries. The 8-week injectable depot medroxyprogesterone acetate (DMPA) and the 12-week injectable norethisterone oenanthate (NETET, Noristerat, or Norigest) generate relatively high, ovulation-suppressing levels of steroid initially, which fall to mucus-affecting levels as the time for the next injection approaches. Their use is associated with irregular but light bleeding, the positive side-effect of reduced iron loss, and anaemia.

Negative side-effects of steroidal contraceptives

We have stressed the balance between contraceptive efficiency and unwanted side-effects. Some of the unwanted side-effects are social or psychological, such as the need for either a regular bleed or for no bleed at all, or a requirement for highly organized self-administration or the availability of good medical care. Other side-effects can include weight gain, headaches, libido changes, acne etc., each of which will be specific for an individual woman and will vary with different preparations. All of these factors are extremely important, as marrying the physiology, psychology and social condition of the woman to her contraceptive is the key element in providing effective and *acceptable* contraception for her.

However, beyond these elements, there has also been widespread concern and discussion about more severe life-threatening side-effects of steroidal contraceptives. Most studies have been done on COCs, and many of the older studies have used COCs containing much higher doses of steroids than are needed with the relatively new generation of steroid analogues now used. There are two general points that need to be made emphatically. First, the life-threatening risk associated with modern steroidal contraception is less than that from having a baby or driving a car. Second, there is little evidence that steroidal contraceptives *cause* life-threatening conditions;

but they may be co-associated with *other* causal agents to promote *their* effects. As these co-associated factors are defined, the risk can be quantified for each woman. Two main general factors contraindicate steroidal contraception: smoking, especially in older women, and a history of thromboembolic or cardiovascular disease.

There is evidence that the higher doses of oestrogen in some combined oral contraceptives increase clotting factors and blood coagulability, although they also increase fibrinolytic activity but *only* in the absence of smoking (see Chapter 7). Older (but not the newer) progestins reduce levels of high density lipoprotein-cholesterol, a change associated with an increased risk of atherogenesis, but oestrogens tend to counteract this fall. In smokers, there is a significantly increased risk to COC-users of myocardial infarct, hypertension and haemorrhagic stroke, all associated with a high oestrogen component (see Chapter 7). Thus, women who smoke and who have any cardiovascular risk factors are not candidates for use of COCs.

The broad effects of modern steroidal contraceptives on cancers are neutral. Benign breast cancer and carcinomata of endometrium and ovary (both very serious conditions) are *reduced* in pill users. A slight increase in hepatic carcinoma occurs, as does breast cancer, among women who have taken the pill for several years before the age of 25, an outcome perhaps related to the effects of progestagens on cell division in the postpubertal, developing breast. Increased cervical dysplasia is also reported, but may be secondary to increased coital activity without condoms rather than a direct effect. These observations emphasize the value of regular cervical smear and breast examinations in women on steroidal contraceptives.

The latent diabetes revealed during pregnancy (see Chapter 11) can also become evident in users of steroidal contraceptives, who should therefore only use low dose oestrogen preparations under careful supervision.

A plus for most women on steroidal contraception is the reduced incidence of pelvic inflammatory disease, a potent cause of infertility (see later). It is not clear why this is so, as genitourinary infections of the vagina are, if anything, increased. One idea is that infectious agents hitch a ride on spermatozoa, but the hostility of cervical mucus prevents their passage further into the female genital tract.

Post-coital contraception

Post-coital oral oestrogenic contraceptives used to be given in high doses within 72 hours of unprotected intercourse (the inappropriately named '*morning-after pills*') in order to interfere with transport of the conceptus and implantation (see Chapter 7). The high doses given caused nausea and vomiting, and failure of contraceptive action led to

an increased incidence of ectopic pregnancy (see Chapter 7). Therefore, a combined oestrogen–progestagen pill is now given in two doses 12 hours apart; high doses of progestagen only are also used. Alternatively, the insertion of a copper intra-uterine contraceptive device (IUCD) within 5 days of the expected time of ovulation is highly effective as a post-coital contraceptive. A clinical trial of an *anti-oestrogen* is being carried out in India (see also anti-progestin RU486 under 'The future' below).

Steroidal contraceptives for men

The output of gonadotrophin releasing hormone (GnRH) and/or gonadotrophins in the male can be suppressed via negative feedback, just as in the female. Indeed, trials using both progesterone and the continuous administration of a GnRH analogue (Table 2.7) have shown this to be the case. However, androgens are also depressed, reducing masculinizing stimulation and libido, both unacceptable side-effects. To combat these effects, androgens must also be administered. A simpler approach has been to use an oral contraceptive preparation of testosterone itself. Since 1986, more than 700 couples from nine countries have taken part in a clinical trial of the 'male pill'. The preparation is very effective in stopping sperm production in 70–95% of men and, in the remainder, reducing it to as little as 3 000 000/ml of semen, a level of oligospermia regarded as infertile. The contraceptive effect of testosterone is probably mediated by preventing both FSH and LH secretion by the pituitary (see Chapter 5), analogous to the main action of steroidal contraceptives in women. However, as LH normally functions to stimulate testosterone production, presumably the contraceptive effect is mediated largely through a reduced FSH stimulation of the Sertoli cells as well as the absence of pulsatile testosterone (see Chapter 3). Side-effects of the testosterone pill include acne, weight gain and oily skin, which although classed as only 'minor' might affect compliance. There is also some concern that treatment with high doses of testosterone may exacerbate the risk of prostatic hyperplasia, which is quite common in older men. Some men experienced an increase in libido, but this was not particularly common and is unlikely to have been a direct effect (see Chapter 7 for discussion). It is still far from clear whether hormonal contraception for men will have a major impact on fertility control. Pharmaceutical companies seem more interested in developing drugs that increase libido rather than reduce male fertility.

Intra-uterine contraceptive devices

Modern IUCDs are made of copper, the older plastic models being no longer in use. IUCDs function as foreign bodies within the uterus to produce a low-grade, local, chronic inflammatory response, the composition of the uterine luminal fluid resembling a serum transudate containing large numbers of invading leucocytes. There is a broad correlation between the capacity of the IUCD to induce inflammation and its efficiency as a contraceptive. Such a uterine environment inhibits sperm transport, and thereby fertilization; it is also cytotoxic to the developing pre- and post-implantation conceptus, and may inhibit or impair decidualization. In addition, luteal life may be abbreviated due to premature prostaglandin release in some species (see Chapter 5). The use of copper enhances contraceptive effectiveness, because, in addition to copper's inflammatory effects, it also has specific spermotoxic and embryotoxic effects. The copper IUCDs are also smaller and produce fewer side-effects and spontaneous expulsions. These multiple sites of action make the IUCD highly effective, the Copper T 380A being as effective as the combined oral contraceptive (Table 14.2).

For obvious reasons, contraindications to IUCD insertion include a history of pelvic inflammatory disease, heavy or painful periods, and anatomical abnormality of the uterus. Insertion into nulliparous young women who intend to have a family subsequently is not recommended because of the risks of infection, which could lead to infertility. Complications following IUCD insertion are: heavy menstrual or irregular uterine blood loss (*dysmenorrhoea*); uterine pain or muscular spasm; uterine perforation; and an increased tendency to pelvic sepsis. These side-effects are much reduced by use of the smaller copper-containing devices. Unnoticed expulsion of the IUCD can also occur, as can pregnancies with the IUCD still in place. The skills required for insertion, and the desirability of regular monitoring of patients, mean that trained and available medical or paramedical staff are needed. The IUCD is a widely used and very effective contraceptive device, and may be particularly suited to young parous women.

An extra twist to the contraceptive action of the IUCD is to combine it with steroidal contraception by use of the progesterone impregnated IUCD. This approach delivers progesterone directly to its major peripheral site of action. As an alternative, progestagen-impregnated vaginal rings can be used. The effect is exclusively local, no effect on ovulation being possible, but the contraceptive efficiency is not as high (Table 14.6).

The future

A range of variably effective sexual techniques, devices and pharmaceutical preparations now exists to limit reproduction. The use of these, tailored to the needs of the individual or couple concerned, should be adequate for effective control of fertility, high acceptability, and good compliance. The problem is not with the technology but with its use, issues of education, support and motivation being paramount. It is unlikely, given the restrictions on clinical trials imposed in recent years, that a totally new range of contraceptive approaches will be available in the near future. Some progress is being made with immunological approaches. For example, active immunity to the specific β-chain terminal sequence of hCG neutralizes the embryonic hormone but leaves LH unaffected, thereby giving normal (or slightly lengthened) cycles and protection against pregnancy (see Chapter 10). Results of advanced trials using this approach have been encouraging. No side-effects have been noted, and the immunity declines after about 6 months in the absence of a booster injection, so the approach should be reversible. Smaller scale trials are also underway immunizing against ovine FSH.

Alternatively, immunity to antigens on spermatozoa or on the zona pellucida is known, from clinical and experimental studies, to be associated with infertility. Such an approach might therefore be harnessed for contraceptive use; indeed, trials with immunization to zona antigens produced *in vitro* from cloned genes are in prospect.

However, there is a reasonable reluctance to tinker with the body's immune system until more is understood about its natural regulation. Uncontrollable side-effects, such as a wider autoimmune response, might develop after use of a 'contraceptive vaccine', particularly one that utilizes cellular or cell-associated antigens. Moreover, there is wide variation among individuals in immune responsiveness, which makes for a corresponding and unacceptable variation in contraceptive effectiveness.

Finally, the use of anti-progestins, such as RU486 (Table 2.7), as early abortifacients (see below) may be extendible to a contraceptive action. Thus, RU486 functions as a post-coital contraceptive in an emergency, but could also be used regularly each cycle as a *luteal phase contragestive*. Moreover, for reasons that are not entirely clear, daily administration of low doses prevents fertilization in monkeys, and trials are underway to determine whether the same effect is seen in humans.

Abortion

Induced abortion is a legitimate choice for couples or women faced with a pregnancy that is generated under certain conditions (such as rape, mental or physical stress or disorder) or with abnormal outcome (such as a malformed or genetically diseased fetus or a placental tumour). Abortion is also, whether performed legally or illegally, one approach to fertility control in all communities. However, the importance of its role varies. Thus, in many Western European countries, where contraception is readily available, abortion tends to be used only when contraceptive failure occurs, and only more rarely as an alternative to contraception by, for example, poorly educated or inexperienced women. In some Eastern European countries, abortion has been used as a principal method of fertility control. In countries where contraception is proscribed on religious grounds or is not easily available, abortion is frequently resorted to; for example, it has provided a major artificial control on fertility in South America, many Islamic countries, Japan and Spain.

In the first trimester (or 3 months) of pregnancy, abortion is usually performed by either dilating the cervix with metal sounds and scraping out the conceptus with a curette, or, more frequently now in the UK, by use of vacuum aspiration. The latter approach is particularly useful in the first month or so of conception and is frequently resorted to after a missed period when it is euphemized as *menstrual regulation*. The euphemism is useful practically, as legal or religious proscription of abortion often applies only if a firm diagnosis of pregnancy is made. A delayed period can therefore be treated therapeutically as such and remain within the law and ethical teaching or religious belief; hence, this approach is used extensively in South America, Spain and Bangladesh. The failure rate using these approaches is low (1–1.5%) and complications, such as major blood loss, incomplete aspiration or damage to the cervix or uterus, are about 2–3%. The side-effects of post-operative infection and damage to the uterus or cervix may affect subsequent fertility.

The possibility of conducting totally medical outpatient abortions without the need for surgery under aseptic conditions has been realized with the development of orally active anti-progestins (such as RU486), which compete for progesterone receptors. They are more than 98% effective at terminations undertaken within the first 50 days of amenorrhoea, when followed after 48 hours by a low dose of orally active prostaglandin, such as misoprostol. Terminations during the whole of the first trimester are also highly effective. Such anti-progestins may also be useful as 'once-a-month' pills to terminate the luteal phase.

Later in pregnancy, in mid-trimester, premature delivery was induced traditionally by cervical injection of hypertonic saline intra-amniotically, or of soaps, pastes or saline extra-amniotically. These approaches are little used now in the UK. More recently, controlled prostaglandin infusions into the cervical region, intra- or extra-amniotically or intravenously, to induce both cer-

vical softening and myometrial contractility (see Chapter 12) provide an alternative method of termination. However, side-effects can include vomiting, diarrhoea and nausea.

Risks versus effectiveness

The individual requires control over his or her fertility to be 100% effective at each sexual encounter, with zero risk of side-effects. In practice, this requirement is very difficult to achieve for conventional coital techniques with existing methods of fertility control. Thus, resort to multiple, or hierarchical, levels of fertility control tends to occur. Social regulation of sexual encounters is followed by use of natural techniques or caps and condoms. Failure of these approaches leads to use of the post-coital contraception or to menstrual regulation. There should be little need for unwanted pregnancy in more sophisticated societies given the cumulative contraceptive efficiency of these various approaches.

With the use of some contraceptive approaches there may be attendant risks to health, well-being or even life. However, while these risks should not be minimized, their impact is often somewhat exaggerated. For example, many of the contraindications to the use of steroidal contraceptives, such as thromboembolic episodes, cardiovascular problems or latent diabetes, also contraindicate pregnancy or the surgical procedures required for sterilization. Among groups not at risk from steroidal contraceptives, pregnancy itself and abortion probably constitute greater risks to health and life. Thus, the balance of risks must be evaluated for each individual. The optimal situation is a combination of well-educated general practitioners, a readily available range of contraceptives, and easy and rapid access to early abortion. Sadly, this situation is rare anywhere, and in most of the world it is simply not available.

INFERTILITY AND SUBFERTILITY

The incidence of pathological infertility is difficult to determine precisely because control populations, which lack any social or artificial constraints on fertility, are not readily found. However, results from a number of studies on populations of women having frequent, unprotected intercourse suggest that 50% of couples trying to conceive will have done so within 3 months, and 70% within 6 months. *Failure to conceive within 1 year of unprotected intercourse is defined clinically as subfertility.* Subfertility may be contributed to by both partners such that a relatively minor problem for each may become a more serious problem for both. Estimates of

10–15% of couples experiencing difficulty with conception and successful delivery of a healthy child are probably not wildly inaccurate. This figure is likely to rise as couples defer having a family until they are older (see earlier). These figures reveal the magnitude of a problem that has been ignored by much of the medical profession in developed countries until recently. Even now, diagnosis of, and therapy for, infertility is probably, often protracted and stressful, with a need for good counselling support, although recent advances have rendered treatment increasingly effective.

The causes of infertility vary greatly with socioeconomic and geographic factors. The male and female partners contribute more or less equally to the problem (Fig. 14.7). While psychological factors causing sexual dysfunction (such as impotence, vaginal spasm, premature ejaculation and disorders of sexual identity) are common among infertile patients, they may often be responses to, rather than causes of, the infertile state.

Three major classes of disorder account for about 75–80% of all explicable cases of infertility (Fig. 14.7): disorders of the female tract (in particular *blocked tubes*), disorders of ovulation, and seminal inadequacy. Many of the remaining cases are diagnosed as 'unexplained infertility'. In addition, there is evidence that human fertility patterns are influenced by substantial loss of conceptuses in both fertile and subfertile couples.

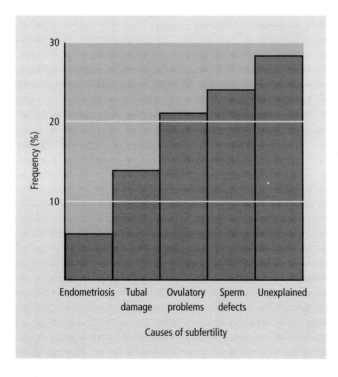

Fig. 14.7 Causes of subfertility in UK couples, expressed as a percentage frequency distribution.

Disorders of the female tract

The diagnosis of *tubal obstruction* is made by visual assessment of the intra-abdominal pelvic organs with a laparoscope and by attempting to insufflate dye from the cervix through the tubes under direct vision. Alternatively, it can be made by X-ray (*hysterosalpingogram*) using a radio-opaque dye. Tubal damage is usually a secondary consequence of pelvic infection, its incidence being elevated after sexually transmitted disease, such as frank or asymptomatic gonorrhoea or chlamydial infection, and following tuberculosis, post-abortal or post-pregnancy sepsis, and in users of the IUCD, especially older varieties that had braided strings attached. Tubal infection leads to loss of cilia on the intraluminal cells, causing impaired oocyte and spermatozoal transport, and to extra-oviducal scarring, leading to adhesions that restrict oviducal movement and oocyte pick up, or may result in physical blockage of the fimbrial ends of the tubes. At present, surgical treatment for this large group of patients has a low chance of success, and more effective therapy is provided by aspiration of oocytes from the ovary and *in vitro fertilization* (IVF) (Fig. 14.8), followed by placement of the conceptuses into the uterus, thereby bypassing the damaged tubes. When IVF technology is to be used, follicular growth is usually controlled exogenously by: (a) shutting down the woman's own hypothalamic activity via several days administration of GnRH analogue (such as buserelin; see Table 2.7 & Chapter 5); (b) then administering an FSH-like preparation; and (c) monitoring follicular growth by ultrasound scanning of the ovary (Fig. 14.9) and/or measurement of urinary oestrogens. Intrafollicular oocytes are recovered when almost fully mature via a probe inserted through the vault of the vagina under continuous monitoring by ultrasound scanning. The probe contains a needle that is inserted into each follicle and through which warm culture medium can be flushed and then collected by aspiration, so carrying the oocyte and its cumulus cells out into a receptacle. Spermatozoa are provided from the male partner by masturbation.

Endometriosis, in which endometrial tissue grows inappropriately in ectopic sites, such as the oviduct, ovary or in the peritoneal cavity, may cause a severe reaction in which the body responds by scarring and adhesion formation. It is often associated with dyspareunia, dysmenorrhoea and subfertility, but its origins and pathogenesis remain unclear.

Disorders of ovulation

This general classification covers a range of disorders. Primary amenorrhoea was discussed earlier. Here we consider only disorders relating to malfunction of the matured reproductive system: namely, absent cycles (*secondary amenorrhoea*) and irregular cycles (*oligomenorrhoea*), both of which are indicative of *anovulatory cycles*. As the

hypothalamus plays such a key role in regulating ovarian function via the pituitary, it is not surprising that these conditions are often associated with stress, obesity, strenuous exercise, anorexia nervosa or use of various drugs,

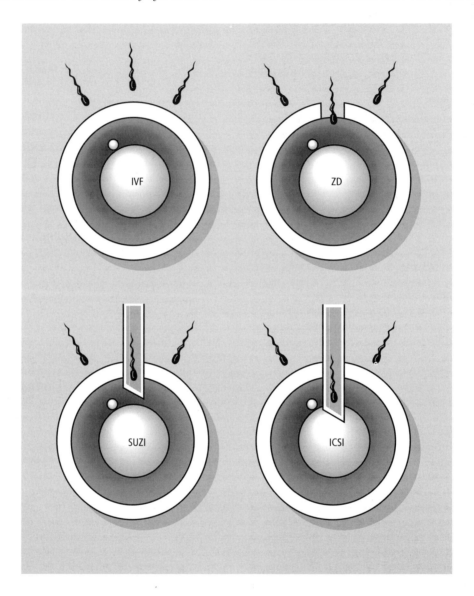

Fig. 14.8 Assisted conception techniques used to overcome sub- or infertility. In IVF, oocytes are recovered laparoscopically from the female partner's ovary (or by donation from another female), usually after hormonally induced stimulation, and spermatozoa are recovered and washed from an ejaculate or post-coital sample from the male partner (or by donation from another male). The two are then mixed *in vitro* for 24 to 48 hours; the oocytes that have been fertilized are identified by the appearance of pronuclei and passage through cleavage to two to four cells, and up to three fertilized zygotes are placed in the uterus via a transcervical catheter. Any fertilized oocytes remaining may be frozen with reasonable success for later use, should this be necessary. After IVF, average pregnancy rates per treatment cycles that go to embryo transfer at established clinics in the UK, are between 15 and 25%. Zona drilling (ZD) is identical to IVF except that a small hole is made in the zona pellucida by local application of a zona-dissolving chemical prior to insemination, and this facilitates the access of spermatozoa to the oocyte. This approach is useful if the seminal quality is poor or if the oocytes are surrounded by particularly tough zonae. The hole may also facilitate the escape of the blastocyst from the zona pellucida prior to attachment and implantation (see Chapter 9). Sub-zonal insemination (SUZI) involves the placement of one or more spermatozoa under the zona pellucida using a micropipette, while intra-cytoplasmic injection (ICSI) involves use of the micropipette to inject a single spermatozoon into the ooplasm. Both these techniques are useful for men with very low sperm numbers, ICSI being useful where there are no motile spermatozoa or even only epididymidal or testicular spermatozoa or spermatids (e.g. in obstructive azoospermia).

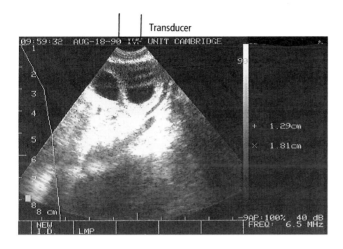

Transducer

Fig. 14.9 Transvaginal ultrasound scan of a human ovary showing two developing follicles, the largest of which has an average diameter of about 1.6 mm.

such as neuroleptics or tranquillizers, and may resolve if the primary cause is removed or its effects alleviated. Indeed, in one study, merely taking women into clinical care and giving placebo treatments resulted in 30% of cases having successful pregnancies! Many other patients, although classified as oligomenorrhoeic or secondarily amenorrhoeic, have never had entirely normal cycles and could therefore represent failures of terminal maturation of the neuroendocrine system at puberty.

The endocrine features of a normal menstrual cycle were described in Chapter 5, and the associated cyclical changes in the woman's anatomy and physiology in Chapter 7. An ideal clinical investigation would examine all of these through at least one cycle, but such a procedure would be time-consuming, costly and inconvenient. In practice, therefore, a more limited range of preliminary tests is applied in an attempt to locate the endocrine defect. In the patient with cycles, a mid-luteal phase measurement of progesterone is the simplest screening test to give an indication as to whether she is ovulating or not. Several underlying causes of ovulation disorders can be identified.

Hyperprolactinaemia

Hyperprolactinaemia is a relatively common cause of menstrual irregularity and ovulation failure (discussed in Chapter 5 in some detail). Diagnosis is by two or more blood samples; ideally, each should be taken at the same time of day under non-stressful conditions, and diagnosed by magnetic resonance imaging (MRI) for evidence of changes in the sella turcica caused by enlarging pituitary tumours, which may press on the optic chiasma. Most cases of hyperprolactinaemia are due to microadenomas, which are unseen with routine imaging

and without effects on the visual system. Hyperprolactinaemia not requiring surgery is treated by bromocriptine therapy.

Hypothalamic–pituitary insufficiency

Deficiency of GnRH reaching the pituitary results in depressed gonadotrophin and oestrogen levels and failure of ovulation with oligo- or amenorrhoea (*hypogonadotrophic hypogonadism*). Ultrasonically, quiescent ovaries and a thin endometrium will be identified. Patients can be treated therapeutically with exogenous gonadotrophins, but the use of a pump to deliver pulsatile infusions of GnRH subcutaneously or intravenously (as described experimentally in Chapters 5 & 6) is a more physiological approach and does not carry the same risk of multiple pregnancy.

Idiopathic ovarian failure

In other cases, gonadotrophin secretion seems to be occurring within the normal range of levels, but is insufficient to support a normal cycle, probably due to end organ insensitivity. In consequence, oestrogen levels fail to rise appropriately and ultrasonography of the ovary reveals antral follicles that fail to mature fully. Most cases will respond to therapy with exogenous gonadotrophins to recruit or maintain follicular growth and oestrogen output. The endogenous LH surge is usually attenuated and, hence, is supplemented or replaced by an injection of hCG (which mimics the ovulating effect of LH). However, the appropriate doses of gonadotrophins can be difficult to gauge and slight overdosing can lead to multiple ovulations and implantations or, in some cases, to ovarian hyperstimulation. Because of these difficulties and the cost of using gonadotrophins, a simpler regimen using oral *clomiphene therapy* tends to be used initially. The mode of action of this drug is not entirely clear. It has anti-oestrogenic properties and is thought to act on the hypothalamus and/or pituitary to compete with endogenous oestrogen and thereby reduce its negative feedback effects. Elevated gonadotrophins would result. However, clomiphene also stimulates aromatase activity in the ovary. As local ovarian oestrogen stimulates granulosa cell proliferation and development of LH receptors on these cells (see Chapter 4), an important additional action of clomiphene might be to stimulate ovarian responsiveness to gonadotrophins. Clomiphene tablets are administered early in the follicular phase usually on days 2–6 or 5–9 of the cycle.

An assessment as to whether treatment with clomiphene or gonadotrophins is likely to be of greatest use is made with the *progesterone challenge test*, which estimates the circulating levels of oestrogen (as a marker of follicular maturation) and thus the likelihood or responsiveness to clomiphene. A synthetic progestin (norethisterone) is administered to the amenorrhoeic or oligomenorrhoeic woman for about 5 days and then stopped. If sufficient levels of natural oestrogen are present to prime the endometrium to respond to the progestin, a withdrawal bleed will be precipitated. Absence of a bleed indicates low endogenous oestrogens and the likelihood that gonadotrophins will be more successful than clomiphene.

Polycystic ovarian syndrome

This syndrome is poorly understood and is associated with tonically elevated LH (but *not* FSH, which distinguishes this syndrome from secondary ovarian failure and (early) menopausal onset, see above). Androgens are mildly elevated in the follicular phase and many small follicles are present ultrasonographically. The primary cause of ovarian failure of this kind is not known, but these ovaries are often exquisitely sensitive to clomiphene and gonadotrophins, and ovarian hyperstimulation is a real risk. In the past, surgical removal of part of the ovary (*wedge resection*) was recommended. This procedure apparently produced an acute reduction in the prevailing high level of follicular circulating androgens, perhaps to reinstate more normal feedback relationships and leave open the possibility of an LH surge and ovulation. However, the precise sequence of events is far from clear, and the procedure has fallen out of favour because of the associated damage to the oviducts and post-operative adhesion formation. More recently, a variant procedure, involving the drilling of small holes in the ovary using diathermy or a laser has achieved remarkably good results. Alternatively, IVF has been used but the quality of oocytes recovered is usually less than found in normal or hypogonadotrophic women.

'Anovulatory' cycles that are endocrinologically 'normal'

Luteinization can occur with the oocyte remaining *in situ*, the so-called *luteinized unruptured follicle syndrome* (LUF), and, in circumstances such as these, the cycle may appear normal but infertile. There is some evidence that oocytes recovered from follicles laparascopically in women classified in this way are deficient when *in vitro* fertilization is attempted.

Abbreviated luteal phase

Some women with evidence of ovulation none the less show slow or reduced rises in progesterone, and this is associated with infertility. Such a pattern is also observed more frequently in women who have undergone gonadotrophin therapy. Whether it follows follicle formation, a deficiency in the maturation of granulosa cells leading to poor luteinization, or whether it is a primary defect of, for example, development of LH or prolactin receptors is unclear (see Chapter 4). However, as it is not yet clear whether, in women, either LH or prolactin are luteotrophic, a deficiency in these cannot be invoked by way of explanation! Treatment with progesterone during the luteal phase is, more often than not, unhelpful.

Oligospermia

Oligospermia (strictly meaning too few spermatozoa) is a term usually expanded to include a wide range of defects in semen quality, such as *asthenozoospermia* (reduced motility) and *teratozoospermia* (abnormal morphologies). Although each of these deficiencies may occur individually, they are usually associated (oligo-astheno-teratozoospermia), reflecting a general deficiency in spermatogenesis (Table 14.7). Total absence of spermatozoa in the ejaculate (*azoospermia*) may be due to deficient production (*aspermatogenesis*) or deficient transport (*obstructive azoospermia*). Deficiencies in the seminal plasma volume or composition usually reflect disease or malfunction of the accessory glands, such as the prostate or seminal vesicle.

A systematic and quantitative assessment of semen quality (either ejaculated or after recovery from the cervix post-coitally) is an essential part of an infertility examination. In addition, the availability of laparascopically recovered human oocytes or of zona-free hamster oocytes on which to assay sperm function diagnostically can be used. It has been proposed recently that sperm counts in men from several developed countries may have declined during the last few decades, the suggestion being that environmental toxins, including some that appear to have oestrogenic metabolites, might have a causal role.

Table 14.7 Characteristics associated with 'normal' and 'subfertile' semen

Criterion	Normal	Subfertile
Volume (ml)	2-5	1
Sperm concentration (no./ml)	$50\text{-}150 \times 10^6$	$<20 \times 10^6$
Total sperm no.	$100\text{-}700 \times 10^6$	$<50 \times 10^6$
Spermatozoa swimming forward vigorously	>60%	<40%
Abnormal spermatozoa	<30%	>60%
Viscosity after liquefaction	low	high
Cellular debris, leucocytes and immature sperm cells	low but variable	high?

Characteristically, hypospermatogenesis is associated with smaller testes (<20 ml volume) of softer consistency, and can result from dietary deficiency, X-irradiation, heating of the testis, exposure to a range of chemicals (notably cadmium, anti-mitotic drugs used in tumour therapy, insecticides and anti-parasitic drugs), and excessive alcohol intake. Removal of the offending agent may restore spermatogenesis from the stem cell population of spermatogonia, if this is undamaged (see Chapter 3). Untreatable forms of hypospermatogenesis include cryptorchid testes (see Chapter 1), genetic abnormalities, such as XXY, XYY and some autosomal translocations (see Chapter 1), germ cell aplasia of unknown cause often with hyalinization of tubules, and as a sequel to severe orchitis or to prolonged and intense drug therapy for tumours. In all these patients (except those with Klinefelter's syndrome), testosterone and LH levels may well be in the normal range. FSH levels, however, tend to be elevated, probably due to the absence of inhibin production (see Chapter 5). In general, elevated FSH is associated with a poor prognosis. Hypospermatogenesis due to a primary neuroendocrine deficit is relatively rare in men and is easily recognized. Treatment may be attempted with exogenous GnRH, gonadotrophins or with bromocriptine, but success is not high.

Obstructive azoospermia is not associated with obvious endocrine disorder, and testes are of normal size and consistency. Obstruction to sperm transport usually occurs in the epididymides, as a congenital disorder or secondary to infection with, for example, gonorrhoea or tuberculosis. Not surprisingly, attempts to bypass the blockage surgically are of limited success, reflecting the important role the epididymis plays in spermatozoal maturation (see Chapter 8).

Abnormal, slow-swimming or dead spermatozoa in the ejaculate, as distinct from low numbers, might also result from suboptimal spermatogenesis and is certainly increased, for example, in cases of *varicocoele* (varicosity of the spermatic vein), genetic abnormality, maintained elevated scrotal temperatures, deficiencies in spermatozoal maturation, genital tract infections, and cytotoxic factors or anti-sperm antibodies in the fluids of the accessory glands.

Both *IVF* and a variant of it, *gamete intra-fallopian transfer* (GIFT), have provided useful routes to circumvent hypospermatogenesis. These therapies avoid the dilution of spermatozoa that would occur during their passage through the female genital tract (see Chapter 8), instead concentrating the small numbers of recovered viable spermatozoa around the oocyte(s) to promote fertilization *in vitro* or in the oviduct (GIFT, in which oocytes and spermatozoa are mixed and then transferred laparoscopically into the oviduct where fertilization occurs *in vivo*).

Recent variants of IVF have allowed successful pregnancies in cases where: there are very few spermatozoa in the ejaculate; all or most of the spermatozoa in the ejaculate are immotile or clumped; no ejaculated spermatozoa are present. In the latter case, spermatozoa or spermatids must be recoverable from the epididymis or testis (obstructive azoospermia). The details of these approaches are summarized in Fig. 14.8. Such procedures can also be used in cases where sperm transport up the female tract is problematic.

Spontaneous pregnancy loss

The earliest sign that implantation is likely to have occurred comes from the detection in the blood, and later in the urine, of hCG during the period 18–30 days after the initiation of the last menstrual flow (see Chapter 10). This observation leads to the diagnosis of a *biochemical pregnancy*, although some tumours may also produce hCG. Definitive evidence of a *clinical pregnancy* is obtained by ultrasonographic investigation from as early as 5 weeks, at which time the presence and number of gestational sacs present can be assessed (see Fig. 14.1). Using an ultrasound probe in the vagina, fetal heart beat can usually be detected by 7 weeks after the last menstrual period. However, fertilization and the early development of the conceptus does not lead inevitably to a sustained pregnancy. Loss of the human conceptus is common and can occur at any stage, either because of its inherent deficiencies or as the result of environmental insult or inadequate maternal support.

The scale of the losses

We pointed out, at the beginning of this chapter, that four out of every five cycles in which frequent, unprotected intercourse occurred none the less failed to yield pregnancies. It seems likely that some of this failure can be accounted for by very early loss of the conceptus. Thus, in one study, human conceptuses were recovered by *uterine lavage* (flushing fluid through the uterus) 4.5 days after the detection of ovulation and the insemination of spermatozoa. Only 20% of the embryos recovered were blastocysts, the rest being retarded or abnormal. Moreover, of those conceptions surviving to the blastocyst stage and signalling their presence by the production of hCG, between 8 and 25% may fail, as this proportion of menstrual cycles is characterized by detectable but transient levels of hCG during the latter part of the (often slightly prolonged) luteal phase. This hCG is assumed to derive from lost peri-implantation conceptuses. Such a loss could arise from production of abnormal conceptuses that develop to the blastocyst and then fail, or from a failure of the uterine–conceptus inter-action or of the corpus luteum to respond adequately to the hCG stimulation. There is some evidence to suggest that the hCG rise is delayed slightly in those cycles destined to fail, which suggests that an embryonic deficiency in its production may be responsible. Of pregnancies that survive further to be detected clinically, some 15–25% are lost subsequently, the vast majority during the first trimester. Most losses come in women with a history of pregnancy loss, the chance of a loss being relatively low in primigravidae or in women with a successful pregnancy behind them. Overall, the cumulative outcome of such losses makes it possible that less than 15–20% of human conceptions survive to a successful birth. Why are the remainder lost?

Abnormal conceptuses

After IVF, and culture of oocytes in the clinical laboratory during therapeutic IVF, only 20–40% result in blastocysts (a similar proportion to that found *in vivo* after lavage, as quoted above). This early developmental failure of the conceptus is associated with abnormalities of chromosomal distribution or number, as many of the conceptuses examined have whole sets of chromosomes missing or duplicated, individual chromosomes gone astray, cells lacking nuclei or having multiple nuclei, or a mixture of cells of differing genetic constitution (*genetic mosaics*).

A high level of genetic abnormality is also detected in recognized clinical pregnancies (in 0.5% of all live births, 5% of stillbirths, and 40–60% of spontaneous abortions, especially those occurring in the first trimester). These figures mean that around 10% of *recognized* pregnancies are identified as being chromosomally abnormal. The type of each genetic abnormality observed, and its approximate incidence in spontaneous abortuses, is recorded in Table 14.8. Three major classes of chromosomal abnormality are represented: *translocations* (i.e. structural rearrangements of chromosomes); *errors of ploidy* (i.e. deletions or duplications of a complete set of haploid chromosomes); and *errors of chromosome number or somy* (i.e. loss or gain of a single sex chromosome or autosome). It is clear from Table 14.8 that some types of abnormality are more common in abortuses and others are compatible with survival to birth. It is also clear that some types of abnormality are missing altogether (e.g. haploids) or are under-represented (e.g. autosomal and sex chromosomal monosomies, which might be expected to occur with the same frequency as trisomies, since when one nucleus gains a chromosome at division the other will lose one!). It is likely that these types of abnormality are lethal very early in development (lack of

Table 14.8 Incidence of chromosomal abnormalities in spontaneously aborted human conceptions

	No. per 100 aborted conceptions	Surviving to birth (%)
Triploidy (Three sets of chromosomes)	12–15	<0.01
Tetraploidy (Four sets of chromosomes)	3–5	<0.01
Sex chromosome trisomies (Three sex chromosomes)	<1	>99
Sex chromosome monosomies (One sex chromosome)	10	<1
Trisomy for one or two autosomes	20–40	3
Monosomy for one or two autosomes	1	None
Structural rearrangement of chromosomes	2–3	35

genetic material being deleterious earlier than excess). Indeed, analysis of early mouse development confirms that most monosomic and haploid conceptuses die at pre-implantation or early post-implantation stages, whereas trisomic, triploid and tetraploid conceptuses survive for longer. It has been calculated from clinical data, and by extrapolation from data derived from comparative studies, that 50% or more of all human conceptions may result in genetically abnormal embryos.

Clearly this massive failure rate cannot be due entirely to constitutional genetic defects in all the germ cells of one or both parents, and many, if not most, genetic abnormalities in fetuses have indeed been shown to arise at, or shortly after, fertilization. For example, disorders of ploidy are likely to result from failure of polar body formation (10–20% of triploids), from polyspermy (80–90% of triploids) or from failure of one early cleavage division (tetraploidy; see Chapter 8). Many of the mono- and trisomic conceptuses also arise from abnormalities of oocyte meiotic divisions. As we saw in Chapter 8, these events are sensitive to a number of environmental perturbations, in particular, the age of the oocyte in hours postovulation, and exposure of the female to alcohol or anaesthesia around the time of ovulation. Other chromosomal abnormalities in the conceptus arise from events in the ovary or testis that affect gametes directly, well in advance of the acute events of fertilization and early cleavage. For example, abnormalities may be induced by exposure to X-irradiation or mutagenic chemicals, and such exposure correlates with a subsequently increased natural abortion rate. Moreover, an increased incidence of fetal abnormality is observed with increasing maternal age, which may reflect the fact that oocytes are maintained in a prolonged dictyate stage (see Chapter 4) during which they are susceptible to such damage. Finally, recent evidence from analysis of failing pre-implantation conceptuses *in vitro* shows that genetic defects occur during early cleavage.

In addition to genetic causes of embryonic and fetal loss, the conceptus may commence development normally but become deformed or incompetent as a result of environmental insult, for example, direct exposure of the fetus to X-irradiation, to certain viruses, such as Rubella (German measles) and cytomegalovirus, and to certain *teratogenic drugs*, such as thalidomide, dilantin and 6-mercaptopurine. These agents are frequently active only at restricted periods of development, most during some point in the first or second trimester. Additionally, women who smoke or who drink heavily put fetal development at risk, especially in the third trimester, by impairing intercirculatory exchange within the placenta (see Chapter 9). Unfortunately, unlike most genetic disorders, these environmental insults to the fetus often result not in spontaneous abortion but in the birth of deformed children. However, fortunately, as the insults are environmental, they are also, in principle, preventable.

Maternal defects

Of the 40–50% of spontaneous abortions that are not clearly ascribable to genetic or induced defects of the conceptus itself, many are of uncertain origin. Clearly ascribable are abortions resulting from anatomical problems, such as cervical incompetence or implantation in eccentric uterine positions or ectopically. In addition, haemolytic diseases of the fetus and neonate account for a diminishing proportion of fetal wastage in Western Europe. Immunological incompatibility of mother and fetus at either the ABO or rhesus blood group loci poses the major problem (see Chapter 11). However, recognition of incompatibility during the first pregnancy, in which sensitization of the mother to these fetal antigens is most likely around parturition, allows prophylactic administration of antibody passively to the mother. The antibody mops up any antigen released by fetal bleeds late in pregnancy and prevents active immunization of the mother.

Paradoxically, incompatibility of mother and fetus at the major histocompatibility antigen loci (HLA) may favour a successful outcome of pregnancy. The reasons for this are obscure. However, compatibility at these loci predisposes the mother to the condition of *pre-eclampsia* (also called *toxaemia* or *gestosis*). Pre-eclampsia is charac-

teristically a disease of first pregnancies and is more common in the last trimester in which maternal diastolic blood pressure is periodically or continuously elevated; oedema and, in severe cases, albuminuria occur. Severe pre-eclampsia can lead to the convulsive state of *eclampsia*. Throughout, both mother and fetus are at risk, but with the improved survival and care of prematurely delivered neonates, fewer losses from eclampsia are occurring in developed countries.

Summary

Over the past 20 years, the plight of the infertile has been taken much more seriously. There is little doubt that this change of attitude stemmed originally from the pioneering work of Bob Edwards, Patrick Steptoe and Jean Purdy in the development of IVF techniques. This work has spawned interest in all forms of infertility, the development of many new techniques, and has opened up the possibility of pre-implantation diagnosis for patients at risk of genetically transmitted disease, thereby avoiding a requirement for late abortion. However, as has happened historically, the infertile are again at risk of becoming the targets of social reaction, as in many countries this work and its application therapeutically are under attack ethically and legally.

REPRODUCTION, SEXUALITY, ETHICS AND THE LAW

Humans, like many other primates and some cetaceans and insectivores, but unlike most other mammals, do not confine their sexual activity to one narrow phase of the cycle at or around the time of ovulation. Neither do humans confine their sexual activity to vaginal penetrative sex with a member of the 'opposite sex'. Thus, for humans, coition and the expression of sexuality has a biological function over and above that of reproduction. The nature of that function in evolutionary terms remains a matter for debate, but the clear biological separability of coital sex from procreation is undeniable. Moreover, through humankind's own talents and capabilities, this separability has been further enhanced. Thus, the development of methods for reducing fertility or overcoming and circumventing infertility has further divorced reproduction from the sexual act. Reproduction may be commenced outside of the body, doctors and scientists becoming essential agents in the process. Oocytes, spermatozoa or embryos may be donated by individuals other than the partners being treated, leading to unusual patterns of 'parenting'; indeed, the very definition of a parent is challenged. Thus, a woman may carry a fetus derived from gametes neither or only one of which came from her or her partner (*gamete or embryo donation*). A woman might carry a fetus derived from IVF with the purpose of giving it at delivery to two other parents, one or both of whom may be genetically related to the child (*surrogacy*). Parents may both be women or both men, only one of them being related genetically to the child. A woman may carry a pregnancy well after her menopause. The possibility of genetic parenthood after death by use of frozen gametes or embryos is real.

This blurring of boundaries between genetic and social parents, between the expression of sexuality and procreation, between conventional and novel patterns of parenting is confusing for many, and challenging for traditional religions and social values. This challenge derives from the fact that religious, social and legal control of sexual and reproductive activities has been a feature of all human societies, as these activities bear heavily on patterns of inheritance, power and the individual freedom of expression, as well as on individual and social health. Thus, these new possibilities appear to threaten many established aspects of social structure. Clear ethical thinking about their impact is required if sensible legal controls on their use are to be enacted. Such legal responses are already occurring with variable outcome in several countries. What are the key ethical issues that have influenced these discussions and decisions? There is insufficient space here to do justice to the arguments, but some of the main points are summarized.

First, it is illegitimate to resort to ethical arguments that equate traditional with natural and novel with unnatural. What is possible biologically is natural and what is happening is therefore natural. Second, there is a presumption that individual liberty to choose how to act sexually and reproductively should be protected, as long as others are not affected adversely. Third, with this liberty goes the responsibility to protect the welfare of the parties practising sexual and reproductive interactions and also of any children that might be born as a result. This means that the adult parties should give adequate and informed consent to their mutual activity, and should not be placed under any duress. The interests and welfare of any child born, and of the 'siblings' of any child born (or not born), must be taken into account. Such a process must consider the prevailing social values and how these will affect both society's perception of the child and the ability of the child to form a clear and positive identity. It will also take into account the genetic health of the child. Fourth, there is a general acceptance, based on biological and ethical arguments, that humans acquire an individual status of personhood progress-

ively, not suddenly. There is no moment at which a human being exists where one did not before, just as there was no moment in evolution at which humankind existed where no humans had existed before. Fifth, and arising from this consideration, the status of the human gametes, pre-implantation conceptus, embryo, and early fetus differs from that of both existing humans and of non-humans. Human material with the potential for development into an individual demands respect for that potential while not having the rights accorded to an existing human; where conflict arises between the needs of a potential mother and those of her developing embryo/fetus, there is a presumption in favour of the mother's rights until such time that the fetus becomes capable of an independent existence. Sixth, there is a presumed right to parenthood. Seventh, there is a presumption that individuals should be treated confidentially and should have control over the reproductive information held about them. Eighth, there is also a presumption that it is better not to withhold from individuals information about their origins, if, for example, they are conceived by techniques of assisted conception, and that genetic information should be stored securely and made available so that genetic incest can be avoided. Ninth, there should be a limit on the numbers of children produced from donated gametes and embryos from one individual, so as to limit the chances of genetic incest occurring subsequently. Finally, there is a presumption that it is better to exist than not to exist.

In the UK, in which much of the pioneering medical and scientific work that has led to this reproductive revolution has occurred, ethical debate and legal action has also come early. There is currently in place, through the Human Fertilisation and Embryology Act of 1990, a Human Fertilisation and Embryology Authority (HFEA) set up to regulate the generation, use and storage of human embryos *in vitro*, the storage and use of human gametes *in vitro* for later therapeutic use, and the use of human embryos and of human fertilization *in vitro* in research. Many of the other issues raised above are also being brought to the HFEA. This body represents an attempt to protect the human individual, society and embryo from exploitation and excess, by applying ethical principles to biological and medical problems through legal processes. It is being observed with interest by many other states as a possible model of enlightened regulation.

CONCLUSION

Our failure to control fertility adequately, concomitant with our success in decreasing neonatal and infant mortality, has, for much of humanity, replaced one tragedy by another. The spectre of deprivation, starvation, mass migration and war is still all too real for much of the world's population. One of the greatest indictments of medical science has been the relative neglect of research in reproduction, human sexuality and the associated clinical disciplines. The social and religious prejudices that have delayed or prevented analysis of these subjects means that much of our knowledge is recent or incomplete. Indeed, many of the pioneers in this area of scientific study are still alive. Echoes of these prejudices recur in debates on the ethics of the so-called 'test tube babies', abortion, and attitudes to sexual behaviour, and the response to AIDS. More importantly, however, are the continuing effects of social and religious prejudice on the effective application of the knowledge that has accrued from such a relatively brief period of rigorous scientific and medical study. Deliberate policies to promote reproduction in some countries fearful of racial imbalance, and the frustration of family planning programmes by religious intolerance in others, is sadly as much a feature of the so-called civilized world as of countries struggling to improve the economic circumstances of their inhabitants. Throughout this book we have refrained from imposing a personal view on our account of the science of reproduction. However, we do hope that study of this book will help students training to take a place in the medical and scientific professions to realize just how pervasive the reproductive process is, both for the individual and through society at large, and how an adequate understanding and control of it represents a crucial element in the survival and well-being of both.

FURTHER READING

Beier HM & Spitz IM (Eds.) (1994) Progesterone antagonists in reproductive medicine and oncology. *Human Reproduction* (Suppl. 1) **9**.

Crosignani PG & Rubin B (1994) Male sterility and subfertility: guidelines for management. *Human Reproduction* **9**, 1260–1264.

Dunstan GR (1990) *The Human Embryo: Aristotle and the Arabic and European Traditions.* University of Exeter Press.

Edwards RG (1980) *Conception in the Human Female.* Academic Press, New York.

Edwards RG (1994) New concepts in fertility control. *Human Reproduction* (Suppl. 2) **9**.

Guillebaud J (1993) *Contraception: Your Questions Answered,* 2nd edition. Churchill Livingstone, Edinburgh.

Hawkins DF & Elder MG (1979) *Human Fertility Control.* Butterworths.

Human Fertilisation and Embryology Authority (1994) *Third Annual Report.*

Kline J, Stein Z & Susser M (1989) *Conception to Birth. Epidemiology of Prenatal Development.* Oxford University Press.

Lenton EA & Woodward AJ (1988) The endocrinology of conception cycles and implantation in women. *Journal of Reproduction and Fertility* (Suppl.) **36**, 1–15.

Mortimer D (1994) The essential partnership between diagnostic andrology and modern assisted reproductive technologies. *Human Reproduction* **9**, 1209–1213.

Potts M & Diggory P (1983) *Textbook of Contraceptive Practice.* Cambridge University Press.

Symposium Report No. 20 (1983) The control of follicle development and ovulation. *Journal of Reproduction and Fertility* **69**, 325–409.

Taylor AS & Braude PR (1994) The role of the GP in the investigation and treatment of subfertility. In: *The Diplomate.* Royal College of Obstetrics and Gynaecology Press, London.

Vom Saal FS & Finch CE (1994) Reproductive Senescence. In: *The Physiology of Reproduction* (Knobil E & Neill JD), Volume 2, pp. 2351–2413. Raven Press, New York.

Index